Biogas in the Circular Economy

Technology, Production, and Applications

Woodhead Series in Bioenergy

The *Woodhead Series in Bioenergy* publishes titles on the topic of bioenergy as it relates to renewable sources of energy. The series notably covers biological resources, chemical and biological processes, and biomass products for energy production, along with the environmental, social, management, and economic aspects of bioenergy.

Series Editor

Prof. Yves Gagnon, Université de Moncton, Moncton, Canada

Previous Volumes in the Series

Kishor Kumar Sadasivuni and Mithra Geetha, Biofuel Cells and Energy Generation, 2025, 9780443216022

Krishna Kumar Jaiswal, Bhaskar Singh and Amit K. Jaiswal, Microalgal Biofuels: Sustainable Production, Conversion, and Life Cycle Assessment, 2025, 9780443241109

Jeyabalan Sangeetha and Devarajan Thangadurai, Microalgal Biomass for Bioenergy Applications, 2024, 9780443139277

Lina María Grajales Agudelo, Juan Carlos Valdés Serra and Erich Collicchio, Agroenergy, 2024, 9780443214301

Stefano Ferrari Interlenghi, José Luiz de Medeiros and Ofélia de Queiroz Fernandes Araújo, Sustainability of Methylic and Ethylic Biodiesel Production Routes, 2024, 9780443219429

Hwai Chyuan Ong, Islam Md Rizwanul Fattah and Indra Mahlia, Waste Valorization for Bioenergy and Bioproducts, 2024, 9780443191718

Woodhead Series in Bioenergy

Biogas in the Circular Economy

Technology, Production, and Applications

Edited by

Hassan El Bari

Abdellatif Barakat

Series Editor

Prof. Yves Gagnon

ELSEVIER

WP
WOODHEAD
PUBLISHING
An imprint of Elsevier

Woodhead Publishing is an imprint of Elsevier
50 Hampshire Street, 5th Floor, Cambridge, MA 02139, United States
125 London Wall, London EC2Y 5AS, United Kingdom

ISBN: 978-0-443-29230-9 (print)
ISBN: 978-0-443-29231-6 (online)

For information on all Woodhead Publishing publications
visit our website at https://www.elsevier.com/books-and-journals

Publisher: Megan Ball
Acquisitions Editor: Edward Payne
Editorial Project Manager: Himani Dwivedi
Production Project Manager: Jayadivya Saiprasad
Cover Designer: Gopalakrishnan Venkatraman

Typeset by MPS Limited, Chennai, India

Contents

List of contributors

Abdellatif Barakat UMR IATE, INRAE, Montpellier University, Institute Agro of Montpellier, Montpellier, France; CAES, University Mohammed VI Polytechnic (UM6P), Benguérir, Morocco

Mirko Barz Faculty 1, Engineering, Hochschule für Technik und Wirtschaft Berlin (HTW Berlin), Berlin, Germany

Rafael Borja Instituto de la Grasa (CSIC), Department of Food Biotechnology, Campus Universidad Pablo de Olavide, Sevilla, Spain

Soumaya Boukrouh African Sustainable Agriculture Research (ASARI), University Mohammed VI Polytechnic (UM6P), Laayoune, Morocco

Yu-Shen Cheng Department of Chemical and Materials Engineering, National Yunlin University of Science and Technology, Douliu, Yunlin, Taiwan

Santi Chuetor Department of Chemical Engineering, Faculty of Engineering, King Mongkut's University of Technology North Bangkok, Bangkok, Thailand; Center of Eco-Materials and Cleaner Technology, King Mongkut's University of Technology, North Bangkok, Bangkok, Thailand

D. de la Lama-Calvente Instituto de la Grasa (CSIC), Department of Food Biotechnology, Campus Universidad Pablo de Olavide, Sevilla, Spain

Ouahid El Asri Laboratory of Microbial Biotechnology and Vegetal Protection, Faculty of Sciences, Ibn Zohr University, Agadir, Morocco

Hassan El Bari Laboratory of Electronic Systems, Information Processing, Mechanics and Energetics, Faculty of Sciences, Ibn Tofail University, Kenitra, Morocco

Saida El Mendili Laboratory of Electronic Systems, Information Processing, Mechanics and Energetics, Faculty of Sciences, Ibn Tofail University, Kenitra, Morocco

Doha Elalami Department of Process Engineering, Mohammadia School of Engineering, Mohammed V University, Rabat, Morocco

Merve Asli Ergun Department of Operations Research, Graduate School of Natural and Applied Sciences, Izmir Democracy University, Izmir, Türkiye

Fernando G. Fermoso Instituto de la Grasa, Spanish National Research Council (CSIC), Pablo de Olavide University Campus, Seville, Spain

M.J. Fernández-Rodríguez Department of Vegetal Biology and Ecology, Faculty of Biology, University of Seville, Seville, Spain

Margarida Gonçalves Chemistry Department, NOVA School of Science and Technology, Universidade NOVA de Lisboa, Caparica, Portugal; MEtRICs – Mechanical Engineering and Resource Sustainability Center, Universidade NOVA de Lisboa, Caparica, Portugal; VALORIZA – Research Centre for Endogenous Resource Valorization, Polytechnic Institute of Portalegre, Portalegre, Portugal

Judith González-Arias Inorganic Chemistry Department and Materials Sciences Institute, University of Seville-CSIC, Seville, Spain; Chemical and Environmental Engineering Department, Technical School of Engineering, University of Seville, Seville, Spain

Sanae Habchi Laboratory of Electronic Systems, Information Processing, Mechanics and Energetics, Faculty of Sciences, Ibn Tofail University, Kenitra, Morocco

Enes Icer Department of Operations Research, Graduate School of Natural and Applied Sciences, Izmir Democracy University, Izmir, Türkiye

Fadoua Karouach African Sustainable Agriculture Research (ASARI), University Mohammed VI Polytechnic (UM6P), Laayoune, Morocco

Tugba Keskin Department of Industrial Engineering, Faculty of Engineering, Izmir Democracy University, Izmir, Turkey; Research and Application Center for Sustainable Environmental Studies, Izmir Democracy University, Izmir, Turkey; Department of Operations Research, Graduate School of Natural and Applied Sciences, Izmir Democracy University, Izmir, Türkiye

Nuno Lapa Chemistry Department, NOVA School of Science and Technology, Universidade NOVA de Lisboa, Caparica, Portugal; LAQV-REQUIMTE – Associated Laboratory for Green Chemistry, Universidade NOVA de Lisboa, Campus de Caparica, Caparica, Portugal

Asnakech Lass-Seyoum Faculty of Engineering – Energy and Information, SC Renewable Energy, HTW Hochschule für Technik und Wirtschaft, Berlin, Germany

J.M. Mancilla-Leyton Department of Vegetal Biology and Ecology, Faculty of Biology, University of Seville, Seville, Spain

Ihssane Mnaouer African Sustainable Agriculture Research (ASARI), University Mohammed VI Polytechnic (UM6P), Laayoune, Morocco

José Paulo Mota Chemistry Department, NOVA School of Science and Technology, Universidade NOVA de Lisboa, Caparica, Portugal; LAQV-REQUIMTE – Associated Laboratory for Green Chemistry, Universidade NOVA de Lisboa, Campus de Caparica, Caparica, Portugal

Bruna Petry Chemistry Department, NOVA School of Science and Technology, Universidade NOVA de Lisboa, Caparica, Portugal; MEtRICs – Mechanical Engineering and Resource Sustainability Center, Universidade NOVA de Lisboa, Caparica, Portugal

Ana Paula Ramos Chemistry Department, NOVA School of Science and Technology, Universidade NOVA de Lisboa, Caparica, Portugal; MEtRICs – Mechanical Engineering and Resource Sustainability Center, Universidade NOVA de Lisboa, Caparica, Portugal

Zinnabu Tassew Redda Addis Ababa University, Institute of Technology, Addis Ababa, Ethiopia

Tomás R. Reina Inorganic Chemistry Department and Materials Sciences Institute, University of Seville-CSIC, Seville, Spain

Fatima Safa Laboratory of Microbial Biotechnology and Vegetal Protection, Faculty of Sciences, Ibn Zohr University, Agadir, Morocco

Pau-Loke Show Department of Chemical Engineering, Khalifa University, Abu Dhabi, United Arab Emirates

Malinee Sriariyanun Biorefinery and Process Automation Engineering Center, Department of Chemical and Process Engineering, The Sirindhorn International Thai-German Graduate School of Engineering, King Mongkut's University of Technology, Bangkok, Thailand

Saida Tayibi AgroBiosciences, College for Sustainable Agriculture and Environmental Sciences, Mohammed VI Polytechnic University, Hay Moulay Rachid, Ben Guerir, Rehamna Province, Morocco

Catarina Viegas MEtRICs – Mechanical Engineering and Resource Sustainability Center, Universidade NOVA de Lisboa, Caparica, Portugal; LEAF– Linking Landscape, Environment, Agriculture and Food, Associate Laboratory TERRA, School of Agriculture, University of Lisbon, Lisbon, Portugal

Hartmut Wesenfeld Faculty II, Mathematics - Physics - Chemistry, BHT Berliner Hochschule für Technik, Berlin, Germany

Carla Wunsch Faculty of Engineering – Energy and Information, SC Renewable Energy, HTW Hochschule für Technik und Wirtschaft, Berlin, Germany

Soraya Zahedi Instituto de la Grasa, Spanish National Research Council (CSIC), Pablo de Olavide University Campus, Seville, Spain

About the editors

Dr. Hassan El Bari has been a Professor at Ibn Tofail University in Morocco since 1988. He obtained his PhD in Materials Science from INPL (France) in 1987 and also has a "Doctorat d'Etat" in Mechanical Engineering from the Mohamed V University in Morocco, received in 1998. He is the President of the Moroccan Association of Solid Waste. He was the Master Degree Coordinator: Waste Management and Recovery (2010–14) and Member of the International Solid Waste Association (2004–14). He has been involved in many funded projects as a coordinator and partner, and has significant experience in teaching solid waste management, biomass and bioenergy, and anaerobic digestion technology. Prof. El Bari is an Associate Editor of the *Cleaner Waste Systems Journal* (Elsevier), Editorial Board Member of the *Applied Science and Engineering Progress* journal, and Guest Lecturer at the University of KwaZulu-Natal, South Africa. He has authored or coauthored several book chapters and edited or coedited three books.

Dr. Abdellatif Barakat is a Director Researcher in Green Chemistry and Valorization of Biomass at INRAE, JRU of Agropolymer Engineering and Emerging Technologies (IATE), Montpellier, France, which focuses on dry fractionation and bioconversion of agro-resources and waste, biofuels, anaerobic digestion, digestate valorization, pyrolysis, biofertilizers, and molecules of interest. His current research interests include valorization of biomass for bioenergy production, pretreatments, extraction of biopolymers, and biofertilizers' formulation. He received his PhD in Chemistry and Physicochemical of Biopolymers from Reims University in 2007 and has previously held several postdoc positions, including at the Laboratory of Biotechnologies of the Environment (LBE, INRA) and at the Institut Charles Gerhardt (MACS, CNRS). Dr. Barakat has published over 120 papers in international journals, 4 patents, and 6 book chapters.

Foreword

The energy sector is undergoing an important transition to meet the world's energy needs, reduce greenhouse gas emissions, support climate change mitigation, and protect our planet for future generations. A major element of the energy transition is the integration of large-scale renewable energy sources in the energy mix of countries, which includes bioenergy.

Bioenergy is the transformation of biomass, derived from recently living organic materials, into electricity, heat, and transportation fuels. Along with wind and solar energy, bioenergy is poised to play a significant role in the energy transition.

In support of the global research community and the commercial bioenergy sector, the Woodhead Series in Bioenergy publishes research- and application-oriented book titles on the topic of bioenergy as it relates to renewable sources of energy. The series covers biological resources, chemical and biological processes, and biomass products for energy production, all while operating on principles of sustainability in bioenergy, namely, applying the environmental, social, managerial, and economic aspects of bioenergy.

Biogas in the Circular Economy: Technology, Production, and Applications, edited by Prof. Hassan El Bari and Dr. Abdellatif Barakat, addresses an important element of the energy transition, that is, the sustainable production of biogas from various biomass feedstocks under a circular economy.

Prof. El Bari, from the Ibn Tofail University in Morocco, has developed broad expertise in solid waste management, biomass, and bioenergy, with experiences in teaching, research, editorship, and in industrial applications. His involvement with Moroccan and international waste management organizations also allows him to influence public policies on waste management and valorization.

Dr. Barakat, from the National Research Institute on Agronomics (*Institut national de la recherche agronomique*, INRAE) in France, brings complementary expertise and experience in the valorization of biomass for bioenergy production.

The result of this partnership is a book that documents the production of biogas under a circular economy framework, with a focus on biomethane and biohydrogen production. Besides the fundamental aspects of biogas production, this book ventures in leading-edge technologies and approaches such as artificial neural networks and genetic algorithms for the optimization of biogas production, innovative biogas digester technology, carbon neutrality, and the valorization of the digestate of anaerobic digestion processes. This book also looks at biomethane and biohydrogen applications in the energy and transport sectors and in the water–energy–food nexus.

With its presentation of the most recent developments and innovations in the biogas sector, *Biogas in the Circular Economy: Technology, Production and Applications* is a valuable reference for those interested in the sustainable valorization of biomass in a circular economy.

Prof. Yves Gagnon P.Eng., D.Sc.
Université de Moncton, Canada
Series Editor, Woodhead Series in Bioenergy

Preface

There is a significant and huge amount of organic waste and residue worldwide. This potential is particularly apparent in the sludge from domestic wastewater treatment facilities, the agricultural and agro-food sectors, and at the municipal level. However, almost all of this waste, particularly in municipal and agricultural cases, is neither recovered nor properly treated and thus constitutes a real menace to our environment. Thus these organic wastes need sustainable solutions for their treatment; otherwise, they become an important source of pollution and greenhouse gas emissions. A circular economy is considered a vision of a sustainable society, requiring increased responsibility leading to sustainable production and the material recovery of end-of-life products. Its ambition is to minimize the generation of waste and recover resources at the end of a product's life by inserting them in the economic chain, allowing a better projection of the environment. In this sense, the anaerobic digestion (AD) of organic wastes is seen as a solution to these issues. In fact, biogas technology is highly recommended since it can be used to reduce organic pollution and produce biomethane, a renewable source of energy that could be used in different ways. The AD method is set to become increasingly important as it is considered an energy-positive process since it produces more energy than it requires. Furthermore, organic waste recovery and recycling allow nutrients to be returned to the soil as fertilizers or soil conditioners. The AD is a key technology for the development of a circular economy. By closing the loops on previously linear processes, AD can address the challenges of waste, energy, and nutrient recycling in a sustainable and circular way. Nevertheless, there are several conditions that any biogas plant needs to meet in order to achieve these objectives of the circular economy; otherwise, the loop is not closed. In addition, the AD technology has a crucial contribution to achieving 6 sustainable development goals (SDGs) among the 17 goals, which are: SDG 2 (Zero Hunger): Biomethane energy for farmers' needs and digestate as a balanced biofertilizer for soil quality improvement. SDG 3 (Good Health and Well-Being): AD reduces organic pollution considerably (Chemical Oxygen Demand (COD) reduction). Cooking with biogas reduces respiratory diseases due to wood smoke, especially among rural women in developing countries. SDG 5 (Gender Equality): By using biogas for cooking instead of wood, the rural women will no longer have to waste time collecting wood, and the children will be able to attend school regularly. SDG

6 (Clean Water and Sanitation): sludge from domestic and industrial wastewater could be converted to biogas and provide clean energy for sanitation purposes. A liquid fraction of digestate could be used for agriculture and irrigation, and the solid fraction could be composted. SDG 7 (Affordable and Clean Energy): We produce clean and green energy from biomethane and biohydrogen for different purposes, mainly electricity, heat, and transport. SDG 13 (Climate Action): AD contributes considerably to reducing greenhouse gas emissions, mainly methane.

This book aims to provide guidance on achieving circular economy goals by closing the loop. This could be a various bioenergy combined system that utilizes different types of biomass conversion, including biochemical and thermochemical processes. Additionally, this book shows case examples of renewable energy combined systems. The primary objective of this book is to present and explain biogas technology within the context of a circular economy, with a focus on achieving zero waste through sustainable valorization of digestate. To this end, this book covers (1) advanced AD and dark fermentation techniques for producing biomethane and biohydrogen, (2) the latest technologies for separating and purifying biogas, (3) management methods and applications of digestate in a circular economy, and (4) the various ways in which biomethane and biohydrogen can be used as renewable gases.

This book consists of 10 chapters. The first three chapters are dedicated to advanced AD technology for biomethane production, including the suitable substrates and combined pretreatments for optimal biogas production, innovative anaerobic digester technologies, and the use of artificial intelligence to model and optimize biogas production. In Chapter 4, the reader can discover the latest technologies of biogas separation and purification, in particular, the methods of absorption, adsorption, and membrane technologies, as well as the recent progress in the technologies for in situ biogas upgrading. While in Chapter 5, we present carbon neutrality and challenges in the biogas economy. Chapter 6 deals with the different management methods and applications of digestate in a circular economy context, namely, the use of digestate as an agricultural biofertilizer as well as its energy recovery by thermochemical conversion. In Chapters 7, 8, and 9, we present the different ways of using biomethane and biohydrogen, particularly its cogeneration, its application in the sector of decarbonized transport, and its use in the framework of the water–energy–food nexus. The last chapter is about the life cycle assessment and cost–benefit analysis of biogas and biohydrogen production.

This book is an important resource for the circular economy sector, biogas technology researchers, scientists, and master's/PhD students. This book can work as a helpful manual for experts, scientists, researchers, students, and the biofuel industry. Its goal is to offer the most recent developments in the biogas sector. It could be useful for researchers in bioprocessing, bioenergy, microbiology materials, chemical engineering, biotechnology, and biogas technology. The general public, especially those with a keen interest in the field of biogas production, would be the book's secondary audience. Furthermore, a wide range of readers, including bioenergy and circular economy experts, policymakers, municipal authorities, and industry professionals, can find this work to be a useful resource and practical guide. In academic aspects, there are several possibilities of using this book in course titles such as (1) Waste management and recovery: municipal and industrial, (2) Wastewater treatment: domestic and industrial cases, (3) Bioenergy engineering, (4) Bioprocess technology, (5) Renewable energy, (6) Biomass energy, (7) Biohydrogen production, and (8) Circular economy.

Hassan El Bari
Abdellatif Barakat

Chapter 1

Suitable feedstocks and advanced combining pretreatment for optimized biomethane and biohydrogen production

Malinee Sriariyanun[1], Yu-Shen Cheng[2], Pau-Loke Show[3], and Hassan El Bari[4]

[1]*Biorefinery and Process Automation Engineering Center, Department of Chemical and Process Engineering, The Sirindhorn International Thai-German Graduate School of Engineering, King Mongkut's University of Technology, Bangkok, Thailand,* [2]*Department of Chemical and Materials Engineering, National Yunlin University of Science and Technology, Douliu, Yunlin, Taiwan,* [3]*Department of Chemical Engineering, Khalifa University, Abu Dhabi, United Arab Emirates,* [4]*Laboratory of Electronic Systems, Information Processing, Mechanics and Energetics, Faculty of Sciences, Ibn Tofail University, Kenitra, Morocco*

1.1 Introduction

In recent years, the imperative for sustainable development has gained unprecedented momentum as societies worldwide grapple with the escalating challenges posed by climate change, resource depletion, and environmental degradation. Central to this discourse is the necessity for innovative solutions that not only mitigate these pressing issues but also foster a transition toward a circular economy—a model designed to minimize waste generation while maximizing resource efficiency (Cheng, 2023; Sriariyanun & Dharmalingam, 2023). Within this context, various biofuel production technologies emerge to address multiple facets of sustainability by transforming biomass feedstocks into biofuels to supply the growth of industries and societies. Lignocellulose biomass, encompassing various forms of plant-derived materials such as wood, agricultural residues, and dedicated energy crops, represents an abundant and underutilized resource (Ubando et al., 2020). Concurrently, agricultural waste streams, including crop residues, animal manure, and food-processing by-products, pose significant environmental challenges when improperly managed

Biogas in the Circular Economy. DOI: https://doi.org/10.1016/B978-0-443-29230-9.00014-0

(Culaba et al., 2022). As general practices in developing countries, agricultural wastes obtained after harvesting seasons are burned on site to prepare the field for the next cultivating season and for harvesting agricultural products, especially sugarcane and corn, leading to uncontrolled release of greenhouse gases and health-hazardous particles, for example, PM2.5 and PM10 (Janta et al., 2020). It is estimated that agricultural activities generate billions of tons of waste annually worldwide. According to the United Nations Food and Agriculture Organization, agricultural waste production is significant, with estimates suggesting that it accounts for approximately one-third of total agricultural production globally (Babu et al., 2022).

Among a variety of biofuel production technologies from lignocellulose biomass, biogas production is considered to be a conventional technology that is well developed and implemented in industries and communities due to its numerous environmental and economic benefits (Patinvoh et al., 2017). As a renewable energy source, biogas is produced from the anaerobic digestion of organic materials such as agricultural waste, manure, municipal waste, and food waste, making it a sustainable alternative to fossil fuels. This process not only generates energy but also aids in effective waste management, reducing the amount of organic waste that would otherwise end up in landfills and contribute to greenhouse gas emissions. Moreover, biogas production results in the generation of nutrient-rich digestate, which can be used as a high-quality fertilizer, thus closing the nutrient loop in agricultural practices (Reshmy et al., 2022). Furthermore, a biogas production facility is a solution to manage the uncontrolled releases of greenhouse gases, especially carbon dioxide (CO_2) and methane (CH_4), produced from agricultural sectors, for example, rice cultivation fields (Thomas et al., 2022). CO_2 is the most prevalent greenhouse gas, about 80% of total greenhouse gases. Although CO_2 is less potent than methane in terms of heat-trapping ability, its high concentration in the atmosphere and long atmospheric lifetime make it the primary driver of human-induced climate change. CH_4, on the contrary, is much more effective at trapping heat, with a global warming potential approximately 25 times greater than that of CO_2 over a 100-year period. However, methane is present in much lower concentrations, accounting for about 11% of total greenhouse gases, and has a shorter atmospheric lifetime of about 12 years. Despite its shorter lifespan, methane's high potency means it plays a crucial role in short-term climate dynamics (Bačėninaitė et al., 2022; Bagi et al., 2017).

Biogas is mainly composed of CH_4, which makes up about 50%–70% of biogas and serves as the main energy carrier molecule. The other components include CO_2 (about 30%–35%), water, hydrogen sulfide (H_2S), ammonia (NH_3), hydrogen (H_2), nitrogen (N_2), and oxygen (O_2), and traces of volatile organic compounds (VOCs) (Angelidaki et al., 2018; Jose et al., 2024). In addition to CH_4, numerous types of gaseous biofuels include synthetic natural gas (SNG), syngas, and biohydrogen, and they are readily produced for commercialized scale currently. Bio-SNG is produced from the gasification of biomass followed by a methanation process, resulting mainly in methane.

Syngas, a mixture of carbon monoxide (CO), hydrogen, carbon dioxide, and small amounts of methane, is also derived from biomass gasification (De Tissera et al., 2019; Fona et al., 2024). Biohydrogen, produced through biological processes like fermentation or photosynthesis by microorganisms, or via biomass gasification and subsequent water-gas shift reaction, primarily consists of hydrogen (H_2) (Machhirake et al., 2024; Teke et al., 2024). These gaseous biofuels are compatible with existing infrastructure from production plants to users in factories and households, suggesting their acceptance in the market needs. Furthermore, all of them could be produced from lignocellulose biomass; therefore, they have the potential to be biofuels to secure energy situation and mitigate the environmental problem in waste management.

This chapter explores the principles and applications of methane and hydrogen production from lignocellulose feedstock as a key enabler of sustainable development and circular economy paradigms. We are focusing on analyzing factors for the selection of suitable lignocellulose feedstocks for methane and hydrogen production. Various pretreatment methods on lignocellulose biomass to enhance the degradation of biomass are explored and factors affecting pretreatment efficiency are investigated. Moreover, through case studies and practical examples, we highlight the diverse applications of advanced combined pretreatment methods across different scales and contexts, showcasing its potential to fit in methane and hydrogen production.

1.2 Lignocellulose biomass and its potential as feedstocks for methane and hydrogen production

Lignocellulosic biomass is composed of complex organic materials that provide structural support to plants. It is considered to be the most abundant biomass on Earth, and it consists of three main components: cellulose, hemicellulose, and lignin (Fig. 1.1) Cellulose forms the backbone of the plant cell wall, primary and secondary walls, and comprises long chains of glucose molecules (C6 sugar or hexose). Hemicellulose is a heteropolymer of various sugars including xylose, mannose, and arabinose, which are pentose (C5 sugar) and hexose sugars. It binds to cellulose with hydrogen bonding and van der Waals interactions and provides additional structural support (Igwebuike et al., 2024; Tan et al., 2024). Lignin, a complex polymer made of phenylpropane units, serves as a cementing material between cellulose and hemicellulose, conferring rigidity and impermeability to the plant cell wall. Lignin binds with carbohydrates (cellulose and hemicellulose) by ether bond, ester bond, and acetal bond (Prates et al., 2017; Prates et al., 2019). Furthermore, cellulose forms crystalline microfibrils that provide structural integrity to plant cell walls. These microfibrils are closely packed and stabilized by hydrogen bonds, rendering cellulose highly resistant to enzymatic hydrolysis Gundupalli, Cheenkachorn, et al. (2023). Hemicellulose comprises a variety of sugars arranged in an amorphous matrix, and it is more promptly degradable than

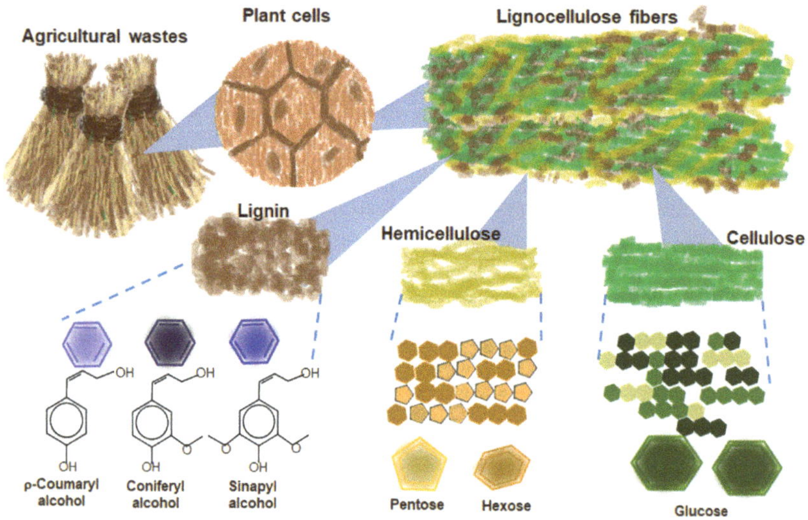

FIGURE 1.1 Compositions of lignocellulose biomass and their molecular arrangement.

cellulose due to its branched structure. Its association with cellulose limits enzymatic access to cellulose. Moreover, lignin's hydrophobic nature and irregular structure further impede the penetration of enzymes, contributing to biomass recalcitrance (Klose & Paës, 2023). This intricate composition makes lignocellulosic biomass highly resistant to enzymatic degradation and necessitates specialized processes, such as pretreatment and enzymatic hydrolysis, to break down its components into fermentable sugars for biofuel production. While cellulose and hemicellulose are fermentable sugars that can be converted into biofuels such as biogas and ethanol, lignin presents a challenge due to its recalcitrant nature, requiring innovative strategies for its valorization in biorefinery processes (Cheng et al., 2020).

The anaerobic digestion process is facilitated by a diverse community of microorganisms, microbial consortiums, that decompose organic material in the absence of oxygen. This process is divided into four phases: hydrolysis, acidogenesis, acetogenesis, and methanogenesis (Fig. 1.2) (Xu et al., 2019). Bacteria primarily perform complex polymeric digestion, breaking down materials like cellulose, protein, and fats, with carbohydrates degrading rapidly within hours, while lignin, fats, and proteins decompose more slowly. Facultative anaerobic microbes play a crucial role by reducing the redox potential through oxygen absorption, favoring the anaerobic digestion process (Adney et al., 1991; Mohd-Nor et al., 2019). Complex organic compounds are converted into simpler forms, yielding volatile fatty acids or organic acids, such as acetic, propionic, butyric, formic, and lactic acids (Xu et al., 2019).

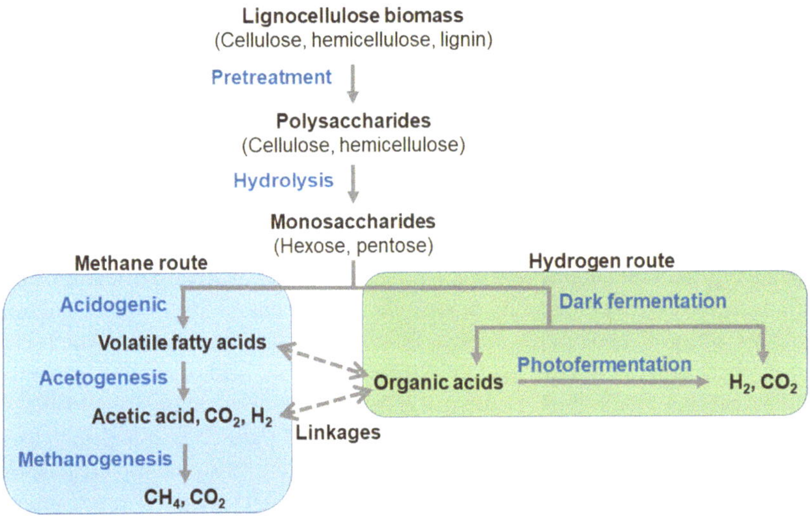

FIGURE 1.2 Methane and hydrogen production from lignocellulose biomass.

However, high concentrations of H^+ ions can impede acetate production by inhibiting the growth of bacteria. During fermentation, sugars, fatty acids, and amino acids are transformed into organic acids and alcohols, with H_2, CO_2, and acetate. Acetogenic and methanogenic microbes coexist during the anaerobic digestion process, with methane produced in the absence of oxygen. In the final stage of methanogenesis, acetotrophic methanogens reduce acetate (CH_3COO^-) into methane, while hydrogenotrophic methanogens convert H_2 and CO_2 into CH_4 (Sawatdeenarunat et al., 2015).

Similar to biomethane production, biohydrogen is produced via fermentation, and it involves a series of biochemical reactions carried out by various microorganisms under anaerobic conditions (Fig. 1.2). The process typically begins with the breakdown of complex organic materials, such as carbohydrates, by hydrolytic microorganisms. These microorganisms, often bacteria or archaea, metabolize sugars through glycolysis, generating intermediate molecules, such as pyruvate and producing hydrogen ions (H^+) and electrons (e^-). Subsequently, hydrogen-producing microorganisms possess hydrogenase enzymes catalyzing the conversion of these H^+ ions and electrons into H_2 through either direct or indirect pathways (Cheng et al., 2011; Singh et al., 2021). In the direct pathway (dark fermentation), hydrogenases directly reduce protons to produce H_2, while in the indirect pathway (photofermentation), formate or other organic compounds serve as intermediates in H_2 production. Additionally, certain microorganisms, such as certain species of Clostridium and Enterobacter, possess specialized enzymes called nitrogenases that can fix atmospheric nitrogen (N_2) and produce ammonia (NH_3), providing an additional source of reducing equivalents for H_2 production (Zhang et al., 2021).

Based on these pathways for methane and hydrogen production, there are sharing steps in both gases, which are hydrolysis of polymeric molecules of biomass, and in the case of lignocellulose, they are cellulose and hemicellulose (Fig. 1.2). However, the lignocellulose properties, including physical and chemical properties, result in ineffective functions of hydrolytic enzymes and microbes. Pretreatment of lignocellulosic biomass plays a crucial role in enhancing the breaking down of the complex and recalcitrant structure of the biomass, thereby increasing its digestibility and accessibility to microbial and enzyme action (Capolupo & Faraco, 2016). Pretreatment methods, which can be physical (e.g., milling or grinding), chemical (e.g., acid or alkali treatment), biological (e.g., microbial or enzymatic treatment), or a combination of these, disrupt this matrix. Physical pretreatment reduces particle size and increases the surface area, while chemical and biological pretreatments breakdown lignin and hemicellulose, exposing the cellulose fibers. A good pretreatment method for lignocellulosic biomass should exhibit several key properties to ensure its efficiency and effectiveness. First, it should efficiently disrupt the lignocellulosic matrix, breaking down lignin and hemicellulose to expose cellulose fibers, thereby increasing the accessibility of enzymes and microorganisms to the biomass (El Gnaoui et al., 2022). Additionally, the method should be economically feasible, with low operational and capital costs, making it viable for large-scale applications. It should use minimal and preferably nontoxic chemicals to reduce handling and disposal issues, thus lowering environmental impact and operational complexities. Furthermore, the pretreatment process should operate under moderate conditions, avoiding extreme temperatures and pressures to save energy and reduce equipment wear. Scalability is another important factor, allowing the method to be applied from laboratory to industrial scales without significant loss of efficiency. Lastly, it should produce minimal inhibitory by-products that could interfere with downstream microbial fermentation processes, ensuring a smooth and efficient conversion of biomass to biogas (Habchi et al., 2022).

Based on the development of pretreatment technology and the suitability of LSB, a large variety of LSB obtained from agricultural waste, agricultural processing plants, dedicated energy crops, municipal waste, husbandry industry, etc., have been demonstrated to be the substrates for methane and hydrogen production from laboratory scale testing, pilot-scale demonstration to industrial application (Table 1.1). In a laboratory-scale demonstration, a variety of LSB have been tested as the substrates ranging from crop residues, wood pellets, fruit-peels, food-processing residues, and husbandry residues. The methane and hydrogen production yield from different types of LSB varies significantly, influenced by the composition and proprieties of the biomass and the specific conditions of the anaerobic digestion process (Lay et al., 2022; Martínez-Gutiérrez, 2018). Corn stover, a commonly used

TABLE 1.1 Lignocellulose compositions in various lignocellulose feedstocks reported for biomethane and biohydrogen production via fermentation processes.

Feedstocks	Composition(%)				References
	Cellulose	Hemicellulose	Lignin	Yield*	
Rice straw	37.19	31.30	10.93	85.3 mL—CH_4/g-VS	Luo et al. (2024)
Tobacco straw	23.08	23.48	19.68	563.2 mL—CH_4/g-VS	Wang, Jin, et al. (2024)
Sugarcane leaves	38.89	20.13	25.11	199.3 mL—CH_4/g-VS	Sitthikitpanya et al. (2024)
Rice straw	32.16	32.34	2.66	40 mL—CH_4/g-DM	Kyawt et al. (2024)
Rice husk	37.93	28.44	11.80	15 mL—CH_4/g-DM	Kyawt et al. (2024)
Corn straw	37.61	30.93	3.99	38 mL—CH_4/g-DM	Kyawt et al. (2024)
Corn cobs	33.06	46.56	2.12	62 mL—CH_4/g-DM	Kyawt et al. (2024)
Sugarcane bagasse	36.23	33.56	9.02	25 mL—CH_4/g-DM	Kyawt et al. (2024)
Almond shell	22.0	21.0	29.2	86.1 mL—CH_4/g-VS	Oliva et al. (2024)
Spent coffee grounds	8.7	34.0	18.7	365.2 mL—CH_4/g-VS	Oliva et al. (2024)
Hazelnut skin	10.1	3.5	44.2	400.4 mL—CH_4/g-VS	Oliva et al. (2024)
Rice straw	43.3	36.37	8.53	318.10 mL—biogas/g-VS	Rani and Dhoble (2023)
Water hyacinth	23.99	32.03	15.57	166.32 mL—biogas/g-TS	Jose et al. (2024)
Water hyacinth	24.0	32.0	16.0	228.06 m—Lbiogas/g-TS	Gundupalli, Tantayotai, et al. (2023)

(*Continued*)

TABLE 1.1 (Continued)

Feedstocks	Composition(%)				References
	Cellulose	Hemicellulose	Lignin	Yield*	
Rice straw	47.44	14.72	19.4	170.92 mL—biogas/g-VS	Tantayotai et al. (2019)
Mixed Vegetable Waste	20.67	16.97	31.70	110.72 mL—biogas/g-VS	Dharmalingam et al. (2023)
Spent mushroom bed	35–40	30–35	10-15	17.92mL—CH_4/g-VS, 56.45 mL—H_2/g-VS	Bertasini et al. (2024)
Willow sawdust	32.7	25.5	26.3	43.2 mL—CH_4/g-VS	Ali et al. (2023)
Various lignocellulose	31.8	22.4	9.30	178 mL—CH_4/g-VS	Sonwai et al. (2023)
Rice straw	35.1	30.2	20.8	345 mL—CH_4/g-DM	Luo et al. (2023)
Atriplex crassifolia	37.0	21.0	0	68.6 mL—CH_4/g-VS, 13.2 mL—H_2/g-VS	Nawaz et al. (2023)
Sugarcane bagasse	32.7	22.4	20.4	103.04 mL—H_2/g-DM	Bu et al. (2021)
Cornstalk	36	22	19	77.73 mL—H_2/g-DM	Cao et al. (2014)
Paper sludge	19.21	6.71	1.87	159 mL—H_2/g-DM	An et al. (2018)
Corncob	38.5	34.2	13.5	1.57 mL—H_2/g-DM	Singhvi and Kim (2022)
Sugarcane bagasse	40.42	23.43	24.89	259.57 mL—H_2/g-TS	Mai et al. (2023)
Energy poplar	47.00	18.90	29.00	22.99 mL—H_2/g-DM	Kucharska et al. (2018)

TABLE 1.1 (Continued)

Feedstocks	Composition(%)				References
	Cellulose	Hemicellulose	Lignin	Yield*	
Birch sawdust	37.07	30.87	22.6	121.4 mL—H_2/g-VS	Sarkar et al. (2022)
Cornstalk	42.6	28.8	20.7	152 mL—H_2/g-DM	Zhao et al. (2014)
Cellulose	100	0	0	66.53 mL—H_2/g-DM	Wang et al. (2009)
Cornstalk	17.9–33.7	14.4–25.2	20.3–42.8	286.1 mL—H_2/g-DM	Zhao, Wang, et al. (2021)
Cassava residues	47.61	16.27	19.63	91 mL—H_2/g-DM	Zhang et al. (2019)
Rice straw	49.27	26.63	19.92	63 mL—H_2/g-DM	Zhang et al. (2019)
Cornstalk	43.87	27.52	23.65	85 mL—H_2/g-DM	Zhang et al. (2019)
Corncob	40.66	31.27	24.38	69 mL—H_2/g-DM	Zhang et al. (2019)
Sugarcane bagasse	56.76	21.05	8.30	109.54 mL—H_2/g-DM	Tian et al. (2015)
Corn-bran residue	10.54	40.57	1.06	338.91 mL—H_2/g-VS	Zhang et al. (2015)
Mixed fruit waste	8.00	30.09	4.50	1.88 mL—CH_4/g-DM, 1.03 mL—H_2/g-DM	Rena et al. (2020)
Mixed vegetable waste	14.50	.75	2.50	.85 mL—CH_4/g-DM, 1.86 mL—H_2/g-DM	Rena et al. (2020)

(Continued)

TABLE 1.1 (Continued)

Feedstocks	Composition(%)				References
	Cellulose	Hemicellulose	Lignin	Yield*	
Sugarcane bagasse	47.40	31.85	10.00	.33 $mLCH_4$/g-DM, .17 mL—H_2/g-DM	Rena et al. (2020)
Rice husk	32.00	21.00	19.00	.99 mL—CH_4/g-DM, .17 mL—H_2/g-DM	Rena et al. (2020)
Wheat straw	38.60	26.00	17.00	1.72 mL—CH_4/g-DM, .47 mL—H_2/g-DM	Rena et al. (2020)
Rice straw	36.14	27	13.73	134.64 mL—H_2/g-DM	Dong et al. (2018)
Corncob	40.20	32.62	10.20	73.4 mL—H_2/g-DM	Fan et al. (2022)
Cornstalk	47.6	35.8	15.9	154.3 mL—H_2/g-DM	Zhao, Wu, et al. (2021)

VS, Volatile solid; *TS*, total solid; *DM*, dried mass.

biomass in the United States, yields between 250 and 350 m^3/t VS. This consists of the leaves, stalks, and cobs left in the field after the corn harvest. Corn stover is abundant due to the large scale of corn farming, and its high cellulose content makes it an excellent raw material for biogas production. Sugarcane bagasse, primarily utilized in Brazil and India, typically produces between 250 and 300 m^3/t VS of biogas. Sugarcane bagasse is the fibrous residue left after extracting juice from sugarcane during sugar production. However, the high lignin content of bagasse poses a challenge, making effective pretreatment essential for optimizing biogas production. Wheat straw, widely used in Europe and parts of North America generates around 200 to 350 m^3/t VS of biogas. Rice straw, prevalent in Asia as a major staple food, especially in China and India, offers a biogas yield in the range of 250 to 300 m^3/t VS. Like other straw types, effective pretreatment is vital to break down the complex lignocellulosic structure to improve biogas yield. Miscanthus, a dedicated energy crop with high cellulose content, yields approximately 200 to 300 m^3/t VS of biogas. Miscanthus, a perennial grass grown specifically for energy production, is notable for its high yield and low input requirements. Switchgrass, another dedicated energy crop native to North America, typically produces between 200 and 350 m^3/t VS of biogas. It is known for its high productivity, adaptability to poor soils, and low maintenance needs, making it a viable feedstock for large-scale ethanol production. Wood residues, including sawdust and wood chips, generally have lower biogas yields, around 150 to 250 m^3/t VS (Kulichkova et al., 2020; Martínez-Gutiérrez, 2018; Pote et al., 2023). The high lignin content in wood residues is a significant barrier to efficient anaerobic digestion, necessitating intensive pretreatment to improve biogas yields. Wood residues, including sawdust, wood chips, and logging leftovers, are significant feedstocks in regions with substantial forestry industries. These residues provide a continuous and sustainable source of lignocellulose biomass for ethanol production, especially in areas with established forestry operations. Effective pretreatment is crucial for maximizing biogas yield from lignocellulosic biomass, as it helps to break down complex polymers and make cellulose and hemicellulose more accessible to anaerobic microorganisms. These yield estimates highlight the potential and challenges of different lignocellulosic biomasses for biogas production, emphasizing the importance of optimizing pretreatment and digestion conditions to achieve the best results.

1.3 Selection criteria for pretreament methods

Pretreatment methods for lignocellulose are essential to break down the complex structure of biomass into fermentable sugars, facilitating the production of biofuels and other biochemicals. These methods aim to make the cellulose and hemicellulose components more accessible to hydrolytic enzymes by disrupting the lignin barrier and altering the crystalline structure of cellulose. The main categories of pretreatment methods include physical, chemical, physico-chemical, and biological techniques, each with distinct advantages and disadvantages (Fig. 1.3). Physical pretreatment involves mechanical processes such as milling, grinding, and irradiation techniques like ultrasound and microwave treatments. These methods primarily reduce the particle size and increase the surface area of the biomass, thereby enhancing enzyme accessibility. Physical pretreatment is usually the first step of biomass processing to reduce the bulky size of biomass to further handling size. The main advantages of physical pretreatment are its simplicity and environmental

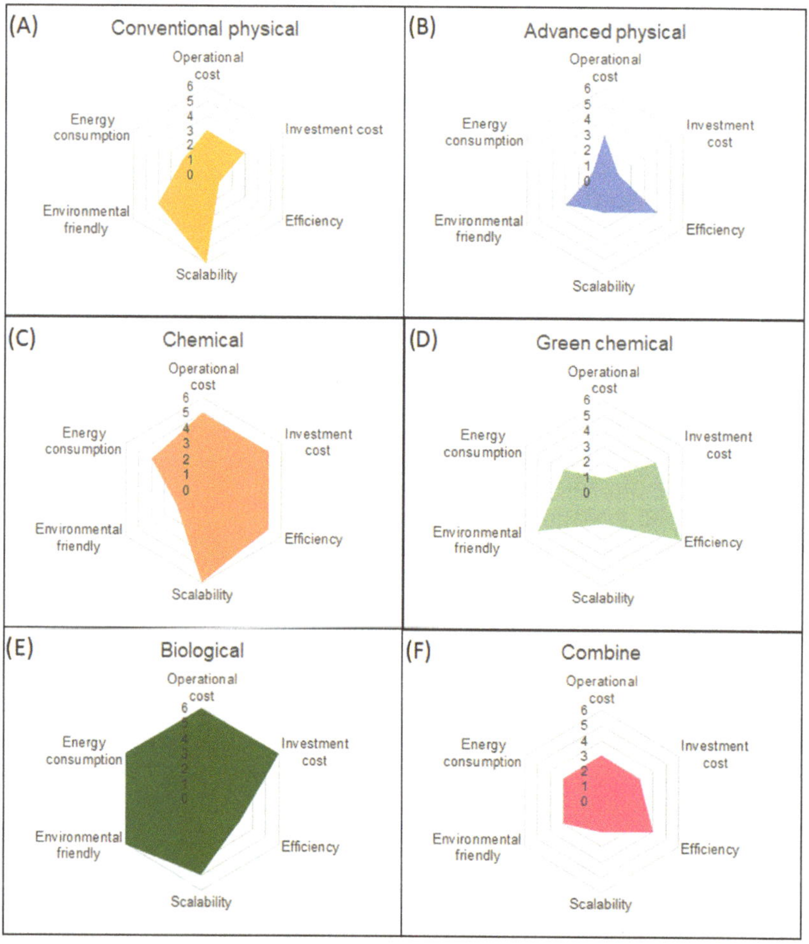

FIGURE 1.3 Characteristics of each pretreatment method (the scoring is given from 1 to 6 to represent from the worst to the best).

friendliness, as it typically does not involve the use of chemicals. Additionally, it can effectively reduce the crystallinity of cellulose, which is beneficial for subsequent enzymatic hydrolysis. However, physical pretreatment methods can be highly energy-intensive and costly, especially when scaled up for industrial applications. Moreover, these methods often do not sufficiently disrupt the lignin structure, necessitating further pretreatment steps to achieve high sugar yields (Mankar et al., 2021; Zhou & Tian, 2022).

Chemical pretreatment methods, such as acid hydrolysis, alkaline hydrolysis, organosolv, and ionic liquids, and deep eutectic solvent (DES), involve the application of chemicals to break down structures of lignin and hemicellulose. Acid hydrolysis uses dilute or concentrated acids to hydrolyze hemicellulose and disrupt lignin, while alkaline pretreatment employs bases like sodium hydroxide to solubilize lignin and hemicellulose. Organosolv pretreatment uses organic solvents to dissolve lignin, and ionic liquids disrupt the hydrogen bonding in cellulose, increasing its digestibility. The primary advantage of chemical pretreatment is its high efficiency in removing lignin and hemicellulose, leading to improved yields of fermentable sugars. Furthermore, the influence of chemical pretreatment is demonstrated to alter the crystallinity and property of cellulose fibrils to be more vulnerable to enzymatic hydrolysis. Additionally, these methods can be tailored to suit various types of biomass. However, the downsides include the high cost of chemicals, potential environmental hazards, and the need for neutralization and recovery processes. The use of harsh chemicals can also lead to the formation of inhibitory compounds that can negatively impact subsequent fermentation processes (Rabemanolontsoa & Saka, 2016; Singh et al., 2018).

Physico-chemical pretreatment, such as steam explosion and ammonia fiber explosion (AFEX), combines physical and chemical methods to enhance biomass digestibility. Steam explosion involves treating biomass with high-pressure steam followed by a sudden decompression, which breaks down the hemicellulose and disrupts the lignin structure. AFEX uses ammonia at high temperatures and pressures to swell the biomass fibers and decrystallize cellulose. These methods are effective in improving enzymatic hydrolysis efficiency and are less damaging to the environment compared with purely chemical methods. However, they require significant energy input and infrastructure investment, and the recovery of chemicals like ammonia can be challenging and costly (Jamaldheen et al., 2022; Kim, 2018). Biological pretreatment uses microorganisms and enzymes to degrade lignin and hemicellulose, making cellulose more accessible for hydrolysis. Fungi such as white-rot and brown-rot fungi are commonly used due to their lignin-degrading capabilities. The main advantage of biological pretreatment is its environmental friendliness, as it operates under mild conditions and does not produce toxic by-products. It also requires lower-energy input compared with physical and chemical methods. However, biological pretreatment is generally slower and less effective in disrupting the lignin structure compared with other methods. Additionally, maintaining optimal conditions for microbial activity can be challenging and requires careful control (Sindhu et al., 2016; Wu et al., 2022).

The choice of pretreatment method for lignocellulose depends on various factors, including the type of biomass, desired sugar yield, formation of inhibitory by-products, environmental impact, energy consumption, scalability, and economic considerations (investment cost and operational cost) (Fig. 1.3). Each method has its unique advantages and disadvantages, and often, a combination of pretreatment techniques is employed to achieve optimal results (Table 1.2). Different types of LSB, such as agricultural residues, forestry wastes, and energy crops, have varying compositions of cellulose, hemicellulose, and lignin (Table 1.1). The choice of pretreatment method must be tailored to the specific characteristics of the biomass being processed. For instance, biomass with high lignin content, like hardwoods, may require more aggressive chemical or physico-chemical pretreatments to effectively disrupt the lignin structure (Ko et al., 2015; Kundu et al., 2021), whereas agricultural residues with lower lignin content might be suitably pretreated using less intensive methods (Paudel et al., 2017). The primary goal of pretreatment is to maximize the yield of fermentable sugars from cellulose and hemicellulose (Table 1.2). The efficiency of different pretreatment methods in achieving this goal can vary significantly. Chemical methods, such as acid and alkaline treatments, are often very effective in breaking down hemicellulose and exposing cellulose, resulting in high sugar yields. Using the different pretreatment methods on the same biomass differently affects the yield of sugar recover (Bondesson et al., 2013; Fernández-Rodríguez et al., 2022). Likewise using the same pretreatment method and the same biomass substrates may also result in different amounts of sugars (Table 1.2). Therefore, optimization of the pretreatment method is one of the keys to maximize the sugar yields. Highly efficient methods are capable of significantly enhancing enzymatic hydrolysis rates, reducing processing times, and increasing overall productivity. For example, steam explosion is known for its ability to efficiently breakdown biomass structures, making it a popular choice despite its high-energy requirements. However, the choice of method must balance the need for high sugar yield with other factors like cost, minimal inhibitory by-products, and environmental impact (Hoang et al., 2023). During pretreatment, the breakdown of lignocellulosic components can lead to the formation of inhibitory compounds such as furfurals, hydroxymethylfurfural, phenolics, and acetic acid (Jönsson & Martín, 2016). These by-products can inhibit microbial fermentation and enzyme activities and reduce the overall efficiency of the biofuel production process. Methods that minimize the formation of such inhibitors or include steps to detoxify the biomass are preferred. For instance, mild acid pretreatment or advanced techniques like ionic liquid and DES pretreatment are designed to reduce the production of inhibitory by-products (Gundupalli et al., 2024; Khan et al., 2024).

TABLE 1.2 Influence on pretreatment on major LSB for sugar productions.

Pretreatment method	Optimal condition	Yield	References
Rice straw			
Combined method: Metal catalyst and UV radiation	.1% (w/v) TiO_2 and an UV-C lamp irradiation at 254 nm for time of 2 h	.203 g/g	Kang and Kim (2012)
Combined method: Steam explosion and alkaline	Steam explosion at 170°C for 10 min, and 2% (w/v) KOH at 121°C for 30 min	.664 g/g	Banoth et al. (2017)
Physical method: Popping pretreatment	15% w/v substrate loading, 1.96 MPa, heating rate of 20°C/min and rapidly open the hatch at 220°C	.567 g/g	Wi et al. (2013)
Chemical method: Soda and sulfate cooking	Pretreatment time of 2 h, 100°C, solid/liquid ratio of 1:10, alkali dosage of 20% of substrat	.489 g/g	Dien et al. (2015)
Chemical method: Alkaline hydrogen peroxide	10% w/w alkaline H_2O_2, 55°C, 2 h	.116 g/g	Deng et al. (2019)
Chemical method: Deep eutectic solvent	Choline chloride and acetic acid (ChCl:AA at ratio of 1:3.59), 126°C and 150 min	.752 g/g	Maibam and Goyal (2022)
Combined method: Deep eutectic solvent, microwave irradiation, and laccase	Choline chloride and formic acid (ChCl-FA), biomass loading 5.51% w/w, temperature 145°C, 8.75 min,	.426 g/g	Sawhney et al. (2023)
Chemical method: Organosolv and diluted acid	75% v/v ethanol with 1% w/w sulfuric acid, 150°C, 60 min	.388 g/g	Amiri et al. (2014)
Biological method: Enzymatic pretreatment	Laccase enzyme 39.40 U/g, 43.70°C, 11.19 h	.193 g/g	Terasawat and Phoolphundh (2021)
Wheat straw			
Chemical method: Ionic liquid	1-ethyl-3-methylimidazolium diethyl phosphate ([EMIM]DEP), 130°C, 30 min	.560 g/g	Li et al. (2009)

(Continued)

TABLE 1.2 (Continued)

Pretreatment method	Optimal condition	Yield	References
Biological method: Ensiling and fungal	Ensiling with 8 mg/g of LACTISIL CCM (*L. buchneri*), 4 weeks. Fungal pretreatment with *Ceriporiopsis subvermispora* CBS 347.63, 30°C, 40 days, 90% humidity	.234 g/g	Thomsen et al. (2016)
Combined method: Ultrasound-assisted ionic liquid	Triethylammonium sulfate ([TEA][HSO_4]), 130°C, biomass loading rate of 20%, with ultrasound treatment at 100 W, 24 kHz, for 30 min	.642 g/g	Ziaei-Rad et al. (2023)
Chemical method: Potassium hydroxide	40% KOH loading, 20°C, 24 h	.727 g/g	Liu et al. (2015)
Combined method: Acetic acid/steam explosion/ supercritical CO_2	50% v/v acetic acid, 50°C, 18 h. Steam explosion at 180°C, 10 min and then pretreated by SC-CO_2 at 180 bar, 180°C, 45 min	.280 g/g	Zabihi et al. (2021)
Chemical method: alkaline peroxide	2.15% w/w H_2O_2, pH 11.5, 35°C, 24 h	.672 g/g	Saha and Cotta (2006)
Chemical method: Ozonation	5.3% w/w of ozone, flow rate of 2 L/min, 120 min	.631 g/g	Bule et al. (2013)
Combined method: Sequential autohydrolysis and alkaline postextraction	Autohydrolysis at 140°C, followed by 6% w/w alkaline postextraction	.495 g/g	Wu et al. (2018)
Chemical method: Aqueous ionic liquid	49.5% w/w of 1-ethyl-3-methylimidazolium acetate, 158°C, 3.6 h	.403 g/g	Fu and Mazza (2011)
Sugarcane bagasse			
Chemical method: Ionic liquid	1-ethyl-3-methylimidazolium acetate ([EMIM]Ac), 120°C, 15% solid loading ratio, 1 h	.564 g/g	Saha et al. (2018)

(Continued)

TABLE 1.2 (Continued)

Pretreatment method	Optimal condition	Yield	References
Physical method: Hydrodynamic cavitation	Pressure drop in the fluid flow in an orifice plate, using 3.5 bar as upstream pressure, 7.5 min,	.460 g/g	Hilares et al. (2019)
Combined method: Vacuum-assisted alkaline pretreatment	2.64% NaOH, 3.34 h, solid loading of 10% w/v, 121°C, 1 h under the vacuum condition	.521 g/g	Lv et al. (2017)
Chemical method: Deep eutectic solvent	Choline chloride and malonic acid (ChCl: MA), 5% solid loading, 12 h, 80°C	.661 g/g	Chourasia et al. (2021)
Chemical method: Organosolv and metal-salt	Tween 80 and $AlCl_3$, 160°C, 10 min, 300 rpm	.362 g/g	Zhang, Chen, et al. (2021)
Physical method: Subcritical CO	5 Mpa of CO_2, 100 min, 160°C	.530 g/g	Zhang and Wu (2013)
Chemical method: Acid-catalyzed organosolv	Ethylene glycol/H_2O-HCl, 130°C, 60 min, 500 rpm	.381 g/g	Wei et al. (2021)
Chemical method: Deep eutectic solvent	Choline chloride/PEG400/ hydroxyethyl sulfonic acid 2.1 h, 150°C	.250 g/g	Huang et al. (2022)
Combined method: Ensiling and hydrothermal	Ensiling with 8 mg/g of LACTISIL CCM (*L. buchneri*), and hydrothermal treatment at 190°C, 10 min	.300 g/g	Ambye-Jensen et al. (2018)

Economic viability and technology level readiness are major considerations in selecting a pretreatment method (Fig. 1.4). This includes the cost of equipment, land investment, utility, chemicals, logistics, energy, and labor, as well as the costs associated with downstream processing and waste management. Methods like biological pretreatment are often less costly in terms of chemical inputs but may require longer processing times and precise control of environmental conditions, which can add to the overall cost. Balancing the upfront and operational costs with the expected benefits is essential for commercial feasibility (Cristóbal et al., 2018). The environmental footprint of

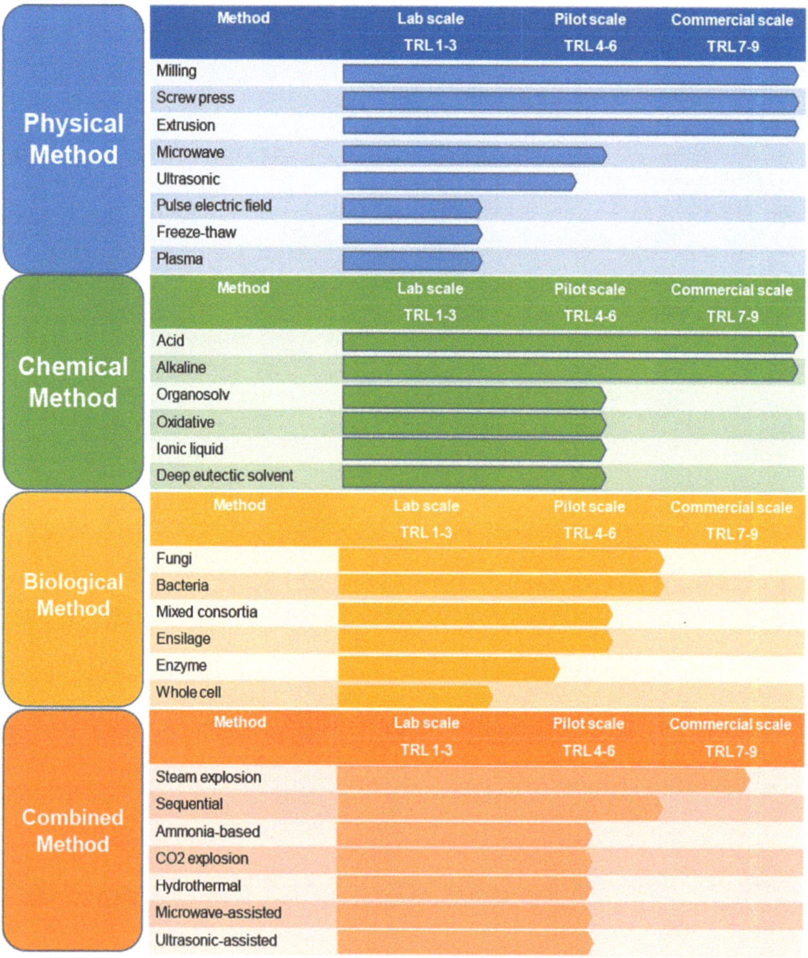

FIGURE 1.4 Pretreatment methods for lignocellulose biorefinery and technology readiness level (TRL).

the pretreatment process is increasingly important in the context of sustainable bioprocessing. Methods that use harsh chemicals, such as concentrated acids or alkalies, can generate hazardous waste and require extensive neutralization and disposal procedures. In contrast, methods like biological pretreatment and certain physical techniques are more environmentally benign, operating under milder conditions and producing fewer toxic by-products. The selection of pretreatment methods must align with environmental regulations and sustainability goals. Also, the scalability of a pretreatment method is crucial for its application in industrial settings. Methods that are effective at the laboratory scale may face challenges when scaled up, including increased energy demands, difficulties in handling large volumes of biomass, and the need for robust and scalable equipment (Igbokwe et al., 2022; Modenbach & Nokes, 2012). Methods like steam explosion and certain chemical treatments have been successfully scaled up, while others, such as some biological pretreatments, may require further development to become commercially viable on a large scale.

1.4 Factors influencing feedstock selection

Selecting the appropriate feedstock for methane and hydrogen production is critical to optimizing the efficiency and sustainability of the process. Several factors influence this selection, including feedstock availability, composition and biodegradability, energy potential, pretreatment requirements, economic considerations, and environmental impact (Aditiya et al., 2016; Raj et al., 2022), which could be categorized to be intrinsic and extrinsic factors. Intrinsic factors are related to the properties of lignocellulose biomass, including compositions, biodegradability, nutrition balance, inhibitors, requirement of pretreatment, and yields of methane and hydrogen. On the contrary, extrinsic factors include the cost of feedstock, logistics and storage, availability and consistency of supplies, competitive market demand, sustainability and regulatory and policy considerations. Intrinsic factors are properties of feedstocks that are not manipulated. Knowing these properties, based on biomass characterization, is necessary to decide for the selection of feedstocks. If the intrinsic properties of any feedstock are not feasible, that feedstocks should be omitted for commercialization. For extrinsic properties, these factors are more likely to be manipulated by interventions of stakeholders such as governmental support and policy. Understanding these factors helps in making informed decisions to achieve a high yield of biogas and ensure the viability of the biogas production facility.

1.4.1 Intrinsic factors

The chemical composition and biodegradability of the feedstock significantly affect biogas production efficiency. Feedstocks rich in carbohydrates, proteins, and lipids generally have higher biogas yields compared with those high in lignocellulosic content. Easily biodegradable materials like food waste and animal manure breakdown faster and more completely in anaerobic digesters, leading to higher biogas production (Shrestha et al., 2023). Conversely, feedstocks with high lignin content, such as woody biomass, are more resistant to microbial degradation and may require pretreatment to enhance their biogas potential (Kundu et al., 2021). High lignin content can hinder biodegradability, affecting yields of methane and hydrogen in biogas. Feedstocks with lower lignin content are preferable. On the contrary, cellulose and hemicellulose are more easily converted into biogas, so higher proportions are beneficial for biogas production. Based on this principle, most studies on 2 G biogas production usually select the types of biomass with high contents of cellulose and hemicellulose. Amon et al. (2007) demonstrated that crude protein, crude fat, cellulose, and hemicellulose have a positive trend in methane production (Amon et al., 2007); however, Gunaseelan et al. (2007) indicated that lignin and cellulose are negatively correlated to methane yield (Gunaseelan, 2007). Additionally, the correlations of methane yields to the composition of nine types of biomass were modeled based on the experimental set up in continuous

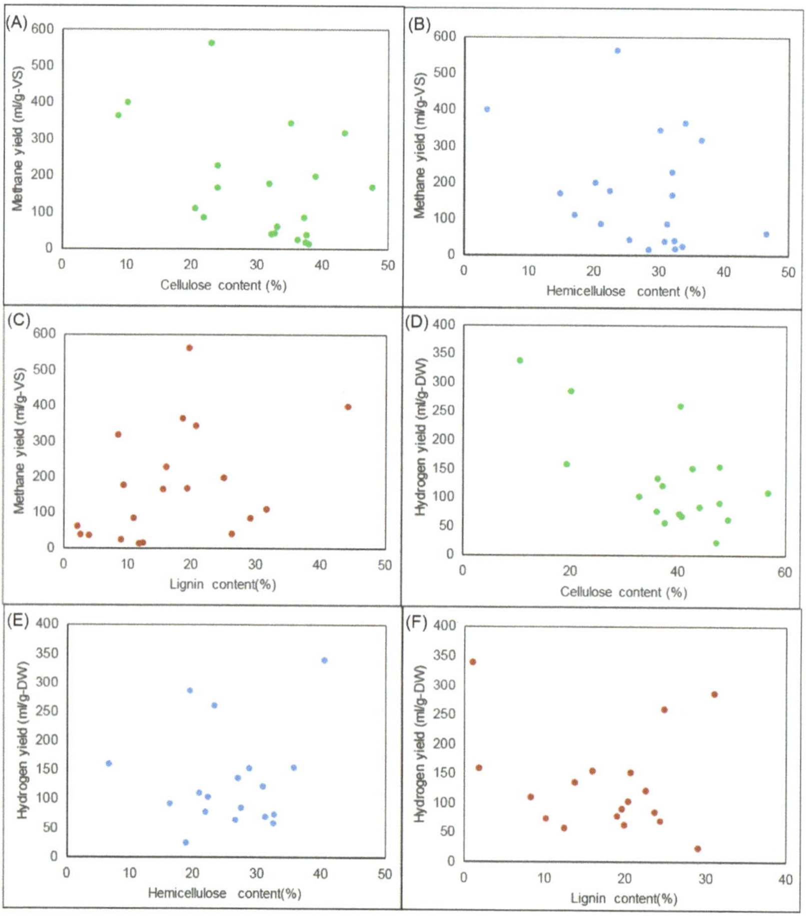

FIGURE 1.5 The correlations of cellulose, hemicellulose, and lignin contents to the yields of methane (A-C) and hydrogen (D-F) obtained from 50 records of publications in SCOPUS database.

anaerobic digestion by using a stepwise regression and found that cellulose affects methane yield when a low organic loading rate is used (Zhao et al., 2020). Therefore, it is presumptive that the correlations of cellulose and hemicellulose contents to methane and hydrogen yields are not in direct proportion and are influenced by other factors (Fig. 1.5). Likewise, some of the biomass with high lignin content produces high biogas yields (Fig. 1.5). These observations suggested that the biogas yields do not only rely on biomass composition but also depend on other chemical and physical properties of biomass, such as cellulose crystallinity, intrinsic inhibitor, and process operation.

The nutrient composition, including carbon, nitrogen, phosphorus, and trace elements, significantly impacts the efficiency and yield of biogas production as they are necessary to support the growth and activities of microbes in digesters. The C/N ratio is a pivotal factor in determining the success of anaerobic digestion. An optimal C/N ratio of around 20–30:1 is typically considered ideal for anaerobic microorganisms. If the C/N ratio is too high, meaning excessive carbon relative to nitrogen, the digestion process slows down due to nitrogen deficiency, limiting microbial growth and activity. Conversely, a low C/N ratio leads to ammonia and acidic accumulation, which can be toxic to the microbial community (Ma et al., 2020). Examples of such plant biomass with high C/N ratio include straw, corn stover, sawdust, wood chips, and rice husk. Straw from cereals like wheat, barley, and rice typically has a C/N ratio of around 50–100:1, making it rich in cellulose and lignin but low in nitrogen. Corn stover, which includes leaves, stalks, and cobs, has a C/N ratio ranging from 40 to 60:1, necessitating nitrogen supplementation for efficient biogas production. Sawdust and wood chips, with C/N ratios often exceeding 100–200:1, contain high lignin content and are deficient in nitrogen. Similarly, rice husk, a by-product of rice milling, has a C/N ratio of approximately 50–60:1 and presents additional complexity due to its silica content. These high C/N ratio materials are often combined with nitrogen-rich substrates, such as manure or food waste, to achieve the optimal nutrient balance for microbial activity, thereby enhancing biogas production. Oppositely, plant biomass with low C/N ratios includes leguminous plants such as clover, alfalfa, and soybean residues, which typically have C/N ratios around 10–20:1 due to their nitrogen-fixing abilities. Vegetable wastes such as lettuce, spinach, and cabbage residues usually have C:N ratios around 10–20:1, providing a good nitrogen source for anaerobic digestion. However, these types of plants with high N-contents are more used to enhance soil fertility than serving as substrates for biogas production (Sun et al., 2021; Zhou & Wen, 2019).

Besides carbon and nitrogen, trace elements such as iron, cobalt, nickel, and selenium are essential for the enzymatic functions of anaerobic microorganisms. These micronutrients act as cofactors for various enzymes involved in the metabolic pathways of biogas production. The availability of these trace elements in lignocellulosic biomass varies widely depending on the type and source of the material. Deficiencies in these elements can lead to suboptimal microbial activity and lower biogas yields. Therefore, supplementing lignocellulosic feedstocks with necessary micronutrients can enhance the performance of the anaerobic digestion process. Phosphorus is another vital nutrient for microbial growth and energy transfer processes as it is one of the main compositions in generic materials. Its availability in lignocellulosic biomass affects the overall microbial metabolism and biogas production efficiency. Some lignocellulosic materials may lack sufficient phosphorus, necessitating supplementation. Additionally, other macronutrients like potassium, magnesium, and sulfur play supportive roles in maintaining microbial health and activity, contributing to more efficient biogas production (Laiq Ur Rehman et al., 2019; Leca et al., 2023).

2Gbiorefinery processes require pretreatment to enhance their digestibility and biogas yield. LSB often need pretreatment to disrupt their rigid structure, making cellulose and hemicellulose more accessible to anaerobic microbes. While pretreatment can significantly improve biogas yields, it also adds to the overall cost and complexity of the biogas production process. The selection of feedstock thus needs to consider the balance between the benefits of higher biogas yield and the additional costs and operational requirements of pretreatment. Furthermore, the amount of biogas produced per unit of feedstock is crucial. Higher yields are more desirable. High-energy feedstocks like fats, oils, and certain energy crops produce more methane per unit of feedstock than lower-energy materials like straw or other fibrous plant residues. Energy crops are specifically cultivated for their high biomass yield and potential to be converted into renewable energy sources. Miscanthus, particularly Miscanthus x giganteus, is a perennial grass known for its substantial biomass yield and ability to thrive on marginal lands with minimal inputs, making it a sustainable choice for energy production (Mironova et al., 2023). Switchgrass (Panicum virgatum), another perennial grass native to North America, is highly adaptable to various soils and climates, and its high cellulose content makes it ideal for biofuel production (Parrish & Fike, 2009). Sorghum (Sorghum bicolor), especially sweet sorghum varieties, are cultivated for their high sugar content, suitable for biofuel production (Hu et al., 2022). Additionally, giant reed (Arundo donax), a tall perennial grass, is well-suited for bioenergy production in warmer climates and can be used for bioethanol, biogas, and solid biofuels (Corno et al., 2014). Evaluation of the methane potential helps in selecting feedstocks that maximize energy output, which is particularly important for the economic feasibility of biogas plants.

1.4.2 Extrinsic factors

The continuous and reliable availability of feedstock is a fundamental factor. Biogas production is often a continuous process requiring a steady supply of organic material in terms of quantity and quality. Local availability of feedstocks such as agricultural residues, animal manure, food waste, and energy crops is crucial to minimize transportation costs and logistical challenges. The density of lignocellulosic biomass varies significantly based on the type and physical state of the material. Loose biomass, such as straw and corn stover, typically has a bulk density ranging from 30 to 100 kg/m^3. For instance, loose straw generally ranges from 30 to 80 kg/m^3, and loose corn stover ranges from 50 to 100 kg/m^3. When biomass is baled, its density increases to about 100 to 200 kg/m^3 for straw and hay and 150 to 200 kg/m^3 for corn stover (Mukherjee et al., 2022; Wang et al., 2021). These variations in density are crucial for the logistics and processing of lignocellulosic biomass in bioenergy applications, influencing storage, transportation costs, and feedstock handling efficiency. Regions with abundant agricultural activity or food-processing industries provide a constant supply of residues and wastes that can be efficiently converted into biogas. Ensuring a consistent feedstock supply is essential to maintaining stable biogas production rates and economic viability. Furthermore, the seasonal variability is another important factor. For example, rice production in Southeast Asia is mostly cultivated twice per year such as planting often begins in March or April, and harvesting occurs around September to November (Ahmad et al., 2023). Sugarcane cultivation usually starts in November or December and its harvesting season is around March to May (Moneda et al., 2022). However, plant operation of biofuel production has to be done in a continuous mode for the whole year. Therefore, the storage infrastructure and operation to maintain the supply for the whole year is one of the main challenges. Nevertheless, biogas production in the form of methane and hydrogen has an advantage over bioethanol due to possible uses of mixed biomass, which is not practical for 2 G bioethanol plants. A recent study investigates versatility to use various biomass types for 2 G ethanol production by mixing each biomass in proportional together. The results showed that the variation in ethanol yield from fixed proportion mixtures of biomass showed an error of no more than 15%, suggesting the potential of mixed biomass in ethanol production (Cheenkachorn et al., 2024).

The cost of feedstock is a critical economic factor. Feedstocks that are readily available as waste products, such as animal manure, food-processing waste, or municipal solid waste, are often low-cost or even come with tipping fees, making them economically attractive. In contrast, energy crops grown specifically for biogas production can be costly, both in terms of land use and cultivation expenses. The cost of lignocellulosic biomass ranges from approximately \$30 to \$150 per dry ton in the United States, depending on the type and quality of the biomass, transportation costs, and other factors. On the contrary, the cost of rice straw and sugarcane bagasse in Thailand ranges from \$30 to \$60 per dry ton. However, if the purchasing is performed on a small scale, such as less than one dry ton, the price is higher and may be up to \$150–\$200 (excluding transportation costs) (Domingues et al., 2022). This scenario highlights the importance of scale of production in a selection of biomass for biofuel production too. However, it is essential to note that these prices can fluctuate over time due to changes in market conditions and government policies. Furthermore, agricultural residues like corn stover and wheat straw and energy crops, such as switchgrass, Nepier grass, and Miscanthus, are key feedstocks for husbandry. If demand for these feedstocks increases for biofuel production, competition with animal feed producers intensifies, potentially driving up prices.

The environmental sustainability of using a particular feedstock is another key consideration. Utilizing waste materials for biogas production can help in waste management, reduce greenhouse gas emissions, and recycle nutrients back into the soil as digestate. Conversely, dedicating large areas of land to grow energy crops for biogas can have negative environmental impacts, such as biodiversity loss, soil degradation, and water resource depletion. Therefore, selecting feedstocks with positive environmental profiles contributes to the overall sustainability of biogas production. Policy and regulation significantly influence the selection of specific lignocellulosic biomass for biofuel production. Renewable fuel standards (RFS) and mandates established by governments set targets for the inclusion of renewable fuels in transportation, driving demand for biofuels, such as ethanol mixed with gasoline or biodiesel mixed with diesel. In addition to RFS, governments often offer incentives and subsidies to stimulate the production and adoption of biofuels. These incentives can take various forms, including tax credits, grants, loan guarantees, or direct payments to biofuel producers (Lark et al., 2022). The availability of such financial support can significantly influence the economics of biofuel production, making certain types of lignocellulosic biomass more attractive for investment and utilization in biofuel facilities. In Brazil, sugarcane bagasse serves as a prime example of lignocellulosic biomass utilized for biofuel production driven by government policy and incentives. The abundance and accessibility of sugarcane as a major crop contribute significantly to the availability of bagasse. This abundant resource is readily accessible and economically viable for biofuel production. Brazil's well-established infrastructure for sugarcane cultivation, processing, and transportation further supports the utilization of sugarcane bagasse. The country's extensive network of sugar mills and ethanol plants provides the necessary infrastructure for processing bagasse into biofuels efficiently. Government policies, notably the Proálcool Program initiated in the 1970s and subsequent policies like RenovaBio, have incentivized biofuel production and provided regulatory frameworks to encourage the use of sugarcane bagasse and other biomass feedstocks. These policies have played a crucial role in driving investments and innovation in the biofuel sector, making sugarcane bagasse a key component of Brazil's renewable energy strategy (Gonçalves et al., 2005).

In Europe, particularly in Denmark, the agricultural sector, which is notably focused on wheat cultivation, generates substantial quantities of straw residues. Straw pellets have emerged as a locally sourced feedstock for biofuel production. Denmark's commitment to renewable energy targets, underscored by policies like the Energy Agreement and the Climate Act, emphasizes the role of biomass in transitioning toward a low-carbon economy. This aligns with the suitability of straw pellets as a renewable and carbon-neutral fuel source. Furthermore, Denmark's well-established district heating infrastructure provides a fitting application for straw pellets, offering an efficient and sustainable alternative to fossil fuels in heating systems. The decentralized production of straw pellets, often utilizing local agricultural residues, minimizes transportation emissions and supports local economies (Elsgaard et al., 2013). Technological advancements in pellet production ensure consistent quality and efficiency, enhancing the viability of straw pellets for use in modern biomass boilers and combined heat and power plants. Moreover, environmental regulations may also play a role in the selection of lignocellulosic biomass for biofuel production. Sustainability criteria, such as greenhouse gas emissions reduction targets, requirements for land use practices, or

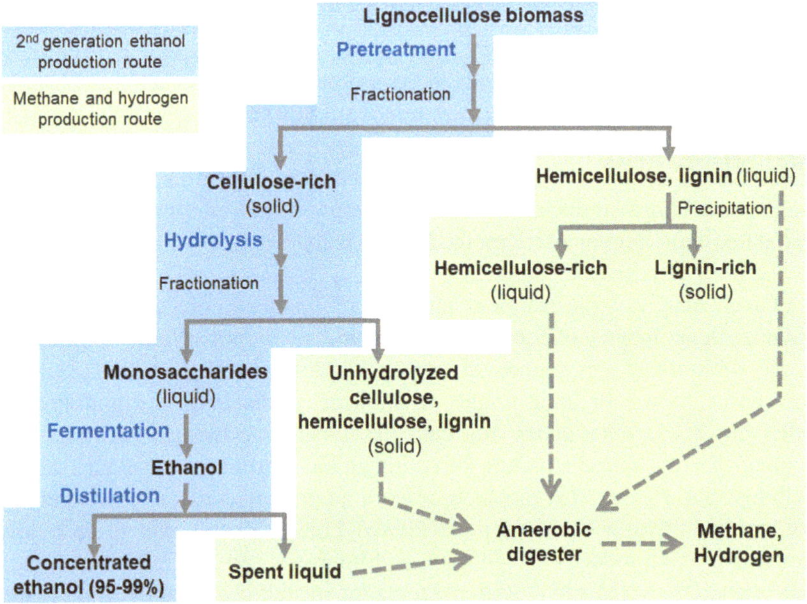

FIGURE 1.6 Production process of second generation ethanol from lignocellulose and integration with anaerobic digester.

carbon credit can guide the choice of feedstocks that minimize environmental impacts and contribute to overall sustainability goals.

Altogether, the selection of feedstock for biogas production involves a complex interplay of various factors. Availability and local sourcing are crucial for logistical and economic reasons. The composition and biodegradability of the feedstock directly affect biogas yield and process efficiency. Energy potential helps prioritize high-yielding materials, while pretreatment requirements and associated costs need careful consideration. Economic factors, including feedstock cost and transportation logistics, alongside environmental impacts, play significant roles in determining the most suitable feedstock for sustainable and efficient biogas production for example the integration of biogas production into second generation of ethanol production (Fig. 1.6).

1.5 Advanced combining pretreatment methods

Conventional methods typically involve basic physical or chemical treatments such as milling, grinding, soaking, or acid hydrolysis, which may not fully breakdown the lignocellulosic structure and often require harsh conditions or extended reaction times. Conventional methods may have limited effectiveness and require additional steps or higher enzyme dosages for efficient conversion. In contrast, advanced pretreatment methods utilize more sophisticated techniques like steam explosion, AFEX, organosolv, alkaline treatment, or deep eutectic solvents (DES). These advanced approaches are designed to efficiently disrupt the lignocellulosic structure, remove lignin, and increase the accessibility of cellulose and hemicellulose for subsequent enzymatic hydrolysis or microbial digestion. Furthermore, advanced methods often operate under milder conditions, reducing energy consumption and environmental impact compared with conventional methods, which may require high-energy inputs or the use of hazardous chemicals.

Green chemical pretreatment methods of lignocellulose are designed to enhance the efficiency of converting this biomass into biofuels and bioproducts while minimizing environmental impact (Saratale et al., 2023). One such method is the use of ionic liquids (ILs), which are salts in a liquid state at relatively low temperatures. ILs can dissolve lignocellulosic biomass, disrupting the crystalline structure of cellulose and partially hydrolyzing hemicellulose and lignin. This results in a more amorphous structure that is easier for enzymes to digest. ILs are considered green solvents due to their low volatility, which reduces air pollution and improves safety. Furthermore, they can often be recycled and reused, reducing the overall chemical footprint of the process (Usmani et al., 2020). Another green pretreatment approach involves the use of DESs, which are mixtures of two or more components that form a eutectic mixture with a melting point lower than that of each component. DESs can effectively solubilize lignin and hemicellulose, thus enhancing the accessibility of cellulose. They are derived from natural and renewable materials, such as choline chloride and urea, which makes them particularly attractive from a sustainability perspective. Their biodegradability and low toxicity further underscore their green credentials (Xu et al., 2020). Organosolv pretreatment is another method, which uses organic solvents, often combined with water and catalysts, to fractionate lignocellulosic biomass into its main components: cellulose, hemicellulose, and lignin. Common solvents include ethanol, acetone, and methanol. Organosolv processes can achieve high degrees of delignification, which enhances the subsequent enzymatic hydrolysis of cellulose. The solvents used are often recoverable and reusable, minimizing waste and reducing environmental impact. Additionally, organosolv processes can produce high-purity lignin as a valuable by-product (Ferreira & Taherzadeh, 2020).

To improve the efficiency of green chemical pretreatment, combinations with other pretreatment methods have been applied and demonstrated, with physical or mechanical pretreatment being usually selected. Zhu et al. (2024) examined the synergistic effects of spiral jet milling and DES pretreatment, using choline chloride and lactic acid (ChCl/LA)as well as potassium carbonate and ethylene glycol (K_2CO_3/EG), on grapevine to improve enzymatic hydrolysis. The optimal pretreatment condition of this combined method was identified to be at 110°C, 4 h, and 20% water content, which resulted in a threefold increase in glucose yield compared with the untreated sample. Furthermore, lignin products obtained from this experiment express antioxidant and antibacterial properties. It is demonstrated here that despite its higher power requirement, JM lower-energy consumption and shorter processing time contribute to its overall cost-effectiveness making it becomes economically viable (Zhu et al., 2024). Canadas et al. (2023) develop the combined pretreatment methods by using DES with a steam explosion, or microwave-assisted, or orbital shaking extraction. The optimal condition was conducted by the combinations of steam explosion, microwave-assisted and DES treatment, in which a maximum yield of sugar at 49.8 g/L was obtained (19.37% increase compared with the untreated sample). Additionally, several phenolic bioactive compounds with antioxidant activities were obtained as by-products (Cañadas et al., 2023). A combination of DES (choline chloride, ethylene glycol, and FeCl3·6H2O) and hydrothermal pretreatment was also demonstrated in several studies. Li et al. (2024) set up hydrothermal pretreatment with seawater and followed by microwave-assisted DES (MD) pretreatment to fractionate value-added products from Pinus massoniana. Several ions in seawater function as Lewis acids to enhance the depolymerization of oligosaccharides, and up to 85% of glucose was obtained after enzymatic saccharification (an eightfold increase compared with untreated sample) (Li et al., 2024). The combined deep eutectic solvent (DES) and hydrothermal pretreatment enhance enzymatic hydrolysis and fractionation of lignocellulose biomass by targeting the molecular arrangement and chemical bonding within the biomass. In the

DES pretreatment phase, the solvent, typically composed of hydrogen bond donors and acceptors, disrupts the robust hydrogen-bonding network between lignin, hemicellulose, and cellulose. This disruption weakens the lignin structure and partially solubilizes hemicellulose, thereby exposing the cellulose fibers. When the biomass subsequently undergoes hydrothermal pretreatment, the high temperature and pressure conditions further hydrolyze hemicellulose into soluble sugars and promote additional lignin removal. This dual action results in significant alterations to the biomass's molecular structure: Cellulose becomes more accessible as the protective lignin and hemicellulose barriers are removed. The increased surface area and porosity of the pretreated biomass facilitate more efficient enzymatic access and action. The enhanced enzymatic hydrolysis results from the reduced steric hindrance and minimized enzyme inhibition due to the lower lignin content. Moreover, the combination of these pretreatments improves the fractionation process, enabling better separation of cellulose, hemicellulose, and lignin for downstream applications (del Mar Contreras-Gámez et al., 2023).

Various types of ILs have been selected to combine with physical pretreatment methods for example microwave and ultrasonic pretreatment. Zhang et al. (2023) applied combined ultrasonic pretreatment and ionic liquid (1-butyl-3-methylimidazolium hydrogen sulfate ([Bmim]HSO_4)) to pretreat giant reeds (Arundo donax L.) for biohydrogen production via photofermentation. The maximum lignin removal was achieved at 23% when the pretreatment was done under 60°C for 1.5 h, which resulted in an enhancement of hydrogen production of 46.4% (Zhang et al., 2023). Another type of ionic liquid, tetrabutylammonium hydroxide ([TBA][OH]), was combined with ultrasonic treatment to pretreat eucalyptus. The reducing sugar yield was improved to 4.5 times to .43 g/g. The improvement of sugar yield after pretreatment was explained by the effect of ultrasonic to reduce crystallinity index of cellulose fibrils and ionic liquid functions in the intervention of hydrogen binding between each cellulose fibril and disruption of lignin molecule (Wang et al., 2018). Similarly, 1-butyl-3-methylimidazolium chloride ([Bmim]Cl) and surfactant (PEG-8000) were used in combination with ultrasonic pretreatment on sugarcane bagasse to enhance the enzymatic hydrolysis efficiency of ionic liquid-tolerant Aspergillus assiutensis VS34, a cellulase-producing fungal strain. Under optimal conditions of combined pretreatment, the highest sugar yield was obtained at .25 g/g biomass, which is a 1.44 fold increase compared with the untreated sample. The results show that the combined pretreatment with the surfactant PEG-8000 and [Bmim]Cl significantly solubilized and removed lignin and acetyl groups, enhancing the effectiveness of A. assiutensis VS34 cellulases (Sharma et al., 2019). Furthermore, other physical pretreatment methods, such as ultra-high-pressure pretreatment (Gao et al., 2018) microwave pretreatment (Wang, Cai, et al., 2024) and hydrothermal pretreatment (Wei et al., 2022), were combined with ionic liquid pretreatment; however, most of these combined methods are still in laboratory-scale demonstration.

In addition to the usage of green chemicals for advanced pretreatment, several mechanical techniques have been developed to combine with other pretreatments to improve their efficiency. Microwave-assisted hydrothermal treatment (MHT) was developed for Paulownia wood to fractionate biomass composition. At the optimal condition, an 80% recovery of xylan as xylooligosaccharides and a 95% recovery of cellulose in the pretreated Paulownia wood biomass demonstrate the selectivity of the treatment. This method was compared with conduction–convection heating autohydrolysis. Additionally, experimental data from the MHT were analyzed using a kinetic model using pseudo-homogeneous reactions based on the Arrhenius equation parameters. Both proposed models fit the experimental data well and can predict the behavior of MHT, offering valuable insights for scaling up the process and advancing its industrial development (Río et al., 2021). Nitrogen-protected microwave-assisted extraction (NPMAE) in combination with soda and acetic acid was developed to extract cellulose from Acer platanoides. At the optimal condition (100 W microwave power, 2.45 GHz frequency, and 120°C for 15 min), 70% of cellulose was recovered from biomass due to the removal of hemicellulose and lignin (Rizwan et al., 2021). Hydrodynamic cavitation (HC)-based pretreatment and oxidation (hydroxyl radicals) created by HC at room conditions was developed in combination with sonochemical effect. This effect is primarily driven by acoustic cavitation, a process where ultrasonic waves create alternating high-pressure and low-pressure cycles in a liquid, leading to the formation, growth, and implosive collapse of gas bubbles. The collapse of these bubbles generates extremely high localized temperatures (up to 5000 K) and pressures (up to 1000 atm), albeit for very short durations and in small regions. These extreme conditions result in the homolytic cleavage of chemical bonds, producing highly reactive radicals such as hydroxyl radicals and hydrogen atoms, which can initiate various chemical reactions, including oxidation, reduction, and polymerization. Additionally, the collapse produces intense localized shock waves and microjets, causing physical changes like erosion of solid surfaces, fragmentation of particles, and enhanced mass transfer (Sun et al., 2022). Electrothermal pretreatment by Ohmic heating was applied on rambutan substrate for cellulose recovery for food applications, like cellulose gels. At a solid-to-liquid ratio of 10% in a water/ethanol (1:1, v/v) mixture as the electrical transmission medium at 60°C for different holding times (15 and 30 min), then the method was combined with sodium chlorite (NaClO2) solution; up to 25% of cellulose was recovered (Torgbo et al., 2022). Induced electric field-assisted alkali pretreatment (IEF-AAP) and hydrothermal-assisted alkali pretreatment (HT-AAP) were applied to wheat straw to enhance enzymatic saccharification. At the optimal pretreatment condition of IEF-AAP in combination with diluted NaOH pretreatment, 78.9 % of lignin and 70.0 of hemicellulose were removed resulting in 79.7% of sugar recovery. The electrical effect promotes the ionization of OH− ions in the alkali solution, enhancing the breakdown of hydrogen and ester bonds to remove lignin and hemicellulose. Simultaneously, the hydrated hydrogen ions (H3O+) produced through electrolysis degrade a portion of the hemicellulose. Moreover, electroporation induced by the electrical effects increases the permeability of the lignocellulosic surface, accelerating the penetration of charged ions (Zhang et al., 2024a). On the contrary, induced electric field-assisted diluted acid pretreatment (IEF-DAP) was tested in wheat straw. Using low-temperature condition (90°C) of IEF-DAP, 73.6% of hemicellulose was removed and 55.8% of enzymatic efficiency was obtained. The effect of IEF pretreatment is likely caused by the magneto-induced electric field promoting the ionization of H+ ions and H3O+ ions, thereby increasing the medium's acidity. Additionally, electroporation disrupts the structure of wheat straw, enhancing the accessibility of cellulose to cellulases (Zhang et al., 2024b). This suggests that IEF is a green and efficient strategy for facilitating the separation of hemicellulose from lignocellulose. These emerging pretreatment technologies have been developed to establish more greener to environment by reduction in toxic or hazardous chemicals, or saving electricity consumption. However, they are in the early stage of demonstration (i.e., TRL1–4); therefore, several aspects should be continued for example, scaling up, economical feasibility, and safety.

1.6 Challenges and future perspectives

The production of methane and hydrogen from lignocellulose presents multifaceted challenges alongside promising future perspectives. Initially, lignocellulose biomass poses inherent complexities due to its heterogeneous composition, containing cellulose, hemicellulose, and lignin. Effective pretreatment methods are crucial to break down this complex matrix, enhancing accessibility to fermentable sugars or facilitating direct conversion to methane and hydrogen. However, existing pretreatment techniques often involve high-energy inputs and the use of harsh chemicals, raising environmental concerns and escalating production costs. Additionally, achieving optimal pretreatment conditions to balance lignin removal, hemicellulose hydrolysis, and cellulose accessibility remains a challenge. Moreover, the scalability of pretreatment processes from laboratory to industrial scale is often hindered by engineering constraints, such as reactor design and heat transfer considerations. Transitioning from laboratory-scale experiments to industrial-scale operations requires meticulous consideration of reactor design parameters, including size, configuration, and material selection. Ensuring efficient mixing, heat transfer, and residence time while maintaining optimal pretreatment conditions becomes increasingly complex at larger scales. Furthermore, the energy consumption of pretreatment methods, such as steam explosion or dilute acid hydrolysis, presents a significant hurdle. High-energy inputs may not be sustainable or economically viable on a larger scale, necessitating innovative engineering solutions to improve energy efficiency and reduce resource consumption. The concern about the need for energy input to the process becomes more important when there is a geopolitical problem during 2022–24, in which the cost of energy rises about 2 times worldwide. Therefore, if the gaining of output energy is not profitable, the technology is not economically viable.

Moreover, the use of harsh chemicals or solvents in certain pretreatment methods introduces environmental concerns and regulatory challenges. Scaling up processes reliant on these chemicals requires stringent measures for handling, disposal, and environmental impact mitigation. Balancing the need for effective pretreatment with sustainable and environmentally friendly practices becomes increasingly intricate at larger scales. Additionally, ensuring consistent product quality and process reliability throughout scaling up presents further obstacles. Addressing these challenges demands interdisciplinary collaboration between researchers, engineers, and industry stakeholders to develop scalable and environmentally responsible lignocellulose pretreatment processes for biomethane and biohydrogen production.

Looking ahead, future perspectives for lignocellulose pretreatment in biomethane and biohydrogen production hold promise with ongoing research and technological advancements. Innovations in pretreatment technologies, such as novel solvent systems, ILs, and advanced enzymatic methods, offer opportunities to improve efficiency while minimizing environmental impacts [164–166]. Nevertheless, most of the research on formulation and testing the performance of green chemicals or special hydrolytic enzyme cocktails are currently demonstrated in laboratory-scale to pilot-scale stages, which results in lacking comprehensive feasibility studies. Without robust information on process efficiency, yield, scalability, and production costs, it becomes difficult to evaluate the feasibility of large-scale implementation. The more precise and accurate modeling of chemical costs and electricity consumption of laboratory scale and commercial scale needs to be concerned about the scale of the production regime. For example, an ionic liquid EMIM-AC with laboratory grade (50 mL, 97% purity) supplied by Sigma-Aldrich costs ~\$880 (or \$17,600/L), while 95% purity EMIM-AC costs ~\$1450/kg (BLDpharm, USA). Similarly, even just conventional pretreatment chemicals, like NaOH, for laboratory-grade use, it costs \$192/kg (97% purity supplied by Sigma-Aldrich). When the purchasing scale is more than 100 kg, the price of bulk NaOH pellets can be reduced to less than \$.5/kg. Furthermore, feasibility studies often overlook critical aspects of the biomass supply chain, such as feedstock sourcing, logistics, and preprocessing, leading to incomplete assessments of overall feasibility. For example, in most developing countries, agricultural fields are mostly distributed as small- to medium-sized farms, whereas in developing countries, more large-scale farming is available. Therefore, the biomass collection and logistic costs to processing facility/factory are added up for small- to medium-sized farms. Also, the cultivation practices, like monoculture and mixed culture, are another factor. In most developed countries, monoculture is more practical management, while in developing countries, mixing of plant biomass in the same field is more likely to be common. Therefore, if a facility is designed to be strict on one particular type of biomass, there will be a possibility of not having sufficient biomass to feed the process continuously. Addressing these gaps in feasibility studies is essential for driving progress in lignocellulosic biomass conversion technologies and realizing their potential for sustainable methane and hydrogen production. Additionally, exploring the synergistic effects of combined pretreatment methods and integrating with emerging conversion technologies like gasification and pyrolysis could unlock new pathways for lignocellulosic biomass valorization, paving the way for sustainable and efficient biomethane and biohydrogen production.

1.7 Conclusion

Lignocellulosic biomass emerges as a promising feedstock for biomethane and biohydrogen production, given its abundance, renewability, and minimal competition with food crops. To optimize methane and hydrogen yields, selection criteria for pretreatment methods must carefully weigh factors such as feedstock composition, efficiency, environmental impact, and economic feasibility. Advanced combining pretreatment methods, which integrate thermochemical, chemical, and biological approaches, offer significant potential for enhancing biomass conversion efficiency and reducing processing costs. Nevertheless, challenges like reactor design constraints, energy consumption, chemical usage, and environmental impact necessitate concerted efforts to develop innovative, sustainable pretreatment technologies. Collaboration between researchers, engineers, and industry stakeholders is pivotal in overcoming these hurdles and realizing the full potential of lignocellulosic biomass in driving the transition toward a greener, more renewable energy future centered around biomethane and biohydrogen production within integrated biorefinery frameworks.

Acknowledgment

The authors would like to thank National Science, Research and Innovation Fund, and King Mongkut's University of Technology North Bangkok (Grant Contract No. KMUTNB-FF-68-B-01) for financial support during this work.

AI disclosure

During the preparation of this work, the author(s) used ChatGPT in order to for grammar correction and polishing writing. After using this tool/service, the author(s) reviewed and edited the content as needed and take(s) full responsibility for the content of the publication.

References

Aditiya, H. B., Chong, W. T., Mahlia, T. M., Sebayang, A. H., Berawi, M. A., & Nur, H. (2016). Second generation bioethanol potential from selected Malaysia's biodiversity biomasses: A review. *Waste Management (New York, N.Y.), 47*(Pt A), 46–61. https://doi.org/10.1016/j.wasman.2015.07.031 Epub 2015 Aug 5. PMID: 26253329.

Adney, W. S., Rivard, C. J., Ming, S. A., & Himmel, M. E. (1991). Anaerobic digestion of lignocellulosic biomass and wastes: Cellulases and related enzymes. *Applied Biochemistry and Biotechnology, 30*(2), 165–183. https://doi.org/10.1007/BF02921684 PMID: 1952931.

Ahmad, A., Zoli, M., Latella, C., & Bacenetti, J. (2023). Rice cultivation and processing: Highlights from a life cycle thinking perspective. *The Science of the Total Environment, 871*, 162079. https://doi.org/10.1016/j.scitotenv.2023.162079 Epub 2023 Feb 7. PMID: 36754312.

Ali, S. S., Jiao, H., El-Sapagh, S., & Sun, J. (2023). Biodegradation of willow sawdust by novel cellulase-producing bacterial consortium from wood-feeding termites for enhancing methane production. *Bioresource Technology, 383*, 129232. https://doi.org/10.1016/j.biortech.2023.129232 Epub 2023 May 26. PMID: 37244303.

Ambye-Jensen, M., Balzarotti, R., Thomsen, S. T., Fonseca, C., & Kádár, Z. (2018). Combined ensiling and hydrothermal processing as efficient pretreatment of sugarcane bagasse for 2G bioethanol production. *Biotechnology for Biofuels, 11*(1), 336. https://doi.org/10.1186/s13068-018-1338-y, http://www.biotechnologyforbiofuels.com/.

Amiri, H., Karimi, K., & Zilouei, H. (2014). Organosolv pretreatment of rice straw for efficient acetone, butanol, and ethanol production. *Bioresource Technology, 152*, 450–456. https://doi.org/10.1016/j.biortech.2013.11.038, http://www.elsevier.com/locate/biortech.

Amon, T., Amon, B., Kryvoruchko, V., Zollitsch, W., Mayer, K., & Gruber, L. (2007). Biogas production from maize and dairy cattle manure-Influence of biomass composition on the methane yield. *Agriculture, Ecosystems and Environment, 118*(1-4), 173–182. https://doi.org/10.1016/j.agee.2006.05.007.

An, Q., Wang, J. L., Wang, Y. T., Lin, Z. L., & Zhu, M. J. (2018). Investigation on hydrogen production from paper sludge without inoculation and its enhancement by Clostridium thermocellum. *Bioresource Technology, 263*, 120–127. https://doi.org/10.1016/j.biortech.2018.04.105, http://www.elsevier.com/locate/biortech.

Angelidaki, I., Treu, L., Tsapekos, P., Luo, G., Campanaro, S., Wenzel, H., & Kougias, P. G. (2018). Biogas upgrading and utilization: Current status and perspectives. *Biotechnology Advances, 36*(2), 452–466. https://doi.org/10.1016/j.biotechadv.2018.01.011, http://www.elsevier.com/inca/publications/store/5/2/5/4/5/5/index.htt.

Babu, S., Rathore, S. S., Singh, R., Kumar, S., Singh, V. K., Yadav, S. K., Yadav, V., Raj, R., Yadav, D., Shekhawat, K., & Wani, O. A. (2022). Exploring agricultural waste biomass for energy, food and feed production and pollution mitigation: A review. *Bioresource Technology, 360*, 127566. https://doi.org/10.1016/j.biortech.2022.127566, http://www.elsevier.com/locate/biortech.

Bagi, Z., Ács, N., Böjti, T., Kakuk, B., Rákhely, G., Strang, O., Szuhaj, M., Wirth, R., & Kovács, K. L. (2017). Biomethane: The energy storage, platform chemical and greenhouse gas mitigation target. *Anaerobe, 46*, 13–22. https://doi.org/10.1016/j.anaerobe.2017.03.001, http://www.elsevier.com/inca/publications/store/6/2/2/7/8/0/index.htt.

Banoth, C., Sunkar, B., Tondamanati, P. R., & Bhukya, B. (2017). Improved physicochemical pretreatment and enzymatic hydrolysis of rice straw for bioethanol production by yeast fermentation. *3 Biotech, 7*(5), 334. https://doi.org/10.1007/s13205-017-0980-6, http://www.springerlink.com/content/2190-572x/.

Bačėninaitė, D., Džermeikaitė, K., & Antanaitis, R. (2022). Global warming and dairy cattle: How to control and reduce methane emission. *Animals (Basel), 12*(19), 2687. https://doi.org/10.3390/ani12192687 PMID: 36230428.

Bertasini, D., Battista, F., Mancini, R., Frison, N., & Bolzonella, D. (2024). Hydrogen and methane production through two stage anaerobic digestion of straw residues. *Environmental Research, 247*, 118101. https://doi.org/10.1016/j.envres.2024.118101, https://www.sciencedirect.com/science/journal/00139351.

Bondesson, P. M., Galbe, M., & Zacchi, G. (2013). Ethanol and biogas production after steam pretreatment of corn stover with or without the addition of sulphuric acid. *Biotechnology for Biofuels, 6*(1), 11. https://doi.org/10.1186/1754-6834-6-11 PMID: 23356481.

Bu, J., Wang, Y. T., Deng, M. C., & Zhu, M. J. (2021). Enhanced enzymatic hydrolysis and hydrogen production of sugarcane bagasse pretreated by peroxyformic acid. *Bioresource Technology, 326*, 124751. https://doi.org/10.1016/j.biortech.2021.124751 Epub 2021 Jan 27. PMID: 33535152.

Bule, M. V., Gao, A. H., Hiscox, B., & Chen, S. (2013). Structural modification of lignin and characterization of pretreated wheat straw by ozonation. *Journal of Agricultural and Food Chemistry, 61*(16), 3916–3925. https://doi.org/10.1021/jf4001988.

Cao, G. L., Zhao, L., Wang, A. J., Wang, Z. Y., & Ren, N. Q. (2014). Single-step bioconversion of lignocellulose to hydrogen using novel moderately thermophilic bacteria. *Biotechnology for Biofuels, 7*(1), 82. https://doi.org/10.1186/1754-6834-7-82, http://www.biotechnologyforbiofuels.com/.

Capolupo, L., & Faraco, V. (2016). Green methods of lignocellulose pretreatment for biorefinery development. *Applied Microbiology and Biotechnology, 100*(22), 9451–9467. https://doi.org/10.1007/s00253-016-7884-y, http://link.springer.de/link/service/journals/00253/index.htm.

Cañadas, R., Martín-Sampedro, R., González-Miquel, M., González, E. J., Ballesteros, I., Eugenio, M. E., & Ibarra, D. (2023). Green solvents extraction-based detoxification to enhance the enzymatic hydrolysis of steam-exploded lignocellulosic biomass and recover bioactive compounds. *Journal of Environmental Management, 344*, 118448. https://doi.org/10.1016/j.jenvman.2023.118448, https://www.sciencedirect.com/journal/journal-of-environmental-management.

Cheenkachorn, K., Mensah, R. Q., Dharmalingam, B., Gundupalli, M. P., Rattanaporn, K., Tantayotai, P., Show, P. L., & Sriariyanun, M. (2024). The versatility of mixed lignocellulose feedstocks for bioethanol production: An experimental study and empirical prediction. *BioEnergy Research, 17*(2), 1004–1014. https://doi.org/10.1007/s12155-023-10705-4.

Cheng, C. L., Lo, Y. C., Lee, K. S., Lee, D. J., Lin, C. Y., & Chang, J. S. (2011). Biohydrogen production from lignocellulosic feedstock. *Bioresource Technology, 102*(18), 8514–8523. https://doi.org/10.1016/j.biortech.2011.04.059.

Cheng, Y. S. (2023). Insect biorefinery: Sustainable application of insects in circular economy. *Applied Science and Engineering Progress, 16*, 6772. https://doi.org/10.1016/j.envres.2023.115284.

Cheng, Y. S., Mutrakulcharoen, P., Chuetor, S., Cheenkachorn, K., Tantayotai, P., Panakkal, E. J., & Sriariyanun, M. (2020). Recent situation and progress in biorefining process of lignocellulosic biomass: Toward green economy. *Applied Science and Engineering Progress, 13*(4), 299–311. https://doi.org/10.14416/J.ASEP.2020.08.002, http://ojs.kmutnb.ac.th/index.php/ijst/article/view/3788.

Chourasia, V. R., Pandey, A., Pant, K. K., & Henry, R. J. (2021). Improving enzymatic digestibility of sugarcane bagasse from different varieties of sugarcane using deep eutectic solvent pretreatment. *Bioresource Technology, 337*, 125480. https://doi.org/10.1016/j.biortech.2021.125480, http://www.elsevier.com/locate/biortech.

Corno, L., Pilu, R., & Adani, F. (2014). Arundo donax L.: A non-food crop for bioenergy and bio-compound production. *Biotechnology Advances, 32*(8), 1535–1549. https://doi.org/10.1016/j.biotechadv.2014.10.006, http://www.elsevier.com/inca/publications/store/5/2/5/4/5/5/index.htt.

Cristóbal, J., Caldeira, C., Corrado, S., & Sala, S. (2018). Techno-economic and profitability analysis of food waste biorefineries at European level. *Bioresource Technology, 259*, 244–252. https://doi.org/10.1016/j.biortech.2018.03.016, http://www.elsevier.com/locate/biortech.

Culaba, A. B., Mayol, A. P., Juan, J. L. G. S., Vinoya, C. L., Concepcion, R. S., Bandala, A. A., Vicerra, R. R. P., Ubando, A. T., Chen, W. H., & Chang, J. S. (2022). Smart sustainable biorefineries for lignocellulosic biomass. *Bioresource Technology, 344*(Pt B). https://doi.org/10.1016/j.biortech.2021.126215, http://www.elsevier.com/locate/biortech 126215.

Deng, D., Duan, X. X., Lu, J. J., Qin, Y. H., & Wang, C. W. (2019). Electrogenerated alkaline hydrogen peroxide for rice straw pretreatment to enhance enzymatic hydrolysis. *Bioresource Technology, 292*, 122077. https://doi.org/10.1016/j.biortech.2019.122077, http://www.elsevier.com/locate/biortech.

del Mar Contreras-Gámez, M., Galán-Martín, Á., Seixas, N., da Costa Lopes, A. M., Silvestre, A., & Castro, E. (2023). Deep eutectic solvents for improved biomass pretreatment: Current status and future prospective towards sustainable processes. *Bioresource Technology, 369*, 128396. https://doi.org/10.1016/j.biortech.2022.128396, http://www.elsevier.com/locate/biortech.

De Tissera, S., Köpke, M., Simpson, S. D., Humphreys, C., Minton, N. P., & Dürre, P. (2019). Syngas biorefinery and syngas utilization. *Advances in Biochemical Engineering/Biotechnology, 166*, 247–280. https://doi.org/10.1007/10_2017_5, http://www.springer.com/series/10.

Dharmalingam, B., Tantayotai, P., Panakkal, E. J., Cheenkachorn, K., Kirdponpattara, S., Gundupalli, M. P., Cheng, Y. S., & Sriariyanun, M. (2023). Organic acid pretreatments and optimization techniques for mixed vegetable waste biomass conversion into biofuel production. *Bioenergy Research, 16*(3), 1667–1682. https://doi.org/10.1007/s12155-022-10517-y, https://www.springer.com/journal/12155.

Dien, L. Q., Phuong, N. T. M., Hoa, D. T., & Huy Hoang, P. (2015). Efficient pretreatment of Vietnamese rice straw by soda and sulfate cooking methods for enzymatic saccharification. *Applied Biochemistry and Biotechnology, 175*(3), 1536–1547. https://doi.org/10.1007/s12010-014-1359-3, http://www.springer.com/humana+press/journal/12010.

Domingues, J. P., Pelletier, C., & Brunelle, T. (2022). Cost of ligno-cellulosic biomass production for bioenergy: A review in 45 countries. *Biomass and Bioenergy, 165*, 106583. https://doi.org/10.1016/j.biombioe.2022.106583.

Dong, L., Cao, G., Zhao, L., Liu, B., & Ren, N. (2018). Alkali/urea pretreatment of rice straw at low temperature for enhanced biological hydrogen production. *Bioresource Technology, 267*, 71–76. https://doi.org/10.1016/j.biortech.2018.05.055, http://www.elsevier.com/locate/biortech.

Elsgaard, L., Olesen, J. E., Hermansen, J. E., Kristensen, I. T., & Børgesen, C. D. (2013). Regional greenhouse gas emissions from cultivation of winter wheat and winter rapeseed for biofuels in Denmark. *Acta Agriculturae Scandinavica Section B: Soil and Plant Science, 63*(3), 219–230. https://doi.org/10.1080/09064710.2012.751451.

El Gnaoui, Y., Frimane, A., Lahboubi, N., Herrmann, C., Barz, M., & EL Bari, H. (2022). Biological pre-hydrolysis and thermal pretreatment applied for anaerobic digestion improvement: Kinetic study and statistical variable selection. *Cleaner Waste Systems, 2*, 100005. https://doi.org/10.1016/j.clwas.2022.100005.

Fan, X., Li, Y., Luo, Z., Jiao, Y., Ai, F., Zhang, H., Zhu, S., Zhang, Q., & Zhang, Z. (2022). Surfactant assisted microwave irradiation pretreatment of corncob: Effect on hydrogen production capacity, energy consumption and physiochemical structure. *Bioresource Technology, 357*, 127302. https://doi.org/10.1016/j.biortech.2022.127302, http://www.elsevier.com/locate/biortech.

Fernández-Rodríguez, M. J., Mushtaq, M., Tian, L., Jiménez-Rodríguez, A., Rincón, B., Gilroyed, B. H., & Borja, R. (2022). Evaluation and modelling of methane production from corn stover pretreated with various physicochemical techniques. *Waste Management and Research, 40*(6), 698–705. https://doi.org/10.1177/0734242X211038185, https://journals.sagepub.com/home/WMR.

Ferreira, J. A., & Taherzadeh, M. J. (2020). Improving the economy of lignocellulose-based biorefineries with organosolv pretreatment. *Bioresource Technology, 299*, 122695. https://doi.org/10.1016/j.biortech.2019.122695, http://www.elsevier.com/locate/biortech.

Fona, Z., Irvan, I., Tambun, R., Fatimah, F., Setiawan, A., & Adriana, A. (2024). Review on Advance Catalyst for Biomass Gasification. *Applied Science and Engineering Progress, 17*(2), 7295. https://doi.org/10.14416/j.asep.2024.01.001, https://ojs.kmutnb.ac.th/index.php/ijst/article/view/7295.

Fu, D., & Mazza, G. (2011). Optimization of processing conditions for the pretreatment of wheat straw using aqueous ionic liquid. *Bioresource Technology, 102*(17), 8003–8010. https://doi.org/10.1016/j.biortech.2011.06.023.

Gao, J., Zheng, C., Tan, T., Liu, S., & Ji, H. (2018). Enhanced saccharification of rice straw using combined ultra-high pressure and ionic liquid microemulsion pretreatments. *3 Biotech, 8*(4), 208. https://doi.org/10.1007/s13205-018-1216-0, http://www.springerlink.com/content/2190-572x/.

Gonçalves, A. R., Benar, P., Costa, S. M., Ruzene, D. S., Moriya, R. Y., Luz, S. M., & Ferretti, L. P. (2005). Integrated processes for use of pulps and lignins obtained from sugarcane bagasse and straw: A review of recent efforts in Brazil. *Applied Biochemistry and Biotechnology - Part A Enzyme Engineering and Biotechnology, 123*(1-3), 821–826. https://doi.org/10.1385/abab:123:1-3:0821, http://www.humanapress.com/JournalTOC.pasp?issn=0273-2289.

Gunaseelan, V. N. (2007). Regression models of ultimate methane yields of fruits and vegetable solid wastes, sorghum and napiergrass on chemical composition. *Bioresource Technology, 98*(6), 1270–1277. https://doi.org/10.1016/j.biortech.2006.05.014.

Gundupalli, M. P., Bano, K., Panda, T. K., Sriariyanun, M., & Bhattacharyya, D. (2024). Understanding the effect of low-concentrated protic ionic liquids (PILs) on coconut (Cocos nucifera) residues. *Biomass Conversion and Biorefinery, 14*(3), 3275–3291. https://doi.org/10.1007/s13399-022-02572-4, https://www.springer.com/journal/13399.

Gundupalli, M. P., Cheenkachorn, K., Chuetor, S., Kirdponpattara, S., Gundupalli, S. P., Show, P. L., & Sriariyanun, M. (2023). Assessment of pure, mixed and diluted deep eutectic solvents on Napier grass (Cenchrus purpureus): Compositional and characterization studies of cellulose, hemicellulose and lignin. *Carbohydrate Polymers, 306*, 120599. https://doi.org/10.1016/j.

carbpol.2023.120599, http://www.elsevier.com/wps/find/journaldescription.cws_home/405871/description#description.

Gundupalli, M. P., Tantayotai, P., Chuetor, S., Cheenkachorn, K., Joshi, S., Bhattacharyya, D., & Sriariyanun, M. (2023). Improvement of water hyacinth bioconversion by different organic and mineral acid pretreatment and the effect of post-pretreatment washing. *Bioenergy Research, 16*(3), 1718–1732. https://doi.org/10.1007/s12155-022-10528-9, https://www.springer.com/journal/12155.

Habchi, S., Lahboubi, N., Karouach, F., Naim, I., Lahlou, Y., Bakraoui, M., Sallek, B., & Bari, H. E. (2022). Effect of thermal pretreatment on the kinetic parameters of anaerobic digestion from recycled pulp and paper sludge. *Ecological Engineering and Environmental Technology, 23*(1), 192–201. https://doi.org/10.12912/27197050/143568, http://www.ecoeet.com/Archive.

Hoang, A. T., Nguyen, X. P., Duong, X. Q., Ağbulut, Ü., Len, C., Nguyen, P. Q. P., Kchaou, M., & Chen, W. H. (2023). Steam explosion as sustainable biomass pretreatment technique for biofuel production: Characteristics and challenges. *Bioresource Technology, 385*, 129398. https://doi.org/10.1016/j.biortech.2023.129398, http://www.elsevier.com/locate/biortech.

Hu, W., Zhou, L., & Chen, J. H. (2022). Conversion sweet sorghum biomass to produce value-added products. *Biotechnology for Biofuels and Bioproducts, 15*(1), 72. https://doi.org/10.1186/s13068-022-02170-6, https://biotechnologyforbiofuels.biomedcentral.com/.

Huang, Y., Xu, Y., Zhu, Y., Huang, R., Kuang, Y., Wang, J., Xiao, W., Lin, J., & Liu, Z. (2022). Improved glucose yield and concentration of sugarcane bagasse by the pretreatment with ternary deep eutectic solvents and recovery of the pretreated liquid. *Bioresource Technology, 366*, 128186. https://doi.org/10.1016/j.biortech.2022.128186, http://www.elsevier.com/locate/biortech.

Igbokwe, V. C., Ezugworie, F. N., Onwosi, C. O., Aliyu, G. O., & Obi, C. J. (2022). Biochemical biorefinery: A low-cost and non-waste concept for promoting sustainable circular bioeconomy. *Journal of Environmental Management, 305*, 114333. https://doi.org/10.1016/j.jenvman.2021.114333, https://www.sciencedirect.com/journal/journal-of-environmental-management.

Igwebuike, C. M., Awad, S., & Andrès, Y. (2024). Renewable energy potential: Second-generation biomass as feedstock for bioethanol production. *Molecules (Basel, Switzerland), 29*(7), 1619. https://doi.org/10.3390/molecules29071619, http://www.mdpi.com/journal/molecules.

Jamaldheen, S. B., Kurade, M. B., Basak, B., Yoo, C. G., Oh, K. K., Jeon, B. H., & Kim, T. H. (2022). A review on physico-chemical delignification as a pretreatment of lignocellulosic biomass for enhanced bioconversion. *Bioresource Technology, 346*, 126591. https://doi.org/10.1016/j.biortech.2021.126591, http://www.elsevier.com/locate/biortech.

Janta, R., Sekiguchi, K., Yamaguchi, R., Sopajaree, K., Plubin, B., & Chetiyanukornkul, T. (2020). Spatial and temporal variations of atmospheric pm10 and air pollutants concentration in upper Northern Thailand during 2006–2016. *Applied Science and Engineering Progress, 13*(3), 256–267. https://doi.org/10.14416/J.ASEP.2020.03.007, http://ojs.kmutnb.ac.th/index.php/ijst/article/view/2604.

Jose, D., Tawai, A., Divakaran, D., Sriariyanun, M., Phakeenuya, V., Cheng, Y. S., & Tantayotai, P. (2024). Influence of acetic acid pretreatment and its residue on bioethanol and biogas production from water hyacinth. *Applied Science and Engineering Progress, 17*(3). https://doi.org/10.14416/j.asep.2024.02.001.

Jönsson, L. J., & Martín, C. (2016). Pretreatment of lignocellulose: Formation of inhibitory by-products and strategies for minimizing their effects. *Bioresource Technology, 199*,

103–112. https://doi.org/10.1016/j.biortech.2015.10.009, http://www.elsevier.com/locate/biortech.

Kang, H. K., & Kim, D. (2012). Efficient bioconversion of rice straw to ethanol with TiO 2/UV pretreatment. *Bioprocess and Biosystems Engineering, 35*(1-2), 43–48. https://doi.org/10.1007/s00449-011-0589-9.

Khan, M. A., Lee, S. H., & Sriariyanun, M. (2024). Deep eutectic solvent as a tailor-made chemical for pretreatment in a lignocellulose biorefinery. *Applied Science and Engineering Progress, 17*(3). https://doi.org/10.14416/j.asep.2024.03.003.

Kim, D. (2018). Physico-chemical conversion of lignocellulose: Inhibitor effects and detoxification strategies: A mini review. *Molecules (Basel, Switzerland), 23*(2), 309. https://doi.org/10.3390/molecules23020309.

Klose, H., & Paës, G. (2023). Editorial: Understanding plant cell wall recalcitrance for efficient lignocellulose processing. *Frontiers in Plant Science, 14*, 1284658. https://doi.org/10.3389/fpls.2023.1284658, https://www.frontiersin.org/journals/plant-science.

Ko, J. K., Kim, Y., Ximenes, E., & Ladisch, M. R. (2015). Effect of liquid hot water pretreatment severity on properties of hardwood lignin and enzymatic hydrolysis of cellulose. *Biotechnology and Bioengineering, 112*(2), 252–262. https://doi.org/10.1002/bit.25349, http://onlinelibrary.wiley.com/journal/10.1002/(ISSN)1097-0290.

Kucharska, K., Łukajtis, R., Słupek, E., Cieśliński, H., Rybarczyk, P., & Kamiński, M. (2018). Hydrogen production from energy poplar preceded by MEA pre-treatment and enzymatic hydrolysis. *Molecules (Basel, Switzerland), 23*(11), 3029. https://doi.org/10.3390/molecules23113029.

Kulichkova, G. I., Ivanova, T. S., Köttner, M., Volodko, O. I., Spivak, S. I., Tsygankov, S. P., & Blume, Y. B. (2020). Plant feedstocks and their biogas production potentials. *Open Agriculture Journal, 14*(1), 219–234. https://doi.org/10.2174/1874331502014010219, https://benthamopen.com/contents/pdf/TOASJ/TOASJ-14-219.pdf.

Kundu, C., Samudrala, S. P., Kibria, M. A., & Bhattacharya, S. (2021). One-step peracetic acid pretreatment of hardwood and softwood biomass for platform chemicals production. *Scientific Reports, 11*(1), 11183. https://doi.org/10.1038/s41598-021-90667-9, http://www.nature.com/srep/index.html.

Kyawt, Y. Y., Aung, M., Xu, Y., Zhou, Y., Li, Y., Sun, Z., Zhu, W., & Cheng, Y. (2024). Methane production and lignocellulosic degradation of wastes from rice, corn and sugarcane by natural anaerobic fungi-methanogens co-culture. *World Journal of Microbiology and Biotechnology, 40*(4), 109. https://doi.org/10.1007/s11274-024-03938-8, https://www.springer.com/journal/11274.

Laiq Ur Rehman, M., Iqbal, A., Chang, C. C., Li, W., & Ju, M. (2019). Anaerobic digestion. *Water Environment Research, 91*(10), 1253–1271. https://doi.org/10.1002/wer.1219, https://onlinelibrary.wiley.com/loi/15547531.

Lark, T. J., Hendricks, N. P., Smith, A., Pates, N., Spawn-Lee, S. A., Bougie, M., Booth, E. G., Kucharik, C. J., & Gibbs, H. K. (2022). Environmental outcomes of the US Renewable Fuel Standard. *Proceedings of the National Academy of Sciences of the United States of America, 119*(9). https://doi.org/10.1073/pnas.2101084119, https://www.pnas.org/content/119/9/e2101084119.

Lay, C. H., Dharmaraja, J., Shobana, S., Arvindnarayan, S., Krishna Priya, R., Jeyakumar, R. B., Saratale, R. G., Park, Y. K., Kumar, V., & Kumar, G. (2022). Lignocellulose biohydrogen towards net zero emission: A review on recent developments. *Bioresource Technology, 364*, 128084. https://doi.org/10.1016/j.biortech.2022.128084, http://www.elsevier.com/locate/biortech.

Leca, E., Zennaro, B., Hamelin, J., Carrère, H., & Sambusiti, C. (2023). Use of additives to improve collective biogas plant performances: A comprehensive review. *Biotechnology Advances, 65*, 108129. https://doi.org/10.1016/j.biotechadv.2023.108129, http://www.elsevier.com/inca/publications/store/5/2/5/4/5/5/index.htt.

Li, F., Li, Q., Lv, J., Huang, M., Ling, Z., Meng, Y., Chen, F., & Ji, Z. (2024). A novel seawater hydrothermal-deep eutectic solvent pretreatment enhances the production of fermentable sugars and tailored lignin nanospheres from Pinus massoniana. *International Journal of Biological Macromolecules, 267*(Pt 1), 131596. https://doi.org/10.1016/j.ijbiomac.2024.131596, https://www.sciencedirect.com/science/journal/01418130.

Li, Q., He, Y. C., Xian, M., Jun, G., Xu, X., Yang, J. M., & Li, L. Z. (2009). Improving enzymatic hydrolysis of wheat straw using ionic liquid 1-ethyl-3-methyl imidazolium diethyl phosphate pretreatment. *Bioresource Technology, 100*(14), 3570–3575. https://doi.org/10.1016/j.biortech.2009.02.040.

Liu, X., Zicari, S. M., Liu, G., Li, Y., & Zhang, R. (2015). Pretreatment of wheat straw with potassium hydroxide for increasing enzymatic and microbial degradability. *Bioresource Technology, 185*, 150–157. https://doi.org/10.1016/j.biortech.2015.02.047, http://www.elsevier.com/locate/biortech.

Luo, W., Tian, H., Tan, W., & Tan, Q. (2024). Effect of hydrothermal-acid pretreatment on methane yield and microbial community in anaerobic digestion of rice straw. *Bioresource Technology, 402*, 130765. https://doi.org/10.1016/j.biortech.2024.130765.

Luo, X., Liu, Y., Lei, L., Shen, J., Zhang, Q., Wang, Y., Ruan, R., & Cui, X. (2023). Co-ensiling of rice straw and distillers grains to increase methane production and maximise energy output. *Bioresource Technology, 386*, 129496. https://doi.org/10.1016/j.biortech.2023.129496, http://www.elsevier.com/locate/biortech.

Lv, X., Xiong, C., Li, S., Chen, X., Xiao, W., Zhang, D., Li, J., Gong, Y., Lin, J., & Liu, Z. (2017). Vacuum-assisted alkaline pretreatment as an innovative approach for enhancing fermentable sugar yield and decreasing inhibitor production of sugarcane bagasse. *Bioresource Technology, 239*, 402–411. https://doi.org/10.1016/j.biortech.2017.04.053, http://www.elsevier.com/locate/biortech.

Ma, G., Ndegwa, P., Harrison, J. H., & Chen, Y. (2020). Methane yields during anaerobic co-digestion of animal manure with other feedstocks: A meta-analysis. *Science of the Total Environment, 728*, 138224. https://doi.org/10.1016/j.scitotenv.2020.138224, http://www.elsevier.com/locate/scitotenv.

Machhirake, N. P., Vanapalli, K. R., Kumar, S., & Mohanty, B. (2024). Biohydrogen from waste feedstocks: An energy opportunity for decarbonization in developing countries. *Environmental Research, 252*(Pt 4), 119028. https://doi.org/10.1016/j.envres.2024.119028 Epub ahead of print. PMID: 38685297.

Mai, J., Hu, B. B., & Zhu, M. J. (2023). Metabolic division of labor between acetivibrio thermocellus DSM 1313 and thermoanaerobacterium thermosaccharolyticum MJ1 enhanced hydrogen production from lignocellulose. *Bioresource Technology, 390*, 129871. https://doi.org/10.1016/j.biortech.2023.129871, http://www.elsevier.com/locate/biortech.

Maibam, P. D., & Goyal, A. (2022). Approach to an efficient pretreatment method for rice straw by deep eutectic solvent for high saccharification efficiency. *Bioresource Technology, 351*, 127057. https://doi.org/10.1016/j.biortech.2022.127057, http://www.elsevier.com/locate/biortech.

Mankar, A. R., Pandey, A., Modak, A., & Pant, K. K. (2021). Pretreatment of lignocellulosic biomass: A review on recent advances. *Bioresource Technology, 334*, 125235. https://doi.org/10.1016/j.biortech.2021.125235, http://www.elsevier.com/locate/biortech.

Martínez-Gutiérrez, E. (2018). Biogas production from different lignocellulosic biomass sources: Advances and perspectives. 3 Biotech, 8(5), 233. https://doi.org/10.1007/s13205-018-1257-4. http://www.springerlink.com/content/2190-572x.

Mironova, G. F., Budaeva, V. V., Skiba, E. A., Gismatulina, Y. A., Kashcheyeva, E. I., & Sakovich, G. V. (2023). Recent advances in miscanthus macromolecule conversion: A brief overview. *Federation International Journal of Molecular Sciences, 24*(16), 13001. https://doi.org/10.3390/ijms241613001, http://www.mdpi.com/journal/ijms.

Modenbach, A. A., & Nokes, S. E. (2012). The use of high-solids loadings in biomass pretreatment-a review. *Biotechnology and Bioengineering, 109*(6), 1430–1442. https://doi.org/10.1002/bit.24464.

Mohd-Nor, D., Ramli, N., Sharuddin, S. S., Hassan, M. A., Mustapha, N. A., Ariffin, H., Sakai, K., Tashiro, Y., Shirai, Y., & Maeda, T. (2019). Dynamics of microbial populations responsible for biodegradation during the full-scale treatment of palm oil mill effluent. *Microbes and Environments, 34*(2), 121–128. https://doi.org/10.1264/jsme2.ME18104, https://www.jstage.jst.go.jp/article/jsme2/34/2/34_ME18104/_pdf.

Moneda, A. P. C., de Carvalho, L. A. L., Teheran-Sierra, L. G., Funnicelli, M. I. G., & Pinheiro, D. G. (2022). Sugarcane cultivation practices modulate rhizosphere microbial community composition and structure. *Scientific Reports, 12*(1), 19174. https://doi.org/10.1038/s41598-022-23562-6, https://www.nature.com/srep/.

Mukherjee, A., Patra, B. R., Podder, J., & Dalai, A. K. (2022). Synthesis of biochar from lignocellulosic biomass for diverse industrial applications and energy harvesting: Effects of pyrolysis conditions on the physicochemical properties of biochar. *Frontiers in Materials, 9*, 870184. https://doi.org/10.3389/fmats.2022.870184, http://journal.frontiersin.org/journal/materials.

Nawaz, A., Aamir, F., Huang, R., I.u. Haq, F., Wu, M., Munir, R., Chaudhary, A., Rafique, K., & Jiang (2023). Co-production of biohydrogen and biomethane utilizing halophytic biomass atriplexcrassifolia by two-stage anaerobic fermentation process. *Frontiers in Chemistry, 11*, 1233494. https://doi.org/10.3389/fchem.2023.1233494, http://journal.frontiersin.org/journal/chemistry.

Oliva, A., Tan, L. C., Papirio, S., Esposito, G., & Lens, P. N. L. (2024). Use of N-Methylmorpholine N-oxide (NMMO) pretreatment to enhance the bioconversion of lignocellulosic residues to methane. *Biomass Conversion and Biorefinery, 14*(10), 11113–11130. https://doi.org/10.1007/s13399-022-03173-x, https://www.springer.com/journal/13399.

Parrish, D. J., & Fike, J. H. (2009). Selecting, establishing, and managing switchgrass (Panicum virgatum) for biofuels. *Methods in molecular biology (Clifton, N.J.), 581*, 27–40. https://doi.org/10.1007/978-1-60761-214-8_2.

Patinvoh, R. J., Osadolor, O. A., Chandolias, K., Sárvári Horváth, I., & Taherzadeh, M. J. (2017). Innovative pretreatment strategies for biogas production. *Bioresource Technology, 224*, 13–24. https://doi.org/10.1016/j.biortech.2016.11.083, http://www.elsevier.com/locate/biortech.

Paudel, S. R., Banjara, S. P., Choi, O. K., Park, K. Y., Kim, Y. M., & Lee, J. W. (2017). Pretreatment of agricultural biomass for anaerobic digestion: Current state and challenges. *Bioresource Technology, 245*(Pt A), 1194–1205. https://doi.org/10.1016/j.biortech.2017.08.182, http://www.elsevier.com/locate/biortech.

Pote, X. R., Malekar, V., Iyer, S., Sharma, K., Kharate, S., & Balbudhe, I. (2023). Analysis of various feedstocks's used for biogas generation. *InE3S Web of Conferences, 405*, 04032. https://doi.org/10.1051/e3sconf/202340504032 EDP Sciences.

Prates, Crowley, M. B., Skaf, M. O., Beckham, G. T., Gall, D. L., Ralph, Donohue, & Noguera, D. R. (2017). Biochemical transformation of lignin for deriving valued commodities from lignocellulose. *Current Opinion in Biotechnology, 123*(48). https://doi.org/10.1021/acs.jpcb.9b06243.Epub.

Prates, E. T., Crowley, M. F., Skaf, M. S., & Beckham, G. T. (2019). Catalytic mechanism of aryl-ether bond cleavage in lignin by LigF and LigG. *The Journal of Physical Chemistry. B, 123*(48), 10142–10151. https://doi.org/10.1021/acs.jpcb.9b06243 Epub 2019 Nov 19. PMID: 31687816.

Rabemanolontsoa, H., & Saka, S. (2016). Various pretreatments of lignocellulosics. *Bioresource Technology, 199*, 83–91. https://doi.org/10.1016/j.biortech.2015.08.029, http://www.elsevier.com/locate/biortech.

Raj, T., Chandrasekhar, K., Kumar, A. N., Rajesh Banu, J., Yoon, J. J., Bhatia, S. K., Yang, Y. H., Varjani, S., & Kim, S. H. (2022). Recent advances in commercial biorefineries for lignocellulosic ethanol production: Current status, challenges and future perspectives. *Bioresource Technology, 344*(Pt B), 126292. https://doi.org/10.1016/j.biortech.2021.126292, http://www.elsevier.com/locate/biortech.

Rani, J., & Dhoble, A. S. (2023). Effect of fungal pretreatment by pycnoporus sanguineus and trichoderma longibrachiatum on the anaerobic digestion of rice straw. *Bioresource Technology, 387*, 129503. https://doi.org/10.1016/j.biortech.2023.129503, http://www.elsevier.com/locate/biortech.

Rena, Zacharia, K. M. B., Yadav, S., Machhirake, N. P., Kim, S. H., Lee, B. D., Jeong, H., Singh, L., Kumar, S., & Kumar, R. (2020). Bio-hydrogen and bio-methane potential analysis for production of bio-hythane using various agricultural residues. *Bioresource Technology, 309*, 123297. https://doi.org/10.1016/j.biortech.2020.123297, http://www.elsevier.com/locate/biortech.

Reshmy, R., Philip, E., Madhavan, A., Sirohi, R., Pugazhendhi, A., Binod, P., Awasthi, M. K., Vivek, N., Kumar, V., & Sindhu, R. (2022). Lignocellulose in future biorefineries: Strategies for cost-effective production of biomaterials and bioenergy. *Bioresource Technology, 344*(Pt B), 126241. https://doi.org/10.1016/j.biortech.2021.126241.

Rizwan, M., Gilani, S. R., Durrani, A. I., & Naseem, S. (2021). Low temperature green extraction of acer platanoides cellulose using nitrogen protected microwave assisted extraction (NPMAE) technique. *Carbohydrate Polymers, 272*, 118465. https://doi.org/10.1016/j.carbpol.2021.118465, http://www.elsevier.com/wps/find/journaldescription.cws_home/405871/description#description.

del Río, P. G., Gullón, B., Romaní, A., & Garrote, G. (2021). Fast-growing Paulownia wood fractionation by microwave-assisted hydrothermal treatment: A kinetic assessment. *Bioresource Technology, 338*, 125535. https://doi.org/10.1016/j.biortech.2021.125535, http://www.elsevier.com/locate/biortech.

Saha, B. C., & Cotta, M. A. (2006). Ethanol production from alkaline peroxide pretreated enzymatically saccharified wheat straw. *Biotechnology Progress, 22*(2), 449–453. https://doi.org/10.1021/bp050310r.

Saha, K., Dwibedi, P., Ghosh, A., Sikder, J., Chakraborty, S., & Curcio, S. (2018). Extraction of lignin, structural characterization and bioconversion of sugarcane bagasse after ionic liquid assisted pretreatment. *3 Biotech, 8*(8), 374. https://doi.org/10.1007/s13205-018-1399-4, http://www.springerlink.com/content/2190-572x/.

Saratale, R. G., Ponnusamy, V., Piechota, G., Igliński, B., Shobana, S., Park, J. H., Saratale, G. D., Shin, H. S., Banu, J. R., Kumar, V., & Kumar, G. (2023). Green chemical and hybrid enzymatic pretreatments for lignocellulosic biorefineries: Mechanism and challenges. *Bioresource Technology, 387*, 129560. https://doi.org/10.1016/j.biortech.2023.129560, http://www.elsevier.com/locate/biortech.

Sarkar, O., Rova, U., Christakopoulos, P., & Matsakas, L. (2022). Organosolv pretreated birch sawdust for the production of green hydrogen and renewable chemicals in an integrated biorefinery approach. *Bioresource Technology, 344*(Pt A), 126164. https://doi.org/10.1016/j.biortech.2021.126164.

Sawatdeenarunat, C., Surendra, K. C., Takara, D., Oechsner, H., & Khanal, S. K. (2015). Anaerobic digestion of lignocellulosic biomass: Challenges and opportunities. *Bioresource Technology, 178*, 178–186. https://doi.org/10.1016/j.biortech.2014.09.103, http://www.elsevier.com/locate/biortech.

Sawhney, D., Vaid, S., Bangotra, R., Sharma, S., Dutt, H. C., Kapoor, N., Mahajan, R., & Bajaj, B. K. (2023). Proficient bioconversion of rice straw biomass to bioethanol using a novel combinatorial pretreatment approach based on deep eutectic solvent, microwave irradiation and laccase. *Bioresource Technology, 375*, 128791. https://doi.org/10.1016/j.biortech.2023.128791, http://www.elsevier.com/locate/biortech.

Sharma, V., Nargotra, P., & Bajaj, B. K. (2019). Ultrasound and surfactant assisted ionic liquid pretreatment of sugarcane bagasse for enhancing saccharification using enzymes from an ionic liquid tolerant Aspergillus assiutensis VS34. *Bioresource Technology, 285*, 121319. https://doi.org/10.1016/j.biortech.2019.121319, http://www.elsevier.com/locate/biortech.

Shrestha, S., Pandey, R., Aryal, N., & Lohani, S. P. (2023). Recent advances in co-digestion conjugates for anaerobic digestion of food waste. *Journal of Environmental Management, 345*, 118785. https://doi.org/10.1016/j.jenvman.2023.118785, https://www.sciencedirect.com/journal/journal-of-environmental-management.

Sindhu, R., Binod, P., & Pandey, A. (2016). Biological pretreatment of lignocellulosic biomass - An overview. *Bioresource Technology, 199*, 76–82. https://doi.org/10.1016/j.biortech.2015.08.030, http://www.elsevier.com/locate/biortech.

Singh, J. K., Vyas, P., Dubey, A., Upadhyaya, C. P., Kothari, R., Tyagi, V. V., & Kumar, A. (2018). Assessment of different pretreatment technologies for efficient bioconversion of lignocellulose to ethanol. *Frontiers in Bioscience - Scholar, 10*(2), 350–371. https://doi.org/10.2741/s521, https://www.bioscience.org/2018/v10s/af/521/list.htm.

Singh, T., Alhazmi, A., Mohammad, A., Srivastava, N., Haque, S., Sharma, S., Singh, R., Yoon, T., & Gupta, V. K. (2021). Integrated biohydrogen production via lignocellulosic waste: Opportunity, challenges & future prospects. *Bioresource Technology, 338*, 125511. https://doi.org/10.1016/j.biortech.2021.125511, http://www.elsevier.com/locate/biortech.

Singhvi, M., & Kim, B. S. (2022). Green hydrogen production through consolidated bioprocessing of lignocellulosic biomass using nanobiotechnology approach. *Bioresource Technology, 365*, 128108. https://doi.org/10.1016/j.biortech.2022.128108, http://www.elsevier.com/locate/biortech.

Sitthikitpanya, N., Ponuansri, C., Jomnonkhaow, U., Wongfaed, N., & Reungsang, A. (2024). Unlocking the potential of sugarcane leaf waste for sustainable methane production: Insights from microbial pre-hydrolysis and reactor optimization. *Heliyon, 10*(3), e25787. https://doi.org/10.1016/j.heliyon.2024.e25787.

Sonwai, A., Pholchan, P., & Tippayawong, N. (2023). Machine learning approach for determining and optimizing influential factors of biogas production from lignocellulosic biomass. *Bioresource Technology, 383*, 129235. https://doi.org/10.1016/j.biortech.2023.129235, http://www.elsevier.com/locate/biortech.

Sriariyanun, M., & Dharmalingam, B. (2023). From waste to wealth: Challenges in producing value-added biochemicals from lignocellulose biorefinery. *Journal of Applied Science and Emerging Technology, 22*(3), e900001. https://doi.org/10.14416/jaset.kmutnb.2023.03.001.

Sun, J., Zhang, L., & Loh, K. C. (2021). Review and perspectives of enhanced volatile fatty acids production from acidogenic fermentation of lignocellulosic biomass wastes. *Bioresources and Bioprocessing, 8*(1), 68. https://doi.org/10.1186/s40643-021-00420-3, https://bioresources bioprocessing.springeropen.com.

Sun, X., Liu, S., Zhang, X., Tao, Y., Boczkaj, G., Yoon, J. Y., & Xuan, X. (2022). Recent advances in hydrodynamic cavitation-based pretreatments of lignocellulosic biomass for valorization. *Bioresource Technology, 345*, 126251. https://doi.org/10.1016/j.biortech.2021.126251, http://www.elsevier.com/locate/biortech.

Tan, Y. Y., Raman, A. A. A., Abidin, M. I. I. Z., & Buthiyappan, A. (2024). A review on sustainable management of biomass: physicochemical modification and its application for the removal of recalcitrant pollutants—Challenges, opportunities, and future directions. *Environmental Science and Pollution Research, 31*(25), 1–40. https://doi.org/10.1007/s11356-024-33375-x, https://www.springer.com/journal/11356.

Tantayotai, P., Rattanaporn, K., Tepaamorndech, S., Cheenkachorn, K., & Sriariyanun, M. (2019). Analysis of an Ionic liquid and salt tolerant microbial consortium which is useful for enhancement of enzymatic hydrolysis and biogas production. *Waste and Biomass Valorization, 10*(6), 1481–1491. https://doi.org/10.1007/s12649-017-0186-5, http://www.springer.com/engineering/journal/12649.

Teke, G. M., Cho, B. A., Bosman, C. E., Mapholi, Z., Zhang, D., & Pott, R. W. M. (2024). Towards industrial biological hydrogen production: A review. *World Journal of Microbiology and Biotechnology, 40*(1), 37. https://doi.org/10.1007/s11274-023-03845-4, https://www.springer.com/journal/11274.

Terasawat, A., & Phoolphundh, S. (2021). Simultaneous biological pretreatment and saccharification of rice straw by ligninolytic enzymes from panus neostrigosus i9 and commercial cellulase. *Journal of Fungi, 7*(10), 853. https://doi.org/10.3390/jof7100853, https://www.mdpi.com/2309-608X/7/10/853/pdf.

Hilares, R. T., Dionízio, R. M., Prado, C. A., Ahmed, M. A., da Silva, S. S., & Santos, J. C. (2019). Pretreatment of sugarcane bagasse using hydrodynamic cavitation technology: Semi-continuous and continuous process. *Bioresource Technology, 290*, 121777. https://doi.org/10.1016/j.biortech.2019.121777, http://www.elsevier.com/locate/biortech.

Thomas, A. S. S., Pongprayoon, W., Cheenkachorn, K., & Sriariyanun, M. (2022). Plant-microbe interactions - Insights and views for applications in sustainable agriculture. *Applied Science and Engineering Progress, 15*(1). https://doi.org/10.14416/j.asep.2021.07.008, http://ojs.kmutnb.ac.th/index.php/ijst/article/view/5286/3675.

Thomsen, S. T., Londoño, J. E. G., Ambye-Jensen, M., Heiske, S., Kádár, Z., & Meyer, A. S. (2016). Combination of ensiling and fungal delignification as effective wheat straw pretreatment. *Biotechnology for Biofuels, 9*(1), 16. https://doi.org/10.1186/s13068-016-0437-x, http://www.biotechnologyforbiofuels.com/.

Tian, Q. Q., Liang, L., & Zhu, M. J. (2015). Enhanced biohydrogen production from sugarcane bagasse by Clostridium thermocellum supplemented with CaCO3. *Bioresource Technology, 197*, 422–428. https://doi.org/10.1016/j.biortech.2015.08.111, http://www.elsevier.com/locate/biortech.

Torgbo, S., Sukyai, P., Khantayanuwong, S., Puangsin, B., Srichola, P., Sukatta, U., Kamonpatana, P., Beaumont, M., & Rosenau, T. (2022). Assessment of electrothermal pretreatment of rambutan (Nephelium lappaceum L.) peels for producing cellulose fibers.

ACS Omega, 7(44), 39975–39984. https://doi.org/10.1021/acsomega.2c04551, http://pubs.acs.org/journal/acsodf.

Ubando, A. T., Felix, C. B., & Chen, W. H. (2020). Biorefineries in circular bioeconomy: A comprehensive review. *Bioresource Technology, 299*, 122585. https://doi.org/10.1016/j.biortech.2019.122585, http://www.elsevier.com/locate/biortech.

Usmani, Z., Sharma, M., Gupta, P., Karpichev, Y., Gathergood, N., Bhat, R., & Gupta, V. K. (2020). Ionic liquid based pretreatment of lignocellulosic biomass for enhanced bioconversion. *Bioresource Technology, 304*. https://doi.org/10.1016/j.biortech.2020.123003, http://www.elsevier.com/locate/biortech 123003.

Wang, A., Gao, L., Ren, N., Xu, J., & Liu, C. (2009). Bio-hydrogen production from cellulose by sequential co-culture of cellulosic hydrogen bacteria of Enterococcus gallinarum G1 and Ethanoigenens harbinense B49. *Biotechnology Letters, 31*(9), 1321–1326. https://doi.org/10.1007/s10529-009-0028-z.

Wang, F., Ouyang, D., Zhou, Z., Page, S. J., Liu, D., & Zhao, X. (2021). Lignocellulosic biomass as sustainable feedstock and materials for power generation and energy storage. *Journal of Energy Chemistry, 57*, 247–280. https://doi.org/10.1016/j.jechem.2020.08.060, http://elsevier.com/journals/journal-of-energy-chemistry/2095–4956.

Wang, L., Jin, W., Cai, F., Song, C., Jin, Y., Liu, G., & Chen, C. (2024). Performance and mechanism of various microaerobic pretreatments on anaerobic digestion of tobacco straw. *Bioresource Technology, 393*, 130092. https://doi.org/10.1016/j.biortech.2023.130092.

Wang, Y., Cai, D., Jiang, Y., Mei, X., Ren, W., Sun, M., Su, C., Cao, H., Zhang, C., & Qin, P. (2024). Rapid fractionation of corn stover by microwave-assisted protic ionic liquid [TEA][HSO4] for fermentative acetone–butanol–ethanol production. *Biotechnology for Biofuels and Bioproducts, 17*(1), 62. https://doi.org/10.1186/s13068-024-02499-0, https://biotechnologyforbiofuels.biomedcentral.com/.

Wang, Z., Hou, X., Sun, J., Li, M., Chen, Z., & Gao, Z. (2018). Comparison of ultrasound-assisted ionic liquid and alkaline pretreatment of Eucalyptus for enhancing enzymatic saccharification. *Bioresource Technology, 254*, 145–150. https://doi.org/10.1016/j.biortech.2018.01.021, http://www.elsevier.com/locate/biortech.

Wei, W., Wang, B., Wang, X., Ling, R., & Jin, Y. (2021). Comparison of acid and alkali catalyzed ethylene glycol organosolv pretreatment for sugar production from bagasse. *Bioresource Technology, 320*(Pt A), 124293. https://doi.org/10.1016/j.biortech.2020.124293.

Wei, Y., Fakudze, S., Zhang, Y., Song, M., Xue, T., Xie, R., & Chen, J. (2022). Low-temperature hydrothermal liquefaction of pomelo peel for production of 5-hydroxymethylfurfural-rich bio-oil using ionic liquid loaded ZSM-5. *Bioresource Technology, 352*, 127050. https://doi.org/10.1016/j.biortech.2022.127050, http://www.elsevier.com/locate/biortech.

Wi, S. G., Choi, I. S., Kim, K. H., Kim, H. M., & Bae, H. J. (2013). Bioethanol production from rice straw by popping pretreatment. *Biotechnology for Biofuels, 6*(1), 166. https://doi.org/10.1186/1754–6834-6–166.

Wu, X., Huang, C., Zhai, S., Liang, C., Huang, C., Lai, C., & Yong, Q. (2018). Improving enzymatic hydrolysis efficiency of wheat straw through sequential autohydrolysis and alkaline post-extraction. *Bioresource Technology, 251*, 374–380. https://doi.org/10.1016/j.biortech.2017.12.066, http://www.elsevier.com/locate/biortech.

Wu, Z., Peng, K., Zhang, Y., Wang, M., Yong, C., Chen, L., Qu, P., Huang, H., Sun, E., & Pan, M. (2022). Lignocellulose dissociation with biological pretreatment towards the biochemical platform: A review. *Materials Today Bio, 16*, 100445. https://doi.org/10.1016/j.mtbio.2022.100445.

Xu, H., Peng, J., Kong, Y., Liu, Y., Su, Z., Li, B., Song, X., Liu, S., & Tian, W. (2020). Key process parameters for deep eutectic solvents pretreatment of lignocellulosic biomass materials: A review. *Bioresource Technology, 310*, 123416. https://doi.org/10.1016/j.biortech.2020.123416, http://www.elsevier.com/locate/biortech.

Xu, N., Liu, S., Xin, F., Zhou, J., Jia, H., Xu, J., Jiang, M., & Dong, W. (2019). Biomethane production from lignocellulose: Biomass recalcitrance and its impacts on anaerobic digestion. *Frontiers in Bioengineering and Biotechnology, 7*, 191. https://doi.org/10.3389/fbioe.2019.00191, http://journal.frontiersin.org/journal/bioengineering-and-biotechnology#archive.

Zabihi, S., Sharafi, A., Motamedi, H., Esmaeilzadeh, F., & Doherty, W. O. S. (2021). Environmentally friendly acetic acid/steam explosion/supercritical carbon dioxide system for the pre-treatment of wheat straw. *Environmental Science and Pollution Research, 28*(28), 37867–37881. https://doi.org/10.1007/s11356-021-13410-x, https://link.springer.com/journal/11356.

Zhang, C., Wang, G., Ma, S., Huang, H., Ma, Y., & Li, Z. (2021). Enhancing hydrogen productivity of photosynthetic bacteria from the formulated carbon source by mixing xylose with glucose. *Applied Biochemistry and Biotechnology, 193*(12), 3996–4017. https://doi.org/10.1007/s12010-021-03708-4, http://www.springer.com/humana+press/journal/12010.

Zhang, H., Chen, W., Han, X., Zeng, Y., Zhang, J., Gao, Z., & Xie, J. (2021). Intensification of sugar production by using Tween 80 to enhance metal-salt catalyzed pretreatment and enzymatic hydrolysis of sugarcane bagasse. *Bioresource Technology, 339*, 125522. https://doi.org/10.1016/j.biortech.2021.125522, http://www.elsevier.com/locate/biortech.

Zhang, H., & Wu, S. (2013). Subcritical CO2 pretreatment of sugarcane bagasse and its enzymatic hydrolysis for sugar production. *Bioresource Technology, 149*, 546–550. https://doi.org/10.1016/j.biortech.2013.08.159, http://www.elsevier.com/locate/biortech.

Zhang, J., Zhang, J., & Zang, L. (2015). Thermophilic bio-hydrogen production from corn-bran residue pretreated by calcined-lime mud from papermaking process. *Bioresource Technology, 198*, 564–570. https://doi.org/10.1016/j.biortech.2015.09.082, http://www.elsevier.com/locate/biortech.

Zhang, L., Li, Y., Liu, X., Ren, N., & Ding, J. (2019). Lignocellulosic hydrogen production using dark fermentation by Clostridium lentocellum strain Cel10 newly isolated from Ailuropoda melanoleuca excrement. *RSC Advances, 9*(20), 11179–11185. https://doi.org/10.1039/c9ra01158g, http://10.1039/c9ra01158g.

Zhang, L., Shao, G., Jin, Y., Yang, N., & Xu, X. (2024a). Efficient delignification of wheat straw by induced electric field-assisted alkali pretreatment. *Industrial Crops and Products, 214*, 118564. https://doi.org/10.1016/j.indcrop.2024.118564.

Zhang, L., Shao, G., Jin, Y., Yang, N., & Xu, X. (2024b). Efficient hemicellulose removal from lignocellulose by induced electric field-aided dilute acid pretreatment. *International Journal of Biological Macromolecules, 261*(Pt 2), 129839. https://doi.org/10.1016/j.ijbiomac.2024.129839, https://www.sciencedirect.com/science/journal/01418130.

Zhang, Q., Yang, J., Zhang, T., Shui, X., Zhang, H., Chen, Z., He, X., Lei, T., Jiang, D., Elgorban, A. M., Syed, A., & Solanki, M. K. (2023). Pretreatment of Arundo donax L. for photo-fermentative biohydrogen production by ultrasonication and ionic liquid. *Bioresource Technology, 377*. https://doi.org/10.1016/j.biortech.2023.128904, http://www.elsevier.com/locate/biortech 128904.

Zhao, L., Cao, G. L., Wang, A. J., Ren, H. Y., Zhang, K., & Ren, N. Q. (2014). Consolidated bioprocessing performance of thermoanaerobacterium thermosaccharolyticum M18 on fungal pretreated cornstalk for enhanced hydrogen production. *Biotechnology for Biofuels, 7*(1), 178. https://doi.org/10.1186/s13068-014-0178-7, http://www.biotechnologyforbiofuels.com/.

Zhao, L., Wang, Z., Ren, H. Y., Chen, C., Nan, J., Cao, G. L., Yang, S. S., & Ren, N. Q. (2021). Residue cornstalk derived biochar promotes direct bio-hydrogen production from anaerobic fermentation of cornstalk. *Bioresource Technology, 320*(Pt A), 124338. https://doi.org/10.1016/j.biortech.2020.124338, http://www.elsevier.com/locate/biortech.

Zhao, L., Wu, K. K., Chen, C., Ren, H. Y., Wang, Z. H., Nan, J., Yang, S. S., Cao, G. L., & Ren, N. Q. (2021). Role of residue cornstalk derived biochar for the enhanced bio-hydrogen production via simultaneous saccharification and fermentation of cornstalk. *Bioresource Technology, 330*, 125006. https://doi.org/10.1016/j.biortech.2021.125006 Epub 2021 Mar 18. PMID: 33765629.

Zhao, S., Wang, M., Zhou, D., & Pan, S. (2020). Methane yield predictive model based on the composition of biomass: Focus on the anaerobic digestion mode and regression method. *BioResources, 15*(2), 3850–3858. https://doi.org/10.15376/biores.15.2.3850-3858, https://bioresources.cnr.ncsu.edu/wp-content/uploads/2020/04/BioRes_15_2_3850_Zhao_WZP_Methane_Yield_Composition_Biomass_Anaerobic_Digestion_16986.pdf.

Zhou, H., & Wen, Z. (2019). Solid-State anaerobic digestion for waste management and biogas production. *Advances in Biochemical Engineering/Biotechnology, 169*, 147–168. https://doi.org/10.1007/10_2019_86, http://www.springer.com/series/10.

Zhou, M., & Tian, X. (2022). Development of different pretreatments and related technologies for efficient biomass conversion of lignocellulose. *International Journal of Biological Macromolecules, 202*, 256–268. https://doi.org/10.1016/j.ijbiomac.2022.01.036, http://www.elsevier.com/locate/ijbiomac.

Zhu, Y., Yang, T.-X., Li, H., Zhao, Q.-S., & Zhao, B. (2024). Synergism of jet milling and deep eutectic solvent pretreatment on grapevine lignin fractionation and enhancing enzymatic hydrolysis. *International Journal of Biological Macromolecules, 269*(Pt 2), 132144. https://doi.org/10.1016/j.ijbiomac.2024.132144.

Ziaei-Rad, Z., Pazouki, M., Fooladi, J., Azin, M., Gummadi, S. N., & Allahverdi, A. (2023). Investigation of a robust pretreatment technique based on ultrasound-assisted, cost-effective ionic liquid for enhancing saccharification and bioethanol production from wheat straw. *Scientific Reports, 13*(1), 446. https://doi.org/10.1038/s41598-022-27258-9 PMID: 36624114.

Chapter 2

Artificial neural networks and genetic algorithms for modeling and optimizing biogas production

Hassan El Bari[1], Sanae Habchi[1], Saida El Mendili[1], and Malinee Sriariyanun[2]

[1]Laboratory of Electronic Systems, Information Processing, Mechanics and Energetics, Faculty of Sciences, Ibn Tofail University, Kenitra, Morocco, [2]Biorefinery and Process Automation Engineering Center, Department of Chemical and Process Engineering, The Sirindhorn International Thai-German Graduate School of Engineering, King Mongkut's University of Technology, Bangkok, Thailand

List of abbreviations

AARE	Average absolute relative error
ACO	Ant colony optimization
AD	Anaerobic digestion
ADM1	AD Model No. 1
AI	Artificial intelligence
ANFIS	Adaptive neuro-fuzzy inference system
ANN	Artificial neural network
BP	Backpropagation
BP	Biogas production
BPNN	Backpropagation neural network
BY	Biogas yield
COD	Chemical oxygen demand
CSTR	Continuous stirring tank reactor
DE	Differential evolution
DOE	Design of Experiment
DSS	Decision support system
EBP	Error backpropagation
FIS	Fuzzy inference systems
FV	Fractional variance

Biogas in the Circular Economy. DOI: https://doi.org/10.1016/B978-0-443-29230-9.00011-5

GA	Genetic algorithm
GDR	Generalized delta rule
GPR	Gaussian process regression
HRT	Hydraulic retention time
IBP	Incremental back propagation
ID	Input data
KNN	k-nearest neighbors
LM	Levenberg–Marquardt
LMBP	Levenberg–Marquardt backpropagation
MAE	Mean absolute error
ML	Machine learning
MLFF	Multilayer feedforward
MLP	Multilayer perceptron
MLP	Multi-layer perceptron
Neus	Neurons
NN	Neural networks
OD	output data
ORP	Oxidation and reduction potential
PSO	Particle swarm optimization
RF	Random forests
RMSE	Root mean square error
RSM	response surface methodology
SVM	Support vector machines
TF	Transfer function
TOL	Total Organic Loading
TS	Total Solids
VFA	Volatile fatty acids
VS	Volatile Solids
XGBOOST	Extreme gradient boosting

2.1 Introduction

Anaerobic digestion (AD) of organic waste and residues is a biological process that operates in the absence of oxygen, producing methane-rich biogas (50%–75%) and a residue known as digestate, rich in micronutrients such as potassium and phosphorus, which can be used as biofertilizer. This is a dynamic four-stage process (hydrolysis, acidification, acetogenesis, and methanogenesis), involving several chemical reactions and many parameters (Beniche et al., 2021; El Gnaoui et al., 2022; Habchi et al., 2024; Ourradi et al., 2022). It is a well-known method for recycling organic waste on a large scale into digestate, a possible organic fertilizer, and biomethane. AD systems, in contrast to the others two biological processes (fermentation and composting), rely more on the balanced functional structure of a variety of bacteria to break down organic materials. A disruption in this synergistic balance might have a direct impact on the stability of the reaction (Tsui et al., 2023).

Numerous mechanistic or first principles models were created for the purpose of designing and refining AD methods. The cumulative biogas yield and compositions are predicted by the mechanistic models taking into account the conservation of mass and energy. The most used mechanistic AD model developed to date is the AD Model No. 1 (ADM1), which can accurately predict the time-dependent biogas output (Batstone et al., 2002; Gupta et al., 2023). Furthermore, given physicochemical processes and microbiological conversions are complicated, it takes a lot of work to calibrate parameters and may be difficult to fully comprehend potential mechanistic links (Tsui et al., 2023). In addition, because of its computationally costly features, the ADM1 model is not suitable for real-time prediction and control. Moreover, extensive calibration of the ADM1 parameter is not possible in industrial scale with variable feedstock proportions. (Gupta et al., 2023).

In recent years, the machine learning (ML), a system that can autonomously acquire and integrate knowledge, has been progressively applied to environmentally relevant domains like wastewater, air pollution, and solid waste treatments. The basic idea behind ML is to use inductive inference to generalize the links between input and output variables and then use the correlations discovered from empirical data to influence judgments in new situations. Three steps are usually involved in a whole ML process: testing, cross-validation, and training. Training stage: The ML model is trained by varying its parameters (not its hyperparameters) on the basis of the training dataset. Cross-validation stage: By adjusting the model's hyperparameters, the best model is chosen using a validation dataset. Testing step: Using a test dataset, the chosen model's performance is further assessed. The model is then prepared for use in making predictions (Guo et al., 2021). Popular ML algorithms for classification, prediction, and other tasks include support vector machines (SVMs) and artificial neural networks (ANNs) (based on supervised ML). The soft computing methods and ML-based models have emerged as an alternative to AD process modeling. The following are the drawbacks of using ML to model AD processes in contrast to mechanistic models (like ADM1): (1) faster execution times; (2) no requirement for interdisciplinary knowledge of bio-kinetics, microbiomes, and transfer of heat/mass; and (3) do not recalibrate when training on a large dataset. To forecast yield, process stability parameters [like volatile fatty acid (VFA)] and effluent quality indicators [like chemical oxygen demand (COD)], numerous regression, and classification models have been created. These include k-nearest neighbors (KNNs), RF, NN, GPR, SVM, and extreme gradient boosting (XGBOOST) (Gupta et al., 2023). Since the prediction may be based just on the data that is already available, the data-driven ML technique in this situation is less dependent on resolving the interactions included in the ADM1 model (Tsui et al., 2023).

ANN models, in contrast to ADM1, are driven by data and do not require any prior understanding of the kinetics or mechanisms behind microbial digestion. Without prior knowledge of the relationships between the independent input elements and the dependent output variables, the multilayered perceptron that represents the ANN models enables the evaluation of the relationship between them. Because of their quick and accurate computations, ANNs are effective tools for figuring out the relationship between variables in nonlinear and nonstationary systems. As a result, ANNs have been used in AD systems to improve process parameters, estimate the rate of BP or process intermediates, and maximize bioreactor performance (Cruz et al., 2023). A feedforward ANN model translates sets of input data into a suitable set of outputs. A forward-looking ANN is trained using a supervised learning method called backpropagation, which determines the gradient of the loss function in relation to the ANN weights. It is frequently a component of algorithms that modify the weights of the network to maximize performance (i.e., gradient descent method) (Alejo et al., 2018).

The aim of this work is to show the importance of ML modeling on AD performance, mainly by using the GA and ANN algorithms for optimization and prediction biogas production. Particularly, this chapter focuses on the ANN method for biogas prediction using suitable algorithms. This chapter consists of five sections. The first and second sections investigate ML, devoting attention to its different methods, mathematical aspects, and efficiency factors. The third section highlights the different AD modeling approaches (ADM1, RSM, and ANN), including their comparisons in terms of accuracy and efficiency, while the last two sections focus on optimization and prediction of biogas production using GA and ANN methods.

2.2 Machine learning and mathematical algorithms

ML involves methods enabling a machine to evolve through learning and tackling problems beyond traditional algorithmic capabilities. It analyzes complex systems, including data with symbolic or categorical values, graphs, or temporal curves. This contrasts with classic attribute-value pairs, marking a shift toward continuous data analysis (Hashem, 2016). The first stage of analysis is classification, which aims to label each data point by associating it with a class.

Different learning systems exist which are as follows (Hashem, 2016):

1. Supervised learning, where the system learns to classify according to a predetermined classification model.
2. Unsupervised learning or automatic classification, where the system does not have examples of data and the algorithm must discover by itself the more or less hidden structure of the data.
3. Semisupervised learning, carried out probabilistically or not, aiming to reveal the underlying distribution of data examples in their description space.

4. Partially supervised learning, carried out probabilistically or not, when data labeling is partial.
5. Reinforcement learning, where the algorithm learns a behavior given an observation.

The algorithms used enable a computer-driven or computer-assisted system to adapt its analyses and behaviors based on empirical data from databases or sensors. The challenge is that the range of possible behaviors for all potential inputs quickly becomes too complex to describe with current programming languages. Therefore, programs are designed to learn and self-improve their analysis or response systems, which is a form of artificial intelligence. These programs, depending on their level of sophistication, incorporate capabilities in probability and statistics, data processing, and possibly analysis of sensor data, recognition (speech recognition or pattern recognition), data mining, and theoretical computer science (Hashem, 2016).

2.2.1 Support vector machines

SVMs, also known as large margin classifiers, are supervised learning models utilized for document classification. They create models of input–output mappings. SVMs typically operate as linear learning systems, constructing two-class classifiers based on linear combinations of features. The discrimination between positive and negative classes relies on the sign of this linear combination (Xue et al., 2011).

Suppose we have training examples A = $\{(x_1, y_1), (x_2, y_2),..., (x_n, y_n)\}$, where $x_i = (x_{i1}, x_{i2}, x_{ik})$ is an input vector of dimension k, and the target values $y_i \in \{-1,1\}$ (where values 1 and –1 denote the positive and negative classes, respectively). The SVM involves defining a separation hyperplane [a linear function $f(x)$], which separates the positive and negative examples from each other with a maximum margin. The margin is the distance between the hyperplane and the nearest samples.

The function $f(x)$ is defined as follows (Eq. 2.1):

$$f(x) = < w, x > + b \tag{2.1}$$

where $< w, x >$ denotes the Euclidean dot product of x and w (weight vector). b is called bias.

An input vector x_i is assigned to class y_i as follows (2.2):

$$y_i = \begin{Bmatrix} 1 & \text{if } f(x_i) \geq 0 \\ -1 & \text{if } f(x_i) = 0 \end{Bmatrix} \tag{2.2}$$

2.2.2 K-nearest neighbors

While previous learning methods derive models from training data, the KNN classification method stands out as a lazy learning approach. Here no model is constructed during training. Instead the learning phase entails memorizing examples in an optimized manner for efficient retrieval later. The calculations are performed in the classification phase. Let A be the training dataset whose categories are known. When a new data point *t* is presented for classification, the algorithm calculates the distance or similarity to each element in the set A. Then, the *k* closest examples, known as the KNN of *t*, are selected. *t* is then assigned the class most common among its KNN.

The key component of a KNN algorithm is the distance or similarity function, chosen based on the application and data type. Cosine similarity is commonly used for textual documents. The optimal number of neighbors (*k*) is determined through cross-validation to achieve the best accuracy. Generally, $k = 1$ is insufficient due to outliers. Although KNN is simple and effective for text classification, it is time-consuming because each test instance must be compared to all training data, which slows down classification with large datasets.

2.2.3 Decision trees

Decision trees are commonly used for classification and prediction. They serve as decision support tools, representing a series of choices in a tree-like graphical form. The tree's branches lead to leaves, which represent decisions and their possible consequences, including the outcomes of random events. The paths from the root to the leaves signify the classification rules, with the decision-making process concluding at the leaves (Solorzano & Plevris, 2022).

A decision tree is a structure similar to an organizational chart and can be transformed into a set of conditional control rules (if–then). When the variable to be predicted is categorical and its values represent classes, we refer to it as a classification tree. A decision tree consists of three types of nodes: chance nodes, decision nodes, and terminal nodes. A chance node illustrates multiple uncertain outcomes. A decision node represents a decision to be made, and a terminal node represents the final result of a decision path. Decision trees are highly effective tools, which are easy to understand and interpret. However, calculations can become very complex, especially if many values are uncertain and/or if many outcomes are linked.

2.2.4 Regression modeling methods

Regression analyses, like the least squares method, traditionally explore correlations between target outputs and independent variables. However, their vulnerability to outliers is a drawback. Polynomial and logistic regression offer solutions, each with unique processing methods. Polynomial methods address nonlinear data relationships through polynomial equations. Logistic regression suits discrete problems, while ridge regression handles multicollinearity effectively. Kernel smoothing, a nonparametric technique, maps nonlinear data into higher-dimensional spaces. Regression methods, compared with other ML algorithms, are less complex and offer faster computational speeds (Solorzano & Plevris, 2022).

2.2.5 Artificial neural network

This algorithm, popular in the 1980s and 1990s, draws inspiration from the human brain's functioning. It relies on "artificial" neurons for learning, defined as nonlinear, parameterized, bounded-value algebraic functions. In general, an ANN can be segmented into three parts, referred to as layers (Xue et al., 2011):

Input layer: This initial layer receives data from the external environment, typically normalizing inputs using activation functions. This normalization enhances numerical accuracy for subsequent mathematical operations, as discussed earlier.
Hidden layers: These intermediary layers consist of neurons tasked with extracting patterns relevant to the analyzed system. They carry out the bulk of internal processing.
Output layer: Comprising neurons, this final layer generates and presents the network's outputs (such as opinion classification) resulting from the processing conducted by neurons in prior layers.

Originally, ANNs were either "single-layer" or "multilayer" architectures, as illustrated in Figs. 2.1 and 2.2 respectively. In a single-layer network, there is only one input and one output layer, with information flowing unidirectionally from input to output.

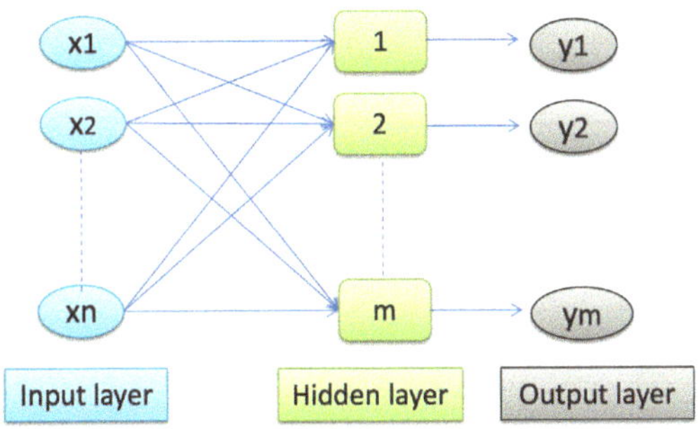

FIGURE 2.1 Single-layer ANN.

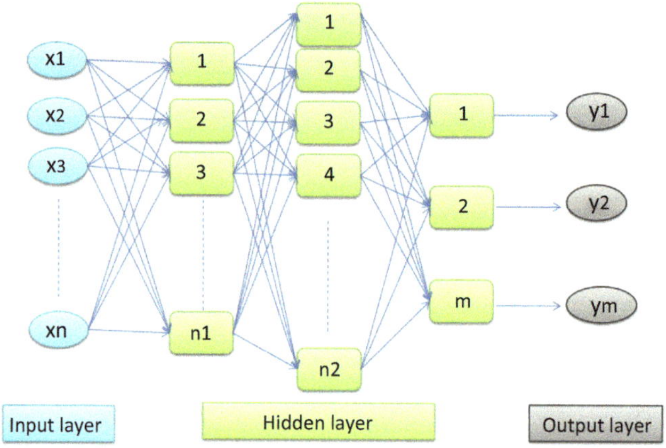

FIGURE 2.2 Multilayer ANN.

As shown in Fig. 2.1, the number of neurons in the first hidden layer may differ from the signals in the input layer. This variation depends on the problem's complexity and the available data quality and quantity. A layer n is composed of a certain number of artificial neurons whose inputs are the outputs of the neurons from layer $n-1$. Neural connection is achieved through unidirectional weighted links. Thus we can interpret a neural network as a directed graph whose nodes are the artificial neurons (Xue et al., 2011).

2.3 Relevance and efficiency factors

In ML methods, the quality of learning and analysis relies on the upfront requirement and the operator's competence in preparing the analysis. It also depends on the complexity of the model and its suitability to the subject matter at hand. Furthermore, the quality of work will also depend on how the results are visually highlighted for the end user (relevant outcomes might be obscured in an overly complex diagram or poorly highlighted by an inappropriate graphical representation). Before this, the quality of work hinges on initial constraining factors related to the current database (Xue et al., 2011):

1. The number and quality of attributes describe these examples, implying that the distance between two numerical data examples (price, size, weight, light intensity, and noise intensity) is straightforward to establish, while the distance between two categorical attributes (color, beauty, and utility) would be more intricate: **(1)** the percentage of populated data and **(2)** the noise, number, and location of questionable (error) values or those naturally deviating from the general distribution pattern of examples in their distribution space will impact the quality of the analysis.

As previously discussed, once a "classifier" is selected and built, assessing its performance and accuracy becomes crucial. Various methods exist for evaluating a "classifier," with classification accuracy serving as the primary metric (Solorzano & Plevris, 2022). This metric indicates the proportion of documents in the test set correctly classified relative to the total of documents number relative to the test set. Subsequent sections will introduce other frequently employed evaluation metrics and techniques for classifiers.

2.3.1 Cross-validation

Cross-validation is invaluable for small datasets, allocating a portion for model testing during the learning phase. Data are split into = subsets, with one serving as the test set and the rest forming the training set (Piche et al., 2000).

This process repeats n times, yielding n accuracies. The average of these n accuracies represents the final estimated accuracy. Commonly used are 10-fold and fivefold cross-validations, suitable for parameter estimation as well.

2.3.2 Precision and recall

Precision and recall are two statistical measurement criteria for evaluating classifiers, also known as positive predictive value and sensitivity. We denote the following:

TP: The number of correctly labeled positive examples (true positives).
FN: The quantity of false negatives or positive cases that were misclassified.
FP: The quantity of instances that were mislabeled as positive (false positives).

TN: The quantity of negative examples that were accurately categorized as true negatives. Precision (p) for a class represents the ratio of true positives to the total elements categorized as positive (2.3):

$$p = \frac{\mathrm{TP}}{\mathrm{TP} + \mathrm{FP}} \tag{2.3}$$

Recall (r) is defined as the ratio of true positives to the total elements belonging to the positive class (2.4).

$$r = \frac{\mathrm{TP}}{\mathrm{TP} + \mathrm{FN}} \tag{2.4}$$

A precision score of 1 for Class C means that every associated element is correctly labeled, while a recall score of 1 indicates that all Class C elements are accurately labeled. However, this does not account for incorrectly labeled elements or the inverse relationship between precision and recall (Juba & Le, 2019).

2.3.3 *F*-score

F-score, or *F1*-score, is a widely used metric that combines both precision and recall to assess a test performance. It provides a single metric for comparing different classifiers (2.5).

$$F = 2 * \frac{p * r}{p + r} \tag{2.5}$$

The *F*-score referred to as the *F1*-score or *F*-measure represents the weighted harmonic mean of precision and recall (Eq. 2.6).

$$F = \frac{2}{\frac{1}{p} + \frac{1}{r}} \tag{2.6}$$

ML revolutionizes industries like manufacturing, retail, healthcare, finance, and more by analyzing extensive data, refining decision-making, and predicting outcomes accurately. Challenges remain, including a lack of expertise in applying ML to business problems, understanding its integration into daily operations, ensuring data quality, and addressing technological gaps in utilizing big data for algorithms.

Data modeling provides a visual aspect for managing data resources and creating fundamental architectures, thus optimizing data usage in applications while reducing processing costs. All modeling techniques have strengths and weaknesses in their approach to the user. Most available techniques are suitable for developer operation, while very few are dedicated to users with limited technical knowledge. Moreover, these techniques produce complex data models in an attempt to cover all aspects, at the expense of simplicity of use of the solution and data readability.

2.4 Anaerobic digestion modeling for biogas production

2.4.1 Anaerobic digestion model 1

Over the years, several models have been developed to describe AD. As shown, most of the time, these models have been built to meet specific needs, almost always related to the type of substrate to be degraded. ADM1 is a structured mathematical model, using COD as a common unit to quantify the substrate to be digested, the digestion products, and the biomass. The model describes the AD process using a continuously fed CSTR reactor as a reference.

The biochemical transformations included in ADM1 are as follows:

1. Disintegration of particles (compound materials),
2. Hydrolysis of particulate COD (proteins, lipids, and carbohydrates),
3. 19 bacterial degradation and lysis processes, and
4. 24 particulate components (composite materials, carbohydrates, proteins, lipids, inerts, and seven bacterial groups).

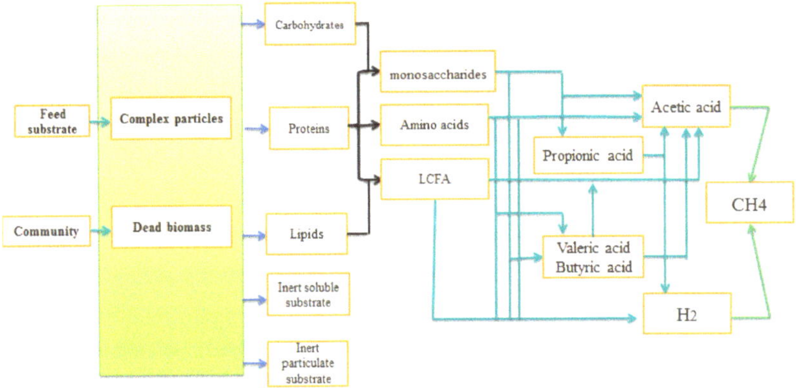

FIGURE 2.3 Metabolic model proposed by ADM1.

ADM1 takes into account five digestion stages. What is new is the introduction of the first stage, which models the disintegration of the particulate substrate into particulate proteins, lipids, and carbohydrates. Fig. 2.3 shows the metabolic steps proposed by ADM1.

The extensive approach of ADM1 uses a variety of kinetic equations. The viability of scaling up to simulate the effects of substrate effects and operating modes on digesters has been demonstrated. However, the insufficient knowledge of potential mechanistic linkages may impede the needs for prediction due to the complexity of physicochemical reactions and microbiological conversions, and it also necessitates a substantial amount of work for parameter calibration (Tsui et al., 2023). Numerous inhibitory balances, six acid–base reactions, three gas transfer reactions, 19 process rate equations, and more than 30 equations of water phase relative to soluble and particulate matter, are all included in the ADM1 model. Additionally, a number of kinetic parameters are needed by ADM1, and each process's rates must be predicted. As a result, the application is restricted to a particular stage of the process due to the complexity of the microbial degrading processes (Beltramo et al., 2019). Numerous research works address ADM1 in great depth in the realm of dynamic AD process modeling. The outcomes of this research work were very transparent, because they also released programs and source code (Heiker et al., 2021). However, because of its computationally intensive nature, the ADM1 model is exceedingly unfeasible to use for real-time prediction and control purposes. Additionally, significant ADM1 parameter calibration is not feasible with large-scale industrial AD processes with fluctuating feedstock contents (Gupta et al., 2023).

Compared with ML and ANN modeling, what are the drawbacks and difficulties encountered in ADM1 approach to modeling biogas production?

ANNs are simpler to form than other models now in use, such as the ADM1. Additionally, a number of kinetic parameters are needed by ADM1, and each process's rates must be predicted. As a result, the application is restricted to a particular stage of the process due to the complexity of the microbial degrading processes (Beltramo et al., 2019).

2.4.2 Response surface methodology

Response surface methodology (RSM) is used to estimate the ideal conditions, analyze the influence of inputs, and create and construct empirical models. The use of second-degree polynomial (quadratic) models to simulate experimental responses was first introduced by Box and Wilson (1951). RSM is made to maximize a response that is influenced by several different factors. With regard to two independent variables (x_1 and x_2) and an output or response variable [y; (2.7)], the two variables (x_1 and x_2) will determine the response (y):

$$y = f(x_1, x_2) + \varepsilon \tag{2.7}$$

The error margin that could be seen in response y is represented by ε. Therefore, the surface shown by (2.8) below is the response surface:

$$\eta = f(x_1, x_2) \tag{2.8}$$

RSM provides a benefit over conventional optimization techniques in terms of the reduced number of experiments and the multivariate interactivities across a response. Two additional benefits of RSM include its ability to identify optimal conditions and quantify the response–input variable relationship by assessing how sensitive these conditions are to variations in the experimental variables. Additionally, RSM allows for the creation of projections, which facilitate the visual interpretation of this relationship. The main advantage of RSM is still the time and money savings from fewer experimental trials.

2.4.3 Artificial neural network

ANN is a mathematical approach rooted in nonlinear regression, belonging to a variety of soft sensor methods. Its widespread adoption across various fields stems from its ability to effectively capture the intricate nonlinear relationships among variables (both input and output) within complex systems, thanks to its adeptness in summing, self-learning, and training functions (Wang et al., 2018).

The ANN, derived from nonlinear regression, stands as one of several mathematical methods within the realm of soft sensor techniques. Its broad application across diverse domains owes to its proficient ability in summing, self-learning, and training, enabling the capture of nonlinear interactions among variables (both multi-input and output) within complex systems (Cruz et al., 2022).

ANNs, also known simply as neural networks, represent highly sophisticated computer modeling techniques utilized within the realm of ML to forecast outcomes of processes. These networks emulate the functioning of biological neural networks found in the human brain, particularly in how signals are transmitted between nodes. Numerous studies have explored the utilization of ANNs to enhance anaerobic reactions. For instance, feedforward backpropagation ANNs are utilized to examine how substrate type and organic loading rate impact methane production, with the goal of attaining methane purity levels within the range of 60%–70% (Chen et al., 2022).

ANN comprises an input layer, one or multiple hidden layers, and an output layer. Within this architecture, each layer consists of nodes, akin to neurons in biological systems, which establish connections with the subsequent layer. During the training phase, the weights assigned to these nodes are subject to adjustment. These weight modifications serve to either amplify or diminish the signal strength originating from the interconnected neuron.

An ANN model undergoes three developmental stages: training, testing, and validation, which necessitates dividing the dataset into three segments. Typically, 70%–80% of the dataset is allocated for training, while the remaining portion is utilized for testing and validation. Training a feedforward neural network involves two primary procedures: the feedforward process and the backpropagation method.

The feedforward technique involves transmitting the input through the neural network's hidden layer before reaching the output layer, where a prediction value is generated utilizing the activation function, bias, and initially assigned weights (Chen et al., 2022). Subsequently, the weights undergo adjustment to reduce the disparity between the experimental and predictive values, derived from the input training data, initiating the backpropagation phase. Specific backpropagation training methods cater to this process (Chen et al., 2022).

2.4.3.1 Optimization of artificial neural network

Optimizing an ANN model involves considering four key components of its architecture. First, the total count of hidden layers is significant. Second, the number of neurons within the input, hidden, and output layers must be determined. Third, selecting the appropriate activation function for each layer is essential. Lastly, the choice of training algorithm significantly impacts performance. Nevertheless, it is crucial to acknowledge that the performance of the model can also be influenced by the quality and quantity of training data. Determining the optimal architecture involves comparing various ANN designs through a trial-and-error method (Chen et al., 2022).

Neural network techniques typically comprise three primary layers: input, hidden, and output layers, arranged sequentially. The hidden layer, often consisting of multiple layers, lies between the input and output layers. Within the hidden layer, every neuron communicates with neurons in adjacent layers. Gradient descent is a technique employed by ANNs to learn the weights that facilitate mapping any input to any output within the network's connections.

2.4.3.2 *Artificial neural network model*

To simulate both linear and nonlinear systems, the ANN structure integrates and interprets the influence of each process element on the output function. Compared with complex models with physical connections, it is considered a straightforward and swift approach for predicting the performance of various operations. ANN does not rely on an underlying physical relationship to establish a mathematical connection between the process response (output data) and the model. Instead, it achieves this by utilizing experimental data gathered at different process parameters, often referred to as input data.

In an ANN, the independent and dependent process parameters are linked by nodes known as neurons. Through the TF, neurons can receive, retain, and adjust one or multiple input signals at different strengths, influenced by the input data, and contribute to the output data. The accuracy of predictions made by the ANN model can be improved by implementing error correction learning strategies (Almomani, 2020)

2.4.3.3 *Artificial neural network design and training—Model evaluation*

For the purpose of generalization, it is crucial to evaluate ANN prediction accuracy following training. As stated in Eqs. (2.9)–(2.11), three metrics were used in this evaluation of the models: correlation coefficient (R-value), mean absolute error (MAE), and root mean square error (RMSE), respectively.

$$\text{RMSE} = \sqrt{\frac{1}{N}\sum_{i=1}^{N}\left[\left(Y_{iexp} - Y_{ipred}\right)\right]^2} \tag{2.9}$$

$$\text{MAE} = \frac{1}{N}\sum_{i=1}^{N}\left|Y_{iexp} - Y_{ipred}\right| \tag{2.10}$$

$$R = \frac{\sum_{i=1}^{N}\left(Y_{iexp} - \overline{Y_{iexp}}\right)\left(Y_{pred} - \overline{Y_{pred}}\right)}{(N-1)\left(\sigma_{\exp} - \sigma_{pred}\right)} \tag{2.11}$$

N is the quantity of experimental data in this case. The values that correspond to experimental values, predicted values, the mean of experimental values, and predicted values are Y_{iexp}, Y_{ipred}, Y_{iexp}, and Y_{ipred}, respectively. The values of $\sigma_{\exp}$ and σ_{pred} correspond to the expected and experimental outcomes (Cruz et al., 2023). ANNs are widely used mathematical techniques for biological system optimization. A system of ANNs was developed to forecast the rate of biogas generation. GAs and an ant colony optimization were used to carry out the variable selection. They decreased the model dimension, found the important process factors, and enhanced the ANN models' ability to predict. Due to the complications of the phenomena involved in the multiple stages of aerobic and anaerobic wastewater treatment, as well as the numerous interaction subprocesses that are operating concurrently at various process scales but have not yet been identified, it is necessary to address concerns regarding the application of conventional modeling in addition to using specialized techniques to identify patterns and correlations within these data sets. Among these are ANNs, which have shown to be effective instruments for modeling and predicting the path of other complicated processes.

Due to the intricate feedback complexity and often unknown interconnections among various processes, especially nonlinear ones across different scales, traditional regression techniques frequently yield unsatisfactory results when attempting to effectively model objects, phenomena, or processes. In such scenarios, an ANN trained solely on a representative dataset of input–output pairs can construct the necessary data processing framework to accurately correlate the input set with related outputs or effects, thereby revealing underlying patterns of interrelationships (Sakiewicz et al., 2020).

Various modeling and computational applications can benefit from different types and topologies of neural networks. The following network types are commonly employed to predict object responses based solely on their sampling under potentially diverse operational conditions: multilayer perceptron (MLP) featuring nonlinear continuous activation functions, typically of the sigmoid type, and linear networks with linear activation functions.

$$f(x) = \frac{1}{1 + e^{-x}} \tag{2.12}$$

$$f(x) = \tanh(\beta x) \tag{2.13}$$

Eq. 2.12 is the unipolar activation function, and Eq. 2.13 is the bipolar activation function.

β is the numeric parameter. It has to be adjusted for the course change of the activation function.

The backpropagation (error) algorithm, or BP algorithm, regulates the weight update strategy inside the multilayer ANN structure by using gradient optimization techniques on a continuous goal (energetic) function that is specified as follows:

$$E(w) = \frac{1}{2} \sum_{k=1}^{M} \left[f\left(\sum_{i=0}^{K} w_{ki}^{(2)} \cdot v_i \right) - d_k \right]^2 \tag{2.14}$$

where M is the number of neurons in the output layer and w is the matrix of weight values. vi is the output signal of the selected ith neuron from the hidden layer, and d_k is the target signal of the kth neuron in the output layer (in supervised training mode).

The conjugate gradient (CG) operates within the selected weight matrix hyperspace, traversing conjugate directions without mutual interference. An even more efficient application compared withthe BP technique is the correlated CG, which identifies the direction of the largest gradient within a chosen weight matrix hyperspace point and navigates conjugate directions without interference (Sakiewicz et al., 2020).

To accomplish this prediction in regression or classification models, one can use recently developed algorithms such as KNN, logistic regression multiclass (GLMNET), RF, SVM, ANN, and RF (Wang et al., 2020). Recently, AI methods, including ANNs, have shown potential in predicting the kinetic behavior of specific engineering processes. ANNs have been highlighted as one of the most effective approaches for accurately forecasting the behavior of both linear and nonlinear processes.

The prediction accuracy for specific output functions is typically influenced by the input data points. Various ML algorithms, including SVM, ANN, KNN, RF, and the nonlinear and intricate interactions in the AD process, are modeled by ANFIS. For example, ANN models have reliably predicted biogas output in well-controlled laboratory-scale research (Chen et al., 2022).

Different kinds of ANNs emerged as the frequently used ML techniques in the AD sector. Three-quarters of all the publications that were reviewed were ANN-based studies, with the remaining percentages being ANFIS (10%), SVM/RF/ GA (8%), and PSO (4%). It is important to note that both PSO and GA are well-known optimization techniques that are typically coupled to create hybrid algorithms. Of those, 15% of the examined studies had optimization of some kind, and most of them combined ANN, GA, and PSO. Furthermore, 14% of the total was accounted for by additional models, including XGBoost, KNN, ACO, and multiclass logistic regression (GMLET).

ANNs, a type of ML, have been used more frequently in recent years to simulate and optimize processes in place of mechanistic models. ANNs have been used in many research to optimize AD, with high R^2 values of up to 0.998 (Chen et al., 2022). Similar to this, Sathish and Vivekanandan also employed ANN to maximize the AD process by utilizing four input variables: substrate concentration, pH level, agitation time, and temperature. They are able to estimate the production of biogas with a satisfactory accuracy of $R^2 = 0.998$ (Sathish & Vivekanandan, 2016).

Powerful technologies such as ANN and ANFIS can be harnessed to attain high yields of value-added products during process development. When integrated with evolutionary algorithms like PSO, ACO, or GAs, the utilization of ANN and ANFIS tools has yielded impressive outcomes. These approaches facilitate the establishment of optimal operating conditions for the efficient production of enzymes, lignin, organic acids, bioethanol, and biogas (Pradhan et al., 2022).

In fact, in recent years, researchers have increasingly focused on artificial intelligence-based nonconventional empirical modeling methodologies, such as fuzzy logic, ANNs, and ANFIS, to model and optimize processes for lignocellulosic biomass (LCB) valorization. ANNs are mathematical techniques that process data, mimicking the structure of real neurons. A fundamental component of the ANN model is the neuron, which receives multiple sources of input data, combines them, and then subjects them to a nonlinear process before producing results. Two key advantages of ANNs are their ability to learn from examples and identify patterns in datasets without making assumptions about the nature of the data or its relationships (Pradhan et al., 2022).

ANNs have garnered significant attention in recent years as versatile tools for mapping, grouping, regression, classification, and predictive modeling. Due to their adaptability, ANNs are well-suited for handling diverse data types and addressing nonlinear problems effectively. Consequently, they have been extensively employed to solve modeling and optimization challenges across various fields, including food science, medicine, biotechnology, and more. To enhance the predictive capabilities of ANN models and streamline the optimization process, ANNs have been occasionally coupled with metaheuristic search algorithms such as ant colony optimization (ACO), particle swarm optimization (PSO), and genetic algorithms (GAs). This fusion of ANNs with metaheuristic algorithms enables more efficient exploration of complex solution spaces and often leads to improved model performance (Pradhan et al., 2022).

ANFIS represents a hybrid modeling approach that amalgamates the strengths of fuzzy inference systems (FIS) with ANNs. ANFIS combines the learning capabilities of ANNs with the reasoning capabilities of FIS, resulting in a powerful modeling tool.

However, before employing these modeling tools in a biomass conversion process, several factors should be considered. These include the quantity and quality of available data, the data partitioning strategy for model training, validation, and testing, the architecture of the ANFIS model (including the number of hidden layers and neurons in each layer), the selection of training algorithms, and the choice of transfer functions. These factors significantly influence the accuracy and reliability of the ANFIS tool for modeling biomass conversion processes. Therefore, careful consideration and optimization of these factors are essential for achieving accurate predictions and optimal performance (Pradhan et al., 2022).

GAs have proven highly effective for solving complex and nonlinear problems, making them particularly useful in biological development processes. GAs offer flexibility and adaptability across various models, as noted by Cruz et al. (2023).

ANNs are rooted in the biomimicry of information processing and signal transmission akin to neurons in the brain's higher layers. ANNs, composed of hidden and output layers, learn data patterns by adjusting neuron weights iteratively during model creation. Activation functions such as sigmoid, softmax, and hyperbolic tangent functions have been developed to enhance the efficiency of operations on weighted neuron sums, as highlighted by Karlik and Olgac (2011)

ANNs are frequently chosen for modeling nonlinear relationships due to their capability to achieve remarkable prediction accuracy, even with high-dimensional inputs, as noted by Tsui et al. (2023). Several deep learning architectures, including convolutional networks, spatiotemporal graphs, recurrent neural networks, long short-term memory, and convolutional neural networks, offer computationally efficient solutions for handling time-series applications in biorefinery systems.

However, ANNs often suffer from a lack of interpretability and are characterized by their black-box nature. To address this limitation, model-independent interpretive ML approaches like global surrogate models and feature permutation techniques can offer more user-friendly alternatives.

It is worth noting that creating precise neural networks typically demands large quantities of data, which in turn leads to higher computing costs (Tsui et al., 2023).

2.5 Artificial neural networks for biogas prediction

ANNs have been employed in numerous research endeavors to forecast the generation of biogas in AD systems, both during their initial phases and throughout sustained operation. Additionally, ANN has been utilized to approximate the composition of biogas produced (Almomani, 2020). Dahunsi et al. (2016) utilized ANNs to forecast the biogas production from AD, considering factors such as TOL, HRT, TS, VS, and pH. Holubar et al. developed an ANN model aimed at forecasting the potential biogas output, methane production rate, and gas composition throughout lignocellulosic AD. The model considered pH, COD, concentrations of VFAs, and VS as determining factors (Holubar et al., 2002). As mentioned earlier, process modeling serves as a valuable tool for forecasting and elucidating the behavior of biological processes. ANN has the capacity to be transformed into process models and proves effective in this regard, primarily because they can encapsulate the nonlinear relationships among variables (multi-input/output) within intricate systems (Antwi et al., 2017).

2.5.1 Multilayer full feedforward network

A multilayer feedforward neural network is made up of interconnected perceptrons. Data and computations flow from inputs to outputs in a

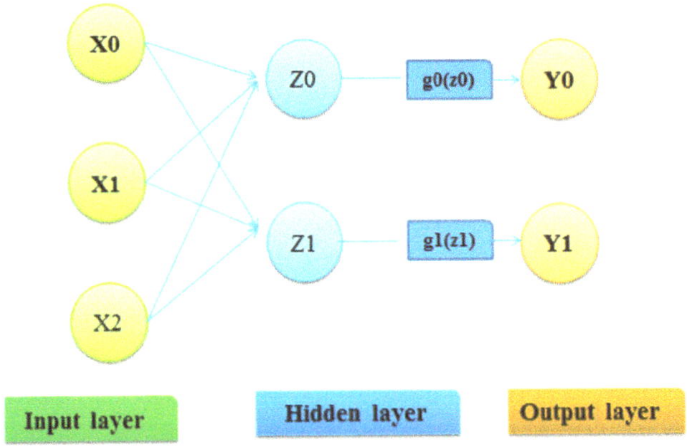

FIGURE 2.4 A single-layer feedforward neural network.

unidirectional manner. Such a neural network's layer count is = to the number of perceptrons in it. Fig. 2.4 shows an example of this architecture in a network. Because only activation calculations are performed by the output layer, this structure is technically referred to as a one-layer feedforward network with two outputs.

The inputs of this single-layer feedforward neural network are directly connected to the Z_1 and Z_2 perceptrons in the output layer. These output perceptrons produce the outputs Y_1 and Y_2, respectively, by using activation functions, denoted as g_1 and g_2:

$$Z_1 = \sum_{i=1}^{3} W_{1,i}x_i - \mu_1 \text{ and } Z_2 = \sum_{i=1}^{3} W_{2,i}x_i - \mu_2 \tag{2.15}$$

$$Y_1 = g_1(Z_1) = g_1\left(\sum_{i=1}^{3} W_{1,i}x_i - \mu_1\right) \tag{2.16}$$

$$Y_2 = g_2(Z_2) = g_2\left(\sum_{i=1}^{3} W_{2,i}x_i - \mu_2\right) \tag{2.17}$$

A linear regression model is identical to a single-layer neural network where the activation functions g_1 and g_2 are identity activation functions. Similarly, the single-layer neural network is the same as logistic regression, provided g_1 and g_2 are logistic activation functions. Single-layer neural networks are rarely utilized in place of these conventional regression techniques because of the clear relationship that exists between single-layer neural networks and linear/logistic regression models.

2.6 Neural network error calculations

Error computations are, in fact, quite important to neural network training. Scholars have investigated a range of error computation techniques in an effort to find ones that have quick training times appropriate for the network's intended use. Error estimations are generally not the same, mostly due to the particular application requirements of the network. The most widely used error function for predicting jobs is the sum-of-squared errors, or one of its scaled versions. This is similar to how linear regression uses the minimal least squares optimization criterion. Similar to least squares, the sum-of-squared errors are calculated by measuring the squared difference between each training instance's predictions made by the network and the associated target or observed value. The equation is formalized as half of the conventional least squares error:

$$E = \frac{1}{2} \sum_{i=1}^{N} \sum_{j=1}^{C} (t_{ij} - \hat{t}_{ij})^2 \tag{2.18}$$

where t_{ij} is the observed output for the i_{th} training case and the j_{th} network output. The number of network outputs is C, while the total number of training cases is N, and $\hat{t}_{ij}$ is the network's forecast for that case.

Fitting a different network for each predicted variable is standard procedure. It is advised to set the number of outputs, C, to 1 when forecasting with a multilayer feedforward neural network. It is common practice to allocate one output per class in classification assignments with more than two classes; that is, C = number of classes.

The sum-of-squared mistakes in ordinary least squares are not multiplied by half. This has no effect on the answer in the end, but it significantly lessens the computational strain during training. Furthermore, the sum-of-squared errors likewise rise with the number of training patterns. As a result, using the RMS error rather than the unscaled sum-of-squared errors is frequently advantageous:

$$E^{\mathrm{RMS}} = \frac{\sum_{i=1}^{N} \sum_{j=1}^{C} (t_{ij} - \hat{t}_{ij})^2}{\sum_{i=1}^{N} \sum_{j=1}^{C} (t_{ij} - \bar{t})^2} \tag{2.19}$$

where $\bar{t}$ is the average output:

$$\bar{t} = \frac{\sum_{i=1}^{N} \sum_{j=1}^{C} t_{ij}}{N.\,C} \tag{2.20}$$

In contrast to the unscaled sum-of-squared errors, the E^{RMS} grows with decreasing N. The network forecasts its training targets more accurately when the E^{RMS} value is smaller. When $E^{\mathrm{RMS}} = 0$, all training targets are accurately predicted by the network. On the contrary, $E^{\mathrm{RMS}} = 1$ indicates that the network's forecasts are comparable to if every estimate is equal to the mean of the training targets.

$$E^{\mathrm{RMS}} = \frac{2}{\bar{t}}.\ E \tag{2.21}$$

2.6.1 Incremental back propagation

In the 21st century, the emergence of ML has introduced novel approaches for addressing the complexity of unknown process mechanisms. This has mitigated the necessity for exhaustive studies of AD mechanisms to enhance feedback control of the process (Ge et al., 2023). Data-driven research methodologies, unlike mechanism-based approaches such as the ADM1, circumvent the need to address the numerous complex changes inherent in the AD process. Consequently, they have made greater strides in progress. For instance, Abu Qdais et al. (2010) utilized ANN coupled with GA to forecast and optimize methane production configurations. Antwi et al. utilized a three-layered feedforward backpropagation (BP) ANN that incorporated a multiple nonlinear regression model. Their aim was to estimate methane yield in anaerobic upflow sludge blanket reactors. Their findings suggested the potential for controlling the AD process (Antwi et al., 2017).

In the study by Beltramo et al. (2016), the hidden layer of the neural network played a pivotal role in transforming the input data. During the training process, the input patterns were fed into the network's input layers, and the output was computed in the layer output, as described by Liu et al. (2016). The BP learning rule gave the network a way to modify its weights so that, for each given input pattern in the training set, it could generate outputs that nearly matched the expected output patterns, as outlined by Cheng et al. (2016). The output neuron experienced additional change based on the inputs from the hidden neurons. Equation (2.22) was used to predict the output of the BP-ANN with a single output neuron and a hidden layer:

$$\hat{Y} = f_0\left[\sum_{j=1}^{HN} WO_j * f_h\left(\sum_{i=1}^{m} WH_{ij} * X_{it} + b_j\right) + b_0\right] \tag{2.22}$$

The neural network architecture involves various components and functions. Here is a breakdown of the elements mentioned as follows:

- WH_{ij} : Link weight between the j_{th} hidden neuron and the i_{th} input
- m: Input neurons number
- WO_j : The weight of the connection between the output neuron and the j_{th} hidden neuron.
- f_h : The hidden neurons' activation function.
- f_0 : The output neurons' activation function of.
- b_j: Bias of the j_{th} hidden neuron.

- b_0: Bias of the output neuron.
- X_{it}: Input variable.
- *HN*: Number of hidden neurons.

The activation functions for the hidden and output layers are specified as follows:

- Hidden layer activation function: Tangent sigmoid transfer function (tansig) is represented by Eq. 2.23.
- Output layer activation function: Linear transfer function (purelin) is represented by Eq. 2.24.

These equations provide the mathematical framework for the operations within the hidden and output layers of the neural network.

$$f(x) = \frac{2}{(1 + \exp^{-2x})} - 1 \tag{2.23}$$

$$f(x) = x \tag{2.24}$$

where x is the vector of inputs.

For the ANN model, the original dataset was split into three subsets at random: train, validation, and test. Seventy-eight data points (thus 70% of the original dataset) were assigned to the training set, and 17 data points (15% of the original dataset) were chosen for the validation and testing subsets, according to Antwi et al. (2017). The test set was employed to generate predictions.

Next, two metrics were used to assess the performance of the BP-ANN models: the fractional variance (FV) and the index of agreement (IA), which are represented in Eqs. (2.25) and (2.26), respectively. These metrics shed light on the degree of agreement between the actual and projected values as well as the variability that the model could account for.

$$\mathrm{IA} = 1 - \frac{\sum_{i=1}^{n} (P_i - O_i)^2}{\sum_{i=1}^{n} |P_i - O_m| + |O_i - O_m|)^2} \tag{2.25}$$

$$\mathrm{FV} = \frac{2(\delta_O - \delta_p)}{(\delta_O + \delta_p)} \tag{2.26}$$

In the context of the performance evaluation metrics for the BP-ANN models, the variables are defined as follows:

- O: Experimental data or observed values.
- P: Anticipated values or predicted values generated by the BP-ANN models.
- m: Arithmetic mean of the observed data points.
- δ: Standard deviation of the observed data points.

These variables are used to calculate the IA and FV to assess the accuracy and reliability of the BP-ANN models in predicting the outcomes.

2.7 Artificial neural networks for biogas optimization

The optimization process was conducted by targeting the peak efficiency operating point, aiming to minimize emissions while minimizing efficiency loss. However, the optimization technique employed involved systematically adjusting operating parameters one at a time until the optimal conditions were attained (Alrbai et al., 2022). Traditional fermentation optimization strategies, such as the one variable at a time strategy (OVAT), have encountered significant challenges. This approach, characterized by a one-dimensional search involving consecutive variations in variables, suffers from two notable flaws. First of all, because other variables are arbitrarily kept constant, it ignores the influence of parameter interactions. Second, it becomes intrinsically challenging to get an acceptable optimum in a limited number of experiments (Lotfy et al., 2007). When more than five factors at four levels are taken into account, the complete factorial DOE becomes unfeasible in terms of labor costs, time, and resources (4^5 = 1024 runs). As an alternative, partial factorial designs like RSM are frequently used. They might, however, ignore "less significant" characteristics if they are unaware of the possible interactions between them and the bioprocess. This constraint may result in less-than-ideal outcomes and impede a thorough comprehension of the procedure (Kana et al., 2012). Emerging AI tools such as ANN and GA (Prakasham et al., 2011), fuzzy logic (FL), ant algorithm (AA), and PSO are considered as efficient alternatives in the design of optimal production media and optimal process operating conditions in fermentation research and development (Haider et al., 2008).

It is possible to build bioprocess models using ANNs even in the absence of prior knowledge about the kinetics of metabolic fluxes between the culture environment and the cell. These models only use methods that are driven by data. The MLP is the most widely used architecture among ANNs. The nonlinear interactions between various causal (input) process parameters and their related dependent (output) variables are accurately approximated by it (Nandi et al., 2001). It is possible to secure optimal values of process variables by adjusting the input space of an ANN-based process model once it has been constructed and has demonstrated good generalization capabilities. The model can be optimized to direct the choice of process variables toward reaching desired results by suitably fine-tuning the input space.

2.7.1 Genetic algorithms

Based on the idea of "survival of the fittest," GAs imitate the processes of mutation and selection observed in evolutionary processes. A population of

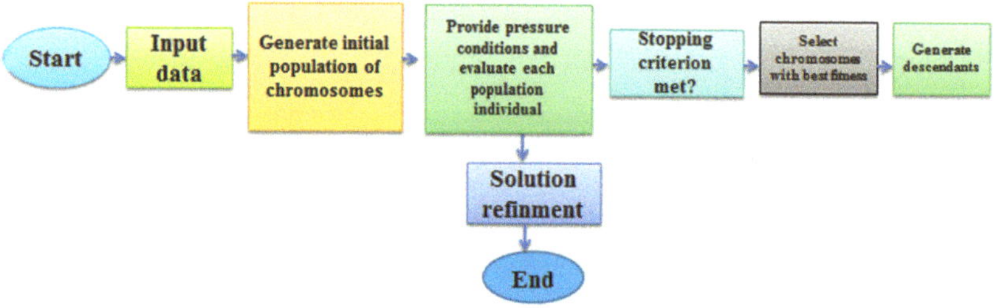

FIGURE 2.5 General flowchart of GA.

possible solutions goes through a number of crossover and mutation changes in these algorithms. Fitter individuals are given preference for the following generation in this selection technique, which drives competition for survival. As the method iteratively converges over subsequent generations, the ideal outcome is the identification of the best individual, which is the optimal solution (Fig. 2.5) (Haider et al., 2008).

GA is used in fermentation process optimization in the following ways: Every fermentation experiment's property such as the medium's composition and physical makeup is represented as "chromosomes," with each concentration of a medium component or physical parameter standing in for a "gene." To establish a generation, a population of chromosomes is created, and their performance is assessed according to predetermined standards. The performance of a chromosome determines its selection and subsequent duplication. After replication, randomly chosen genes within the chromosomes undergo crossover and mutation procedures. Through iterative refinement of the solutions until convergence is attained, this procedure yields a new generation of experiments (Kana et al., 2012).

For the study of Kana et al. (2012), by forecasting the yield of 20% of the validation data that had not been used in the training phase, the trained model was proven correct. The ability to evaluate the model's performance on hypothetical data was guaranteed by this validation procedure. The verified model was then used as an assessment function in the GA optimization procedure. Because of this integration, the GA was able to use the model's predictive power to direct its search for the best answers.

$$f(x) = \frac{1}{1 + e^{-x}} \tag{2.27}$$

All of the media or substrate profiles were represented as chromosomes in the Pro-optimizer program's GA optimization process. With the exception of sawdust, each parameter's search range was set between 5% and 30% in the Pro-optimizer module's configuration. Furthermore, 25%, 20%, 20%, and 60% were chosen as the generation size, parent size, mutation rate, and crossover rate, respectively. Pro-optimizer produced the first generation G0 with these parameters, which included 25 potential media compositions or substrate profiles.

Using the previously built ANN model, the performance value of each substrate profile for the production of biogas was determined. Then, to produce the next generation G1, Pro-optimizer was used a few of the G0 substrate profiles that performed well and used genetic processes, including crossovers and mutations to produce a total of 25 distinct substrate profiles. The ANN model was then used to assess these profiles. The performance value increased as generations went by. This procedure was repeated until the desired medium composition was reached, as shown by a performance value that remained constant for three generations (Kana et al., 2012).

2.7.2 Particle swarm optimization

Using the PSO technique, a collection of particles moves iteratively over the search space in an attempt to find the optimum of the objective function to be minimized. Based on its own experiences (local search) and those of nearby particles (global search), each particle modifies its position. Particles converge into favorable areas of the search space as a result of their cooperative behavior, perhaps finding the best possible solution (Roldán-Blay et al., 2023). The PSO algorithm simulates a swarm's movement of particles. The positions of the particles, which are vectors including the decision variables for the problem, indicate the values they choose at each instant. Typically, the PSO algorithm consists of three steps as follows:

1. **Evaluate fitness:** Find each particle's fitness value at its present location. Calculate the goal function's value at each particle's current position in particular.
2. **Update best individual and global fitness:** For every particle, update its best global and individual fitness. To do this, each particle's present fitness is compared to its best-ever fitness (local best), and the greatest-ever fitness values of all the particles in the swarm are compared (global best).
3. **Update particle velocity and position:** Particle location and velocity are updated for the subsequent iteration depending on the existing positions and velocities of the particles as well as the best positions (individual and global best) discovered thus far. This stage strikes a balance between exploration and exploitation by directing particles toward potential areas of the search space.

Genetic algorithm (GA), multilayered feedforward (MLFF), backpropagation (BP), error backpropagation (EBP), average absolute relative error (AARE), generalized delta rule (GDR), differential evolution (DE), backpropagation neural network (BPNN), biogas yield (BY), biogas production (BP), Levenberg–Marquardt (LM), Levenberg–Marquardt backpropagation

TABLE 2.1 Algorithm types used in ANN for biogas optimization.

References	Scale	ANN algorithm	Input	Output	Main findings
Barik and Murugan (2015)	–	GA, FF	pH, HRT, C/N	BY	The optimized data can be used to monitor a large-scale anaerobic plant.
Dibaba et al. (2016)	–	Coupled ANN-DE, BP, AARE, EBP, GDR	HRT, OLR	BP	ANN-DE result can be inferred that the network response is satisfactory and the simulation results are meaningful.
Sathish and Vivekanandan (2016)	–	BPNN	Temperature (T), pH, substrate concentration (SC) and agitation time (AT)	BY	The prediction accuracy of ANN is better than RSM.
Palaniswamy et al. (2017)	Pilot	BPNN trained with LM	pH, RT, T	BY	The BPNN architecture, trained with LM algorithm, is a technique that can be efficiently used to predict the biogas yield from food waste.
Wang et al. (2018)	Lab	FFNN and LMBP (training)	pH, ORP, and EC	Alkalinity	The soft sensor methods based on ANN with online measured input parameters were used to develop the online monitoring of alkalinity
Najafi and Ardabili (2018)	Lab	ANFIS, MLP	C/N ratio, T, RT	BY	The ANFIS network accurately predicted the output values in both thermophilic and mesophilic environments.

Beltramo et al. (2016)	Pilot	Ant colony optimization algorithm	Amino acids, LCFA, carbohydrates, proteins, and lipids	BY	ANN models have the benefit of being driven by data, so no prior understanding of process dynamics is needed.
Asadi and McPhedran (2021)	–	GA	VFA, TS, FS, pH, inflow rates	BY	The modeling can be used to optimize other MWTPs and possibly other treatment processes, such as AD, resulting in higher biogas production.
Yetilmezsoy et al. (2021)	Real full-scale	ANN, fuzzy, and multiple regression models	Sludge inflow rate, TS, VS, alkalinity, and VFA	BY	ANN model and conventional multiple regression approach, the suggested MISO fuzzy logic-based model produced lesser deviations and displayed improved predictive ability in estimating the biogas production rate.
Asadi and McPhedran (2021)	–	ANN, ANFIS, Regression modeling	VFA, TS, FS, pH, inflow rates	BY	The generated nonlinear regression model with unprocessed inputs performed the best, with values such as $R = 0.81$, RMSE = 0.95, and IA = 0.89.
Long et al. (2021)	Lab	GLMNET, RF, ANN, kNN, SVM, XGBoost	VFA, T, OLR, HRT	BY	The model has the ability to offer guidance for early warning and process control.

(LMBP), oxidation and reduction potential (ORP), adaptive neuro- fuzzy inference system (ANFIS).

Table 2.1 presents the most important ANN algorithms used on AD process for BP. We notice that the main algorithms used are the GA and BPNN. The main input variables are pH, VFA, temperature, and the retention time. Furthermore, we notice that the BY is the chosen output in the quasi-majority cases.

Both RSM and ANN can be used to optimize and model AD processes; they differ in their approach and underlying principles. RSM relies on statistical analysis and experimental design, while ANN utilizes computational models inspired by neural networks to learn and predict complex relationships from data. The study by Sathish and Vivekanandan (2016) shows that RSM and ANN had R^2 values of 0.991 and 0.998, respectively. These strong R^2 values show that the BY from agricultural waste may be efficiently predicted using both models. Because RSM is structured and provides essential insight information about the interactions between various system variables, ANN's prediction accuracy is superior to that of RSM. For another study by Gopal et al. (2021), the response value for biogas generation is extremely close to the experimental result. It was discovered that the ANN R^2 and RSM values were, respectively, 0.999 and 0.9951. Finally, it should be noted that for predicting BY, both models work better. When compared to RSM, the ANN prediction is far more efficient.

2.8 Conclusion

In contrast to ADM1, ANN method for AD is driven by data and does not necessitate any prior understanding of the kinetics or mechanisms of microbial digestion. The ANN is a mathematical methodology based on nonlinear regression, one of the many soft sensor techniques.

Because of their quick and accurate computations, ANNs are valuable tools for figuring out the relationship between variables in nonlinear and nonstationary systems. The multilayered perceptron that depicts the ANN models allows the evaluation of the correlation between the dependent output variables and the independent process input elements without any prior knowledge of those correlations. An appropriate collection of outputs can be generated from sets of input data using a feedforward ANN model. Backpropagation is a supervised learning technique that finds the gradient of the loss function with respect to the ANN's weights to train a forward-looking ANN. Since the ANN technique has a good summation, self-learning, and training function in capturing the nonlinear interactions existing between variables in complex systems, it has been widely applied in a number of applications.

Powerful technologies such as ANN and ANFIS can be harnessed to attain high yields of value-added products during process development. When integrated with evolutionary algorithms like PSO, ACO, or GAs, the utilization of ANN and ANFIS tools has yielded impressive outcomes.

ANNs have been applied to AD systems in this context to optimize digester efficiency, estimate the rate of BP or process intermediates, and improve process parameters. Several studies have shown that the ANN prediction accuracy is largely better than the RSM one.

It was found that the ANN and GA implementations to be defined and applied to model the biogas digestion process based on multiple process variables for evaluation and management of individual process parameters or the process as a whole had been a dependable technique for optimizing and managing AD operations. For predicting BY, both models work better. Nevertheless, the ANN prediction is significantly more accurate than the RSM prediction.

References

Abu Qdais, H., Bani Hani, K., & Shatnawi, N. (2010). Modeling and optimization of biogas production from a waste digester using artificial neural network and genetic algorithm. *Resources, Conservation & Recycling, 54*(6), 359–363. https://doi.org/10.1016/j.resconrec.2009.08.012.

Alejo, L., Atkinson, J., Guzmán-Fierro, V., & Roeckel, M. (2018). Effluent composition prediction of a two-stage anaerobic digestion process: Machine learning and stoichiometry techniques. *Environmental Science and Pollution Research, 25*(21), 21149–21163. https://doi.org/10.1007/s11356-018-2224-7.

Almomani, F. (2020). Prediction of biogas production from chemically treated co-digested agricultural waste using artificial neural network. *Fuel, 280*, 118573. https://doi.org/10.1016/j.fuel.2020.118573.

Alrbai, M., Abubaker, A. M., Ahmad, A. D., Al-Dahidi, S., Ayadi, O., Hjouj, D., & Al-Ghussain, L. (2022). Optimization of energy production from biogas fuel in a closed landfill using artificial neural networks: A case study of Al Ghabawi Landfill, Jordan. *Waste Management (New York, N.Y.), 150*, 218–226. https://doi.org/10.1016/j.wasman.2022.07.011 Alrbai.

Cruz, I. A., Chuenchart, W., Long, F., Surendra, K. C., Andrade, L. R. S., Bilal, M., Liu, H., Figueiredo, R. T., Khanal, S. K., & Ferreira, L. F. R. (2022). Application of machine learning in anaerobic digestion: Perspectives and challenges. *Bioresource Technology, 345*, 126433. https://doi.org/10.1016/j.biortech.2021.126433.

Cruz, I. A., Nascimento, V. R. S., Felisardo, R. J. A., Dos Santos, A. M. G., De Jesus, A. A., De Vasconcelos, B. R, Kumar, V., Cavalcanti, E. B., De Souza, R. L., & Ferreira, L. F. R. (2023). Evaluation of artificial neural network models for predictive monitoring of biogas production from cassava wastewater: A training algorithms approach. *Biomass Bioenergy, 175*(15), 106869. https://doi.org/10.1016/j.biombioe.2023.106869.

Antwi, P., Li, J., Boadi, P. O., Meng, J., Shi, E., Deng, K., & Bondinuba, F. K. (2017). Estimation of biogas and methane yields in an UASB treating potato starch processing wastewater with backpropagation artificial neural network. *Bioresource Technology, 228*, 106–115. https://doi.org/10.1016/j.biortech.2016.12.045.

Asadi, M., & McPhedran, K. (2021). Biogas maximization using data-driven modelling with uncertainty analysis and genetic algorithm for municipal wastewater anaerobic digestion.

Journal of Environmental Management, 293, 112875. https://doi.org/10.1016/j.jenvman.2021.112875.

Barik, D., & Murugan, S. (2015). An artificial neural network and genetic algorithm optimized model for biogas production from co-digestion of seed cake of karanja and cattle dung. *Waste Biomass Valorization, 6*(6), 1015–1027. https://doi.org/10.1007/s12649-015-9392-1.

Batstone, D. J., Keller, J., Angelidaki, I., Kalyuzhnyi, S. V., Pavlostathis, S. G., Rozzi, A., Sanders, W. T. M., Siegrist, H., & Vavilin, V. A. (2002). The IWA anaerobic digestion model no 1 (ADM1). *Water Science and Technology, 45*(10), 65–73. https://doi.org/10.2166/wst.2002.0292.

Beltramo, T., Klocke, M., & Hitzmann, B. (2019). Prediction of the biogas production using GA and ACO input features selection method for ANN model. *Information Processing in Agriculture, 6*, 349–356. https://doi.org/10.1016/j.inpa.2019.01.002.

Beltramo, T., Ranzan, C., Hinrichs, J., & Hitzmann, B. (2016). Artificial neural network prediction of the biogas flow rate optimised with an ant colony algorithm. *Biosystems Engineering, 143*, 68–78. https://doi.org/10.1016/j.biosystemseng.2016.01.006.

Beniche, I., El Bari, H., Siles, J. A., Chica, A. F., & Martín, M.Á. (2021). Methane production by anaerobic co-digestion of mixed agricultural waste: cabbage and cauliflower. *Environmental Technology, 42*, 4550–4558. https://doi.org/10.1080/09593330.2020.1770341.

Box, G. E. P., & Wilson, K. B. (1951). On the Experimental Attainment of Optimum Conditions. *Journal of the Royal Statistical Society: Series B (Methodological), 13*(1), 1–38. https://doi.org/10.1111/j.2517-6161.1951.tb00067.x.

Chen, W.-Y., Chan, Y. J., Lim, J. W., Liew, C. S., Mohamad, M., Ho, C.-D., Usman, A., Lisak, G., Hara, H., & Tan, W.-N. (2022). Artificial neural network (ANN) modelling for biogas production in pre-commercialized integrated anaerobic-aerobic bioreactors (IAAB). *Water, 14*(9), 1410. https://doi.org/10.3390/w14091410.

Cheng, J., Wang, X., Si, T., Zhou, F., Zhou, J., & Cen, K. (2016). Ignition temperature and activation energy of power coal blends predicted with back-propagation neural network models. *Fuel, 173*, 230–238. https://doi.org/10.1016/j.fuel.2016.01.043.

Dahunsi, S. O., Oranusi, S., Owolabi, J. B., & Efeovbokhan, V. E. (2016). Comparative biogas generation from fruit peels of fluted pumpkin (Telfairia occidentalis) and its optimization. *Bioresource Technology, 221*, 517–525. 10.1016/j.biortech.2016.09.065.

Dibaba, O. R., Lahiri, S. K., T'Jonck, S., & Dutta, A. (2016). Experimental and artificial neural network modeling of a upflow anaerobic contactor (UAC) for biogas production from Vinasse. *International Journal of Chemical Reactor Engineering, 14*(6), 1241–1254. https://doi.org/10.1515/ijcre-2016-0025.

El Gnaoui, Y., Frimane, A., Lahboubi, N., Herrmann, C., Barz, M., & El Bari, H. (2022). Biological pre-hydrolysis and thermal pretreatment applied for anaerobic digestion improvement: Kinetic study and statistical variable selection. *Cleaner Waste Systems, 2*(4), 100005. https://doi.org/10.1016/j.clwas.2022.100005.

Ge, Y., Tao, J., Wang, Z., Chen, C., Mu, L., Ruan, H., Rodríguez Yon, Y., Su, H., Yan, B., & Chen, G. (2023). Modification of anaerobic digestion model no.1 with Machine learning models towards applicable and accurate simulation of biomass anaerobic digestion. *Chemical Engineering Journal, 454*, 140369. https://doi.org/10.1016/j.cej.2022.140369.

Gopal, L. C., Govindarajan, M., Kavipriya, M. R., Mahboob, S., Al-Ghanim, K. A., Virik, P., Ahmed, Z., Al-Mulhm, N., Senthilkumaran, V., & Shankar, V. (2021). Optimization strategies for improved biogas production by recycling of waste through response surface methodology and artificial neural network: Sustainable energy perspective research. *Journal of King Saud University - Science, 33*, 101241. https://doi.org/10.1016/j.jksus.2020.101241.

Kana, E. B. G., Oloke, J. K., Lateef, A., & Adesiyan, M. O. (2012). Modeling and optimization of biogas production on saw dust and other co-substrates using Artificial Neural network and Genetic Algorithm. *Renewable Energy, 46*, 276–281. https://doi.org/10.1016/j.renene.2012.03.027.

Guo, HN, Wu, SB, Tian, YJ, Zhang, J, & Liu, HT (2021). Application of machine learning methods for the prediction of organic solid waste treatment and recycling processes: A review. *Bioresour Technol, 319*, 124114.

Gupta, R., Zhang, L., Hou, J., Zhang, Z., Liu, H., You, S., Ok, Y. S., & Li, W. (2023). Review of explainable machine learning for anaerobic digestion. *Bioresource Technology, 369*, 128468. https://doi.org/10.1016/j.biortech.2022.128468.

Habchi, S., Lahboubi, N., Asbik, M., & Bari, H. E. (2024). Enhancing biomethane production from food waste using olive pomace hydrochar: An optimization study. *Environmental Advances, 15*, 100477. https://doi.org/10.1016/j.envadv.2023.100477.

Haider, M. A., Pakshirajan, K., Singh, A., & Chaudhry, S. (2008). Artificial neural network-genetic algorithm approach to optimize media constituents for enhancing lipase production by a soil microorganism. *Applied Biochemistry and Biotechnology, 144*, 225–235. https://doi.org/10.1007/s12010-007-8017-y.

Hashem, H., 2016. Modélisation intégratrice du traitement BigData (phdthesis). Université Paris Saclay (COmUE).

Heiker, M., Kraume, M., Mertins, A., Wawer, T., & Rosenberger, S. (2021). Biogas plants in renewable energy systems—A systematic review of modeling approaches of biogas production. *Applied Sciences, 11*(8), 3361. https://doi.org/10.3390/app11083361.

Holubar, P., Zani, L., Hager, M., Fröschl, W., Radak, Z., & Braun, R. (2002). Advanced controlling of anaerobic digestion by means of hierarchical neural networks. *Water Research, 36*, 2582–2588. 10.1016/S0043-1354(01)00487-0.

Juba, B., & Le, H. S. (2019). Precision-recall versus accuracy and the role of large data sets. In: *Proceedings of the AAAI Conference on Artificial Intelligence, 33*(1), 4039–4048. https://doi.org/10.1609/aaai.v33i01.33014039.

Karlik, B., & Olgac, A. V. (2011). Performance Analysis of Various Activation Functions in Generalized MLP Architectures of Neural Networks. *International Journal of Artificial Intelligence and Expert Systems, 1*(4), 111–122.

Liu, S., Xu, L., & Li, D. (2016). Multi-scale prediction of water temperature using empirical mode decomposition with back-propagation neural networks. *Computers & Electrical Engineering, 49*, 1–8. https://doi.org/10.1016/j.compeleceng.2015.10.003.

Long, F., Wang, L., Cai, W., Lesnik, K., & Liu, H. (2021). Predicting the performance of anaerobic digestion using machine learning algorithms and genomic data. *Water Research, 199*, 117182. https://doi.org/10.1016/j.watres.2021.117182.

Lotfy, W. A., Ghanem, K. M., & El-Helow, E. R. (2007). Citric acid production by a novel *Aspergillus niger* isolate: II. Optimization of process parameters through statistical experimental designs. *Bioresource Technology, 98*, 3470–3477. https://doi.org/10.1016/j.biortech.2006.11.032.

Najafi, B., & Ardabili, S. F. (2018). Application of ANFIS, ANN, and logistic methods in estimating biogas production from spent mushroom compost (SMC). *Resources, Conservation and Recycling, 133*, 169–178. https://doi.org/10.1016/j.resconrec.2018.02.025.

Nandi, S., Ghosh, S., Tambe, S. S., & Kulkarni, B. D. (2001). Artificial neural-network-assisted stochastic process optimization strategies. *AIChE Journal American Institute of Chemical Engineers, 47*, 126–141. https://doi.org/10.1002/aic.690470113.

Ourradi, H., Lahboubi, N., Habchi, S., Hanine, H., & El Bari, H. (2022). Methane production from date seed cake (Phoenix dactylifera L.) using mesophilic fed-batch anaerobic digestion. *Cleaner Waste Systems, 2*, 100009. https://doi.org/10.1016/j.clwas.2022.100009.

Palaniswamy, D., Ramesh, G., Sivasankaran, S., & Kathiravan, N. (2017). Optimising biogas from food waste using a neural network model. In: *Proceedings of Institution of Civil Engineers - Municipal Engineer, 170*(4), 221–229. https://doi.org/10.1680/jmuen.16.00008.

Piche, S., Sayyar-Rodsari, B., Johnson, D., & Gerules, M. (2000). Nonlinear model predictive control using neural networks. *IEEE Control Systems Magazine, 20*, 53–62. https://doi.org/10.1109/37.845038.

Pradhan, D., Jaiswal, S., & Jaiswal, A. K. (2022). Artificial neural networks in valorization process modeling of lignocellulosic biomass. *Biofuels Bioproducts & Biorefining, 16*, 1849–1868. https://doi.org/10.1002/bbb.2417.

Prakasham, R. S., Sathish, T., & Brahmaiah, P. (2011). Imperative role of neural networks coupled genetic algorithm on optimization of biohydrogen yield. *International Journal Hydrogen Energy, 36*, 4332–4339. https://doi.org/10.1016/j.ijhydene.2011.01.031.

Roldán-Blay, C., Escrivá-Escrivá, G., Roldán-Porta, C., & Dasí-Crespo, D. (2023). Optimal sizing and design of renewable power plants in rural microgrids using multi-objective particle swarm optimization and branch and bound methods. *Energy, 284*, 129318. https://doi.org/10.1016/j.energy.2023.129318.

Sakiewicz, P., Piotrowski, K., Ober, J., & Karwot, J. (2020). Innovative artificial neural network approach for integrated biogas – wastewater treatment system modelling: Effect of plant operating parameters on process intensification. *Renewable and Sustainable Energy Reviews, 124*, 109784. https://doi.org/10.1016/j.rser.2020.109784.

Sathish, S., & Vivekanandan, S. (2016). Parametric optimization for floating drum anaerobic bio-digester using Response Surface Methodology and Artificial Neural Network. *Alexandria Engineering Journal, 55*(4), 3297–3307. https://doi.org/10.1016/j.aej.2016.08.010.

Solorzano, G., & Plevris, V. (2022). Computational intelligence methods in simulation and modeling of structures: A state-of-the-art review using bibliometric maps. *Frontiers in Built Environment, 8*, 1–24. https://doi.org/10.3389/fbuil.2022.1049616.

Tsui, T.-H., Van Loosdrecht, M. C. M., Dai, Y., & Tong, Y. W. (2023). Machine learning and circular bioeconomy: Building new resource efficiency from diverse waste streams. *Bioresource Technology, 369*, 128445. https://doi.org/10.1016/j.biortech.2022.128445.

Wang, L, Long, F, Liao, W, & Liu, H (2020). Prediction of anaerobic digestion performance and identification of critical operational parameters using machine learning algorithms. *Bioresour Technol, 298*, 122495.

Wang, X., Bai, X., Li, Z., Zhou, X., Cheng, S., Sun, J., & Liu, T. (2018). Evaluation of artificial neural network models for online monitoring of alkalinity in anaerobic co-digestion system. *Biochemical Engineering Journal, 140*, 85–92. https://doi.org/10.1016/j.bej.2018.09.010.

Xue, H., Chen, S., & Yang, Q. (2011). Structural regularized support vector machine: A framework for structural large margin Classifier. *IEEE Transactions on Neural Networks / a Publication of the IEEE Neural Networks Council, 22*, 573–587. https://doi.org/10.1109/TNN.2011.2108315.

Yetilmezsoy, K., Karakaya, K., Bahramian, M., Abdul-Wahab, S. A., & Goncaloğlu, B.İ. (2021). Black-, gray-, and white-box modeling of biogas production rate from a real-scale anaerobic sludge digestion system in a biological and advanced biological treatment plant. *Neural Computing and Applications, 33*(17), 11043–11066. https://doi.org/10.1007/s00521-020-05562-7.

Chapter 3

Innovative biogas digester technology

Fadoua Karouach[1], Sanae Habchi[2], Ihssane Mnaouer[1], Soumaya Boukrouh[1], and Hassan El Bari[2]

[1]*African Sustainable Agriculture Research (ASARI), University Mohammed VI Polytechnic (UM6P), Laayoune, Morocco,* [2]*Laboratory of Electronic Systems, Information Processing, Mechanics and Energetics, Faculty of Sciences, Ibn Tofail University, Kenitra, Morocco*

3.1 Introduction

The production and utilization of bioenergy are focused on providing an affordable and sustainable supply of energy. Rapid developments and innovations in advanced conversion technologies and the emerging global market for carbon sequestration have been combined to enhance the commercial perspectives of bioenergy. Bioenergy can be derived from a wide variety of organic matter and converted into different energy carriers using various process technologies.

The biogas digester was previously regarded as a technology that was only suitable for rural communities. However, the increase in the price of oil and the introduction of various subsidies, along with the concept of green tariffs for renewable energy, have significantly enhanced the economic viability of biogas technology. The additional revenue gained from the sale of heat energy enables the financial viability of small-scale electricity generation using biogas to be forecast. Additionally, system maintenance and operation costs represent key factors in the long-term success of biogas projects.

Biogas production is a biological process that can cause several problems, some of which can lead to system failure. The variety of feedstocks is one of the most significant challenges in biogas deployment. This can lead to problems with process stability and reduced biogas production, which can be critical for projects that rely on revenue from energy production (Priyanka et al., 2022). For a successful biogas digester implementation, several challenges need to be considered and solved. Modern and advanced biogas technologies can significantly improve the quality of traditional practices. The integration of an innovative decision support system enabled by artificial intelligence (AI) will enable biogas

Biogas in the Circular Economy. DOI: https://doi.org/10.1016/B978-0-443-29230-9.00009-7

plants to optimize their performance in real time. Such an AI system will analyze the data coming from the plant, regardless of its sparsity, to identify trends and correlations that are imperceptible by human (Fajobi et al., 2022; Igielski, 2023; Lamas & Giacaglia, 2020; Lubentsov, et al., 2021). AI can perform complex tasks, enabling a digester technology system that is efficient, accurate, and fast (Tufaner and Demirci, 2020). The use of molecular biology techniques has resulted in considerable improvements in our understanding of the biogas-generating process. By identifying the dominant methanogenic pathway, it is possible to optimize digestion processes by creating ideal conditions for the most efficient pathway. These techniques can be used to improve the efficiency of biogas generation by examining the microbial communities present in biogas systems, thus offering a more thorough understanding of the types of microorganisms present and the factors that influence their activity (Andlar et al., 2021). Moreover, third-generation technology, which uses genetically modified organisms, is an example of a more efficient waste treatment process. Although these modern biotechnological advances hold great promise for improving waste management and energy production, they are still under development.

The incorporation of biogas with other renewable technologies like cogeneration systems, solar, and wind power can enhance its efficiency and attractiveness as a renewable energy source, though some, like wind power, still require subsidies due to their lower efficiency (Jean-Jacques et al., 2012).

This chapter presents a comprehensive review of recent advancements in biogas digester technology. The chapter focuses on cutting-edge innovations, including AI-powered monitoring systems for enhanced efficiency and predictive maintenance, advanced microbial optimization techniques to boost biogas yield and quality, and hybrid solutions that integrate biogas with other renewable energy sources. Furthermore, the chapter highlights recent technological breakthroughs that integrate traditional performance metrics with novel methodologies exploring their potential benefits and limitations. Additionally, it investigates pivotal aspects of scalability and economic viability in biogas digestion technologies, their practical applications, and their role in advancing sustainable energy solutions.

3.2 Biogas digester technologies

3.2.1 AI-powered monitoring systems

The implementation of biogas plants presents many opportunities and challenges. Nevertheless, the absence of sufficient infrastructure and capital has severely impeded the successful establishment of a suitable biogas plant, leading to either no biogas production or a low yield (Ilangovan et al., 2023). The conversion of biomass to bioenergy requires the creation of novel technologies and optimization strategies that can boost energy output under carefully monitored circumstances and with reduced resource consumption. There are fresh chances to create policies that govern the prediction and optimization of the anaerobic digestion (AD) process (Sharma et al., 2023). AI has been effectively integrated into bioenergy production systems. AI is the mechanical ability to conduct actions that mirror human intelligence. It is a field of computer science concerned with the development of intelligent machines. AI-powered monitoring systems are a specific type of technology that utilizes AI to collect, analyze, and interpret data from a variety of sensors and devices within a biogas digester system. Because AI is such a complex field, numerous criteria are used to classify it. Learning tasks and algorithm forms can be used to make various classifications. The primary AI methodologies include the random tree, random forest (RF), artificial neural network (ANN), and the adaptive-network-based fuzzy inference system (ANFIS) (Shahsavar et al., 2021). Deep neural networks are regarded as accurate models for complex fermentation process prediction because of their AI-based capacity to learn from massive amounts of data. Selecting the appropriate hyperparameters, such as learning rate, activation function, optimizer, neural network architecture (number of hidden layers and hidden units), and others, is crucial to enhancing the performance of these models (Sadoune et al., 2023). Table 3.1 summarizes the main machine learning (ML) algorithms cited in the literature, their usage percentage, and their characteristics.

3.3 Integration with other renewable energy sources (hybrid systems)

The world's fossil fuel energy needs will continue to grow until 2030, and the planet will no longer be able to meet them independently. This situation will lead to a rise in CO_2 emissions of 42.3 Gt/year by 2030 and an increase of 57% on the current situation. The use of renewable energy resources has been implemented in various countries, both developed and developing, to support environmentally friendly energy (Imran et al., 2021). A hybrid system is a combination of two or more energy conversion devices or two or more fuels for a single process. The performance of hybrid devices is better than that of a single system, offering great potential for cost-effectiveness and limiting adverse environmental effects. However, the complexity, variability, and high cost of hybrid systems complicate their implementation. Software tools, hybrid algorithms, and AI offer more detailed and cost-effective solutions (Thirunavukkarasu et al., 2023). The combination of solar, wind, hydropower, and biogas represents a viable solution, particularly in regions where access to electricity is limited. These renewable sources are sustainable and plentiful, with solar, wind, and hydro providing consistent energy, and biowaste being continuously available. The utilization of these resources can effectively address future energy demands, benefiting developing villages and remote areas within developed countries.

TABLE 3.1 Various machine learning algorithms used in biogas technology systems adapted from (Cruz 2022).

Machine learning algorithm	Distribution (%)	Characteristics
ANN	33	ANNs are computer models that draw inspiration from the composition and operations of the human brain.
ANFIS	10	ANFIS is a hybrid model that combines the strengths of fuzzy logic and neural networks to handle complex and unpredictable systems.
SVM	8	Supervised learning models, or SVMs, are employed in tasks involving regression and classification. They are renowned for handling both linear and nonlinear data and for being efficient in high-dimensional environments.
RF	8	During training, a large number of decision trees are built using the RF ensemble learning technique, which then outputs the mean prediction (regression) or the mode of the classes (classification) of each individual tree.
GA	8	Genetic algorithms and the process of natural selection serve as the inspiration for genetic algorithms, or GAs. They are frequently employed when more conventional approaches are impracticable to handle difficult optimization problems.
PSO	4	PSO is a population-based optimization method that draws inspiration from fish schools and bird flocking. It is frequently employed to resolve optimization issues when more conventional approaches prove to be unworkable or wasteful.
Hybrid	15	—
Others	15	—

ANN, Artificial neural networks; *ANFIS*, adaptive network-based fuzzy inference system; *SVM*, support vector machine; *GA*, genetic algorithm; *RF*, random forest; *PSO*, particle swarm optimization.

3.3.1 Integration of solar energy and biogas systems

The hybrid solar/biogas system is a promising and innovative technology that has the potential to play an important role in the years to come. It involves merging two technologies to convert solar energy and waste into useful energy. Solar energy harmonizes well with biogas production, particularly during the summer months when solar output is stable. However, the initial cost of solar energy and its discontinuous nature can pose problems, particularly for larger biogas plants. A continuous supply of energy is provided by hybrid solar and biogas systems, which overcome the disadvantages of separate solar configurations and the lack of access to energy at night. This advantage is reinforced by the storage capacity of biogas digesters (Oluwaseun et al., 2018). A hybrid system combining solar power and biogas is still a relatively new idea. As a result, few hybrid systems have yet been brought to market. To assess the viability of these systems, several pilot projects and a significant amount of research and development work have been carried out in this field. In a recent study, researchers used computer simulations to design a hybrid system that integrates biogas generators and solar photovoltaic (PV) panels. The hybrid system was able to meet variable electricity needs, producing 25.2 kW during peak sunlight hours and 9 kW at night. The system maintained a stable frequency (50 Hz) and a high-power factor (.92–.98) (Tamoor et al., 2023). The viability of a hybrid renewable energy system combining solar photovoltaics, biogas, and battery storage to power a commercial platform in Berkane, Morocco, is examined in a recent paper. Using actual data and HOMER software, the model recommends an ideal configuration compared with a stand-alone biogas system. The optimized system achieves a levelized cost of electricity (LCOE) of \$.280/kWh and reduces CO_2 emissions by almost 40%. The study offers a viable path to sustainable development by highlighting the financial and environmental benefits of integrating PV/biogas/battery hybrid systems in Morocco. A study carried out in Jordan investigated an on-grid photovoltaic (PV) biogas hybrid energy system. The study used an improved optimization algorithm to optimize the system's reliability and economic viability, focusing on reducing the total net present cost and increasing the reliability index (RI). The study's findings showed that the strategy used is also successful in producing energy and reducing greenhouse gas (GHG) emissions. Another study in Ghana analyzed the feasibility of a hybrid solar PV/biogas/battery energy system for rural electrification, with the aim of achieving a cost-effective LCOE and mitigating GHG emissions. The results showed that the PV/biogas/battery system outperformed the PV/diesel/battery and diesel systems alone in terms of cost and emissions reduction, with an LCOE of around \$.256/kWh (Flavio et al., 2022). Given the intermittent nature of solar energy, electricity production from solar power over a short period can exceed load demand. In this case, excess electricity from the grid can be used to produce biogas, thanks to electricity-to-gas conversion technologies. The high-temperature solid oxide electrolysis cell is a solar-to-gas technology that converts solar heat and electrical energy into syngas. Syngas from solar-to-gas technologies can produce higher-calorific biogas using CO_2 and H_2 from the water-displacement reaction with methanation. A study by Su et al. proposed a model for a combined cooling, heating, and power system that integrates biogas and solar energy to improve efficiency and reduce environmental impact. The study revealed that by converting solar energy into gas, the system achieves higher electricity production, more efficient refrigeration and reduced natural gas consumption compared with other systems. However, the use of power-to-gas technologies for biogas production still requires a reduction in costs, both onshore and offshore.

3.3.2 Integration of wind energy and biogas systems

Wind can be an attractive energy source, particularly in remote (off-grid) areas where grid access is limited (Reuchlin et al., 2023). Although the use of wind energy for heating is not the most efficient, it can be a good option for surplus wind energy. Wind power's fluctuating nature and storage limitations for further consumption make it ideal for hybrid systems, especially when combined with biogas production. As stored energy, biogas complements wind power by offering near-zero emissions and CO_2 compensation. Combining wind and biogas energy requires efficient management of biogas production. Energy cost analysis and flexible tariffs determine the degree to which wind and biogas energy can be integrated. The storage of wind-generated electricity, possibly using batteries, can further enhance this integration. Advanced control systems and simulations are essential to optimize the interaction between wind and biogas energies economically and efficiently (Olabi et al., 2023). Numerous studies have provided relevant information on combined heat and power systems and the optimization of renewable energy systems. These studies suggest that combining different energy sources can lead to more efficient and cost-effective energy production. A study by Norouzi (2021) examined the combination of an electricity and heat cogeneration system with the engines of a gas turbine and a horizontal-axis wind turbine. According to the study, the suggested cogeneration system has a number of benefits over the independent generating system, including fuel consumption and operating cost reductions of about 60% and 55%, respectively. In the same context, research was carried out into the application of renewable energy sources in wastewater treatment plants (WTP) to achieve energy self-sufficiency. The study used a combination of biogas cogeneration fuel cells, micro-turbines, photovoltaic systems, and wind turbines. This concept effectively meets the energy demands of plants (Helal et al., 2013). Chamazkoti et al. (2024) proposed a hybrid renewable energy system for WTP that uses solar energy, wind power, and biogas to power the plant, assist the grid, and create a clean hydrogen fuel. Combining these renewable sources (hybridization) and connecting to the grid improve stability. Another study investigating the economic feasibility of a hybrid wind and AD bioenergy system for the electrification of a remote region has shown that a hybrid wind/biomass system can meet the demand of a community consuming 60 kW per day at peak times, with a backup storage system. This system is the most economical, with a cost equal to 0.118 US$/kWh, compared with a hybrid biomass/wind/diesel system with 0.202 US$/kWh (Raheem et al., 2014). The study by Monnerie et al. (2014) proposes a two-stage method for producing clean fuels using wind power and biogas. Wind energy is used to produce hydrogen by high-temperature steam electrolysis. The hydrogen is then used to react with the CO_2 in the biogas to produce methane. The concept presented involves a production capacity of 10 MWel, with an annual output of 1104 tons of hydrogen and 5888 Mt/y of methane, achieving an overall efficiency of 38%. This method is promising for the integration of wind resources and the transition to clean energy sources.

The EU's renewable energy policy for 2030 sets targets that will stimulate demand for the integration of wind power and biogas systems into the energy infrastructure, focusing on both electricity and gas.

3.3.3 Integration of hydroelectric power and biogas systems

Biogas extraction from organic matter and other processes offers potential synergy with hydropower in terms of electricity generation and capacity management. The variable electrical energy produced by the hydroelectric plant can be difficult to integrate into the national grid. To overcome this problem, a combination of centralized and distributed solutions is recommended. Centralized solutions focus on maintaining grid stability, whereas distributed solutions address issues related to the management of energy flows in local areas. In addition, flexible technologies such as energy storage are generally more effective in addressing these challenges than traditional grid technologies. By providing base load power from biogas, the hydro plant can then sell excess energy during high electricity demand hours. In the same context, researchers explored the best renewable energy system for an university in Sikkim, India. They analyzed various combinations of solar, wind, biogas, and other emerging renewables using HOMER Pro software. The most cost-effective and sustainable option, with a levelized cost of just $.095/kWh, turned out to be a hybrid system combining solar panels, wind turbines, biogas, syngas (synthetic gas), hydrokinetic (water current) power, and batteries. This approach not only reduces reliance on fossil fuels but also creates local jobs and minimizes land use compared with relying solely on solar or wind (Baruah et al., 2021). Another study explored how to improve electricity access in remote communities using renewable energy. The study area already has a hydropower system, but researchers investigated adding solar, biogas, and battery storage to create a hybrid system. The results showed that combining these renewable sources is feasible and could significantly reduce reliance on traditional energy sources, although the overall cost of energy was \$.67 per kWh (Dzobo, 2022). Studies on hybrid systems involving the integration of biogas production with other renewable energy sources have not been very frequent, due to the relative newness of the biogas industry. As a result, information on the subject is limited and often very general. As biogas becomes more established and begins to play a more important role in energy production, more studies will be carried out on the subject, and more information on the specific technologies and strategies involved will become available.

3.4 Biomass and optimization

3.4.1 Utilization of waste streams for biogas production

In recent years, there has been a shift toward a circular economy model that generates energy from agro-industrial and biomass waste, thereby reducing trash and GHGs. Biogas produced from lignocellulosic biomass (LCB) is a promising clean and sustainable energy source. Several processes are utilized to convert LCB to biogas, including pretreatment, AD, and biogas upgrading to biomethane. According to statistics, fossil fuels account for over 80% of the current global primary energy consumption, and fossil fuel-based energy systems are mainly responsible for more than two-thirds of the global GHG emissions. To meet the purpose of productive biogas systems, which is to provide waste management and modern renewable energy services, sustainable biogas production is essential. While there are various factors that affect sustainability, one essential AD component that should be addressed for enhanced and sustained biogas supply is substrate quality (Tumusiime et al., 2022). Lignocellulosic biomass is an important source of nutrient management in agricultural systems, with a large unmet potential for biogas production. Boukrouh et al. (2023a, 2023b) reported a higher organic matter digestibility of alternative feed resources, Sulla flexuosa hay, and sorghum grains (68.45% and 81.4%, respectively), which could have higher potential for biogas production. AD digestates are extensively used in agricultural systems because of their nutritional makeup and organic matter, making them a significant source of plant nourishment (Karouach et al., 2024). The study examined the production of methane (CH_4) from the AD of lignocellulosic biomass waste, such as rice straw, corn stalk, and grape stem. Increased livestock output generates a significant rise in lignocellulosic biomass and animal waste, which must be managed. AD is the most promising waste valorization technology known. It is a type of material recycling that yields sustainable energy in the form of biogas. Fig. 3.1 presents the main common types of waste used in AD process: agricultural waste, which includes livestock waste and crops; agri-food industry waste; wastewater; and food and organic municipal waste.

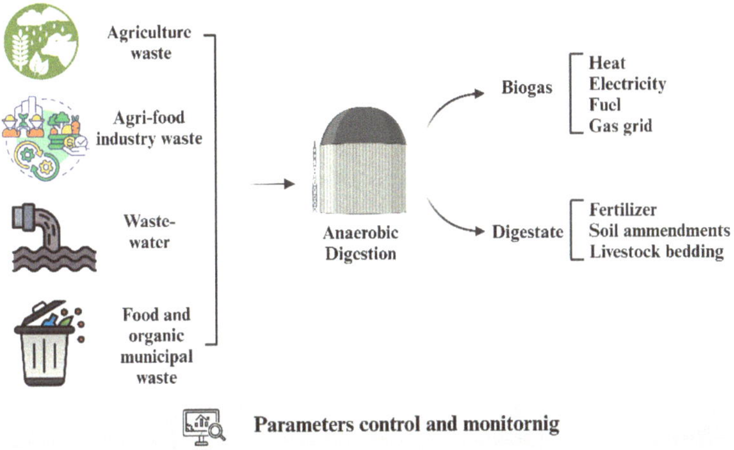

FIGURE 3.1 Recovery of different wastes by generation of biogas and digestate.

3.4.2 Optimization aspects

The optimization of biomass resources for energy production includes strategies that maximize efficiency and sustainability through the whole process. One strategy relies on the diversification of biomass resources. Indeed, the cultivation of microalgae has shown interesting results for energy production due to their high lipid accumulation rates and low carbon dioxide emissions. Cultivating microalgae in different types of wastewaters was reported to improve biomass production and overall energy output with a parallel positive environmental footprint (Zheng et al., 2019). Some authors have used anaerobic co-digestion of two resources. Vijin Prabhu et al. (2021) reported an improvement of AD by 76.6% and reduction in total solids by 41.1% using biomass wastes and cow manure anaerobic co-digestion. Moreover, practices such as physical (drying, size reduction, etc.) and chemical pretreatments prepare the biomass for conversion, increasing the efficiency of the subsequent processes (Lahboubi et al., 2022). Physicochemical pretreatments demonstrated promising results in efficiently breaking down the hemicellulose or lignin polymers inside lignocellulosic-rich biomass before AD (Habchi et al., 2022). Further progress in biomass valorization can be realized by developing integrated approaches and combined systems. For example, integrating biochemical with thermochemical conversion processes can maximize energy output and minimize waste (Okolie et al., 2022).

3.4.3 Advanced microbial optimization techniques

Biomass from organic sources such as organic waste, energy plants, harvest residues, and industrial by-products is generally discharged into the environment where they produce methane after decomposition. As global energy demand is increasing, optimization of biomass waste use has become of vital importance. Reactor feeding type, moisture content, and temperature are among the treatment process characteristics that have an impact on the AD (Karouach et al., 2021). This study observed that methane production was significantly higher in mesophilic digestion (68.52%) compared with thermophilic conditions (5.79%). This study explained the higher CH_4 yield by a greater presence of Methanosarcina (94.42%–94.73%), which is sensitive to thermophilic temperatures and is mainly responsible for acetoclastic methanogenesis. Biogas production through AD by microbial optimization techniques can yield up to 725% improvement in biogas production and +310.6% of methane yield improvement (Suksong et al., 2019). The optimization of these biogas productions will synergistically address environmental sustainability through the disposal of wastes and economic challenges by the generation of heat and electricity through their use as a fuel. Advanced microbial optimization techniques are progressively employed to improve the yield and efficiency of biogas and methane production. An important strategy in this field concerns the use of specialized consortia, which play a crucial role in the AD operation. These consist of microorganism strains that have been either identified or engineered to exhibit enhanced efficiency in substrate degradation and biogas production. Indeed, several authors have claimed the simultaneous use of multispecies consortia compared with single species to improve cellulose and hemicellulose enzymes. This study reported that a synergistic microbial consortium increases significantly cumulative biogas and CH_4 yields by 76.3% and 64.3%, respectively. Similarly, (Kong et al., 2018) reported that during AD, when adding a multispecies lignocellulolytic enzyme (TC-5) in the digestion sludge of wheat straw, a high degradation of 45.7% of wheat straw occurs. They also reported that during 9 days' degradation, the enzyme activity of xylanase, xylan esterase, and β-xylosidase reached the highest levels of 4.23, .15, and .48 U/mg, while cellobiohydrolase, endoglucanase, and β-glucosidase showed the highest specific activity of .36, .22, and .41 U/mg. The improvement in biogas and CH_4 yield was parallel to an improvement in volatile fatty acids, total solids, cellulose, and hemicellulose degradation. Certain methanogenic fungi have been discovered to accelerate the transformation of organic acids into methane, thus enhancing the rate of biogas production. Notably, white- and brown-rot fungi are well known for their ability to degrade lignin and complex polysaccharides such as cellulose and hemicellulose, which are resistant to bacterial decomposition. Pretreatment by *Ceriporiopsis subvermispora* and improved methane yield by up to 154% (Zhao et al., 2014). Mustafa et al. (2016) reported that combining *Pleurotus ostreatus* with milling pretreatment resulted in a 165% of CH_4 yield improvement that reached. To optimize biogas production, some authors exploited the complex microbial community naturally present in the rumen of ruminant animals. Zheng et al. (2019) inoculated with rumen fluid as a pretreatment method and claimed a ratio of 1:5 of rumen microorganisms to the slurry to be optimum for increased biogas and methane yields. Also, another technique involves the optimization of the microbial community structure through co-digestion. This process requires the mixing of different types of organic waste, which provides a more balanced nutrient profile for the microbial population. The synergistic interactions between different microbial species in this mixed substrate environment can lead to enhanced biogas yields. Despite their higher cost, some authors have added synthetic enzymes to improve biogas production, and the results were variable. For example, this study reported an 84% improvement in CH_4 yield by adding lipase from *Candida rugosato* to butter. Also, this study reported an improvement of 90% in CH_4 yield at 1% dose but methane yield decreased by 17% at 2% dose. The use of enzymes mainly speeds up the digestion process, but sometimes without significant improvement of CH_4 yield. This study reported an improvement of .3% to 21.1% in CH_4 yield between days 5 and 20, but there was no significant difference after 60 days of the experiment. Table 3.2 shows some microbial optimization strategies in biogas production.

3.5 Recent developments, scalability, and economic viability

3.5.1 Overview of recent technological advancements

In recent years, significant innovations in biogas digester technology have been driven by the urgent need for sustainable energy solutions and advances in supporting technologies, further supported by rising fossil fuel prices and taxes on energy sources Kumar et al. (2015). These technologies are not only essential for promoting green energy but also for optimizing the efficiency and output of biogas production plants (Kabeyi and Olanrewaju, 2022). Biogas digesters are continually evolving, integrating new features and improvements. Table 3.3 highlights some of the recent innovations in this field, focusing on the specificities and innovations of each technology. The table details various aspects, such as digester volume, production yield, scalability, inoculum source, and associated costs.

3.5.2 Feasibility studies for large-scale biogas digester projects

A feasibility study is crucial for evaluating the economic and technical viability of installing large-scale digesters (Zhu and Zhang, 2020). It involves a detailed techno-economic analysis of different production aspects. Additionally, this study identifies any bottlenecks in the process that could hinder technical feasibility or economic viability.

Before transitioning to large-scale biogas digesters, it is recommended to perform laboratory- and pilot-scale validations. These crucial steps ensure the reliability and efficiency of the process when scaled up.

TABLE 3.2 Microbial optimization strategies in biogas production.

Biomass	Pretreatment	Species	Type	Pretreatment conditions	Incubation conditions	Major results	References
Sawdust waste	Bacteria	*Citrobacter*, *Klebsiella*, *Exiguobacterium*, *Lactococcus*, *Micrococcus*, and the yeasts *Sugiyamaella* and *Vanrija*	Consortia	30°C for 10 days	28 days	+25.6% biogas improvement. +72.6% CH_4 yield improvement.	Ali et al. (2017)
Catalpa sawdust and chlorophenols	Bacteria	CS-5 (*Micrococcus luteus*, *Citrobacter freundii*, *Exiguobacterium acetylicum*, *Acidisoma tundrae*, and *Dyellasp.*) and BC-4 (*Thermoanaerobacterium aciditolerans*, *Ruminococcus flavefaciens*, *Caproiciproducens galcititolivorans*, and *Methanobrevibacter thaueri*)	Consortia	50°C for 7 days	50°C for 40 days	+113.7% of CH_4 yield improvement. −69% and −77% of protein content after 15 days.	Ali et al. (2020)
Wheat straw	Bacteria	TC-5	Consortia	—	7°C–45°C for 12 days	+22.2% and +36.6% CH_4 yield improvement under 37°C and 45°C, respectively.	Kong et al. (2018)

Cotton stalk	Bacteria	MC1	Consortia	50°C for 14 days	35°C for 10 days	+120.4% of biogas yield improvement. +122.4% CH_4 yield improvement.	Yuan et al. (2016) Yuan et al. (2016)
Rotted silage maize straw	Bacteria	MC1	Consortia	—	35°C for 5 days	+74.7% of biogas yield improvement.	Hua et al. (2016)
Oil palm empty fruit bunches (EFB)	Bacteria	*Lachnospiraceae*-rich consortium.	Consortia	40°C for 7 days	40°C for 60 days	+310.6% and +16.4% of CH_4 yield improvement by *Lachnospiraceae*-rich consortium from cow manure and digested sludge, respectively.	Suksong et al. (2019)
Rice straw	Bacteria	Microbial consortiums from horse manure and decomposed wood.	Consortia	—	38°C for 45 days	From horse manure and decomposed wood, respectively: +320.7% and 519.9% of biogas yield improvement. +39.95% and 41.82% of total solids degradation +31.24% and 32.22% of volatile fatty acid degradation	Tantayotai et al. (2017)
Corn straw	Bacteria	(*Saccharomyces cerevisiae* sp., *Coccidioides immitis* sp., and *Hansenula*	Consortia	20°C for 15 days		+33.07% and +75.57% of biogas	Zhong et al. (2011)

(Continued)

TABLE 3.2 (Continued)

Biomass	Pretreatment	Species	Type	Pretreatment conditions	Incubation conditions	Major results	References
		anomala sp.), cellulolytic bacteria (*Bacillus licheniformis* sp., *Pseudomonas* sp., *Bacillus subtilis* sp., and *Pleurotus florida* sp.), and the lactic acid bacteria of *Lactobacillus deiliehii* sp.				and methane yield improvement −34.6% technical digestion time. −5.81% to −25.10% in total lignin, cellulose, and hemicellulose contents.	
Napier grass	Bacteria	*MC1, WSD-5, and XDC-2*	Consortia	50°C for 21 days	35°C for 21 days	+39%, +49%, and +32% of maximum CH_4 yield improvement for MC1, WSD-5, and XDC-2, respectively.	Wen et al. (2015)
Maize silage	Bacteria	16 strains *(of Bacillus, providencia, and Ochrobactrum genera)*	Consortia	—	30°C for 96 h	+38% biogas yield improvement.	Poszytek et al. (2016)
Chicken feather waste	Bacteria	*Bacillus megaterium*	Single strain	—	55°C at 50 days	+155% of CH_4 yield improvement after 1 day.	Forgács et al. (2011)
Straw	Bacteria + fungi	Fungi and fungi + *Bacillus subtilis*	Single strain + mixture	28°C for 14–42 days	37°C for 60 days	+46% and +51% of CH_4 yield improvement for the two treatments, respectively.	Yadav and Paul (2022)

Chestnut leaves and hay	Fungi	*Auricularia auricula-judae*	—	—	37°C for 2–3 weeks	+15% of biogas production improvement.	Mackuľak et al. (2012)
Rice straw	Fungi	*Pleurotus ostreatus* and *Trichoderma reesei*	Single strain	28°C for 10–30 days	45 days at 37°C	+120% of CH_4 yield improvement. −33% of lignin content.	Mustafa et al. (2016)
Yard trimmings	Fungi	*Ceriporiopsis subvermispora*	Single strain	37°C for 40 days	30 days at 28°C	+85%–154% of CH_4 yield improvement. −20% of lignin content.	Zhao et al. (2014)
Sugarcane bagasse	Fungi	*Laetiporus sulphureus,* *Pleurotus ostreatus, Ceriporiopsis subvermispora*	Single strain	—	27°C for 60 days	−48% and −47% lignin and xylan content, respectively following *C. subvermispora* pretreatment.	Machado et al. (2017)
Corn silage and cattail	Fungi	*Piromyces rhizinflata* YM600	Single strain	—	37°C for 60 days	0 CH_4 yield improvement but +62% volatile fatty acid degradation rate improvement.	Nkemka et al. (2015)
Corn silage	Fungi	Trametes versicolor + cow manure	Co-digestion	7 days at 27 °C	37°C for 21 days	+41% of CH_4 yield improvement and improved pH stability.	Tišma et al. (2018)
Wheat and pearl millet	Fungi	*Curvularia lunata*	Single strain	—		+18.7% and +28.6% of biogas	Yadav and Vivekanand (2020)

(Continued)

TABLE 3.2 (Continued)

Biomass	Pretreatment	Species	Type	Pretreatment conditions	Incubation conditions	Major results	References
					28°C–36°C for 14–42 days	production increase for wheat straw and pearl millet straw, respectively. +22.6 and +38.6% of CH_4 yield improvement for wheat straw and pearl millet, respectively. −40.8% and −25% of lignin content for wheat straw and pearl millet straw, respectively.	
Animal manure	Fungi	*Orpinomyces* sp., *Piromyces* sp., *Anaeromyces* sp., and *Neocallimastix frontalis*	Mixture	—	40°C for 40 days	+60% of CH_4 yield improvement.	Yıldırım et al. (2017)
Mushroom residues	Fungi	*Trametes versicolor* and *Pleurotus sajor caju*	Single strain	25°C for 6 weeks	18 days	+74% of VFA production and +58.5% of lignin degradation compared with that of unpretreated compost.	Fang et al. (2020) Fang et al. (2020)

Park wastes + dung manure	Fungi	*Aspergillus terreus* and *Trichoderma viride*	Mixture	25°C for 7 days	35°C for 70 days	+23% and +31% of biogas and CH_4 yield improvement.	Ali and Sun (2015)
Willow sawdust	Fungi	*Leiotrametes menziesii* and *Abortiporus biennis*	Single	25°C for 30 days	35°C for 14–30 days	+31%–43% and +115% of Biochemical methane potential improvement by A. biennis and A. biennis combined with alkaline treatment.	Alexandropoulou et al. (2017)
Rice straw	Fungi	*Pleurotus ostreatus*	Single	28°C for 10–30 days	37°C for 30 days	+56%, +67.6%, and +7.8% of maximum ultimate biogas yield, methane yield, and methane content improvement. +165% CH_4 yield improvement by a combination of fungal and milling pretreatment.	Mustafa et al. (2016)
Maize stover silage	Lactic acid inoculants	*Lactiplantibacillus plantarum,* *Lentilactobacillus buchneri, Limosilactobacillus fermentum*	—	—	37°C for 60 days	+18.8% and +41.0% of CH_4 yield improvement after ensiling and after air exposure, respectively.	Zhang et al. (2023)

(*Continued*)

TABLE 3.2 (Continued)

Biomass	Pretreatment	Species	Type	Pretreatment conditions	Incubation conditions	Major results	References
Pulp and paper sludge	Endoglucanase laccase	Enzyme from *Pleurotus ostreatus*	—	37°C for 4 h	—	+34% CH_4 yield improvement. 12.0% and 22.3% of cellulose and lignin degradation, respectively.	Lin et al. (2017)
Microalgae	Cellulase enzyme mix	—	—	37°C for 6 h	—	+8%–15% CH_4 yield improvement	Passos et al. (2016)
Willow	Laccase	*Trametes versicolor*	—	30°C for 6–24 h	—	+33% CH_4 yield improvement	Schroyen et al. (2015)
Corn cobs	Cellulolytic enzymatic cocktail	Novozymes	—	40°C for 3 h	—	+14% CH_4 yield improvement	Pérez-Rodríguez et al. (2017)
Butter	Lipase	*Candida rugosa*	—	40°C for 16 h	—	+84% CH_4 yield improvement	Domingues et al. (2015)
Agricultural waste (silage, straw, and animal manure)	Cellulase and its mixture with xylanase, and beta-glucanase, and with Xylanase, glucosidase, and endo-pectinase	Cellulase from *Trichoderma citrinoviride*, Xylanase from *Trichoderma citrinoviride*, beta-glucanase from *Trichoderma citrinoviride*	—	—	40°C for 60 days	+0.3% to +21.1% in CH_4 yield improvement between days 5 and 20, but no significant difference after 60 days.	Weide et al. (2020)
Red microalgae	Cellulase protease viscozyme and their mix	—	—	25°C–0°C–55°C, for 9 h (enzyme mix) and 24 h for (cellulase,	—	+70% and +100% of CH_4 yield improvement at mesophilic and thermophilic	Kendir Çakmak and Ugurlu (2020)

				viscozyme, and protease)		conditions, respectively, from *P. cruentum* were achieved after protease pretreatment.	
Microalgae	Mixtures of cellulase, chitinase, protease, xylanase, poligalacturonase, pectinliase, pectinmetilesterase, poligalacturonase	*Trichoderma reesei*, *Trichoderma harzianum*, *Aspergillus oryzae*, *Aspergillus niger*, *Aspergillus aculeatus fungi*	—	37°C for 3 h	—	+90% of CH_4 yield improvement at 1% dose but decreased by 17% for 2% dose.	Avila et al. (2021)
Paper sludge	Rumen fluid	—	—	37°C for 6 h	37°C for 20 days	+725% and +27.7% of CH_4 and biogas production improvement. +24.7%, +22.8%, +4.7%, and +22.9% increase in degradability of TSs, NDF, ADF, and ADL, respectively, during CH_4 fermentation.	Takizawa et al. (2018)
Corn straw	Pig manure + rumen cultures	—	—	39°C for 3 days	34–124 days	+60% and +40% of CH_4 and CO_2 yield improvement	Jin et al. (2018)

(*Continued*)

TABLE 3.2 (Continued)

Biomass	Pretreatment	Species	Type	Pretreatment conditions	Incubation conditions	Major results	References
Brassica napus L.	Cattle rumen fluid	—		35°C for 24 h	32 days	+150% of CH_4 production (due to shifts from bacteroidetes, composed of amylolytic *Prevotella* spp., to Firmicutes, composed of cellulolytic and xylanolytic *Ruminococcus* spp., in only 6 h).	Baba et al. (2017)
Corn straw	Cattle manure	—	—	37°C	3 weeks at 37°C	28.8% and 41% degradation of cellulose and hemicellulose, respectively. +9.8% of CH_4 production improvement.	Nagler et al. (2019)

TABLE 3.3 Overview of innovative anaerobic digestion technologies for enhanced biogas production.

Digester technology	Innovative aspects	Digester volume (m^3)	Production yield	Scalability	Inoculum source	Cost ($)	Refs.
Psychrophilic plug-flow digester with assisted solar heat	Combines psychrophilic AD and solar heating.	4 (2–15 m^3)	4230–14,800 m^3 CH_4/ year	Small scale	Wastewater, cow manure	NS	Ivanovs and Blumberga (2023)
CSIR-IICT anaerobic gas-lift reactor	Employs high rate biomethanation using vegetable waste	28.598	750 units of electricity from 15 tons of waste	Medium scale	Vegetable waste	15,000,000 $ (1.5 crores)	Sandhya Kiran et al. (2023)
Dry anaerobic semi-continuous horizontal reactor	Uses 3D-printed tracers to study solid residence time and flow behaviors.	.5	.179 ± .009 Nm^3 CH_4/ kg VS	NS	Straw-cattle manure	.18$ for tracers (.17 €)	Hernandez-Shek et al. (2023)
Integrated thermophilic anaerobic digester with a power-to-gas system	Enhances biogas production via clean hydrogen addition.	.002	Biogas yield:0.67 NLg-VS-added^{-1}, with 81% ± 1% CH_4	NS	Sewage sludge	NS	Kwang et al. (2023)

(Continued)

TABLE 3.3 (Continued)

Digester technology	Innovative aspects	Digester volume (m^3)	Production yield	Scalability	Inoculum source	Cost ($)	Refs.
Integrated MBR-AD system	Integrates MBR and AD technologies with a zero-waste approach	MBR: 6 AD: 5.4	—	Small-to-medium scale	Wastewater, source-separated biowaste	—	Solomou et al. (2023)
Semi-industrial scale SSAD system (TMethar)	Enhances methane yield and inoculum stabilization through leachate recirculation.	276	.06 m^3 CH_4.kg VS^{-1} removed	Semi-industrial scale	Organic fraction of municipal solid waste	NS	Franca et al. (2022)
Multi-anaerobic reactor (M-SHAR)	Enables selective interaction with materials for direct penetration into microbial biomass and quick cessation of heating by power cutoff.	.1	5% higher COD removal efficiency and 12%–20% higher biogas yields compared with a convection-heated reactor	Small scale	Granular anaerobic sludge sourced from a full-scale dairy effluent treatment plant	NS	Zieliński et al. (2021)
Four stages cascade anaerobic digester	The cascade configuration of CSTRs at low solid retention times significantly improved the	Three reactors of .0022 each, and one of .0154	NS	NS	Waste-activated sludge	NS	Guo et al. (2021)

	enzymatic hydrolysis rate and extend in anaerobic waste-activated sludge digestion.						
Single-stage biohythane reactor	Utilizes a unique single-stage dark fermentation system with entrapped hydrogenic and methanogenic bacteria.	.001	Biohythane: 381–480 mL/L-d CH_4: 395 mL/L-d	NS	Hydrogenic and methanogenic bacteria from digested sludge	NS	Ta et al. (2019)
Farm base biorefinery model (IFBBR)	Combines biogas production, wastewater treatment with microalgae, and bioethanol production from digestate in a multiprocess integration.	.40–23 depending on scale (from 1 to 50 horses)	1.1 of biogas per day per horse, with methane content of 65%–70%	Small-to -arge scales	NS	NS	Bartha et al. (2019)
Domestic anaerobic digester waterless toilet	Integrates a waterless toilet with an anaerobic digester. Simple,	.0069	Specific biogas production of .15 m^3/kgSV,	Laboratory scale	Organic waste from humans	NS-labeled "low cost"	Regattieri et al. (2018)

(*Continued*)

TABLE 3.3 (Continued)

Digester technology	Innovative aspects	Digester volume (m^3)	Production yield	Scalability	Inoculum source	Cost ($)	Refs.
	low-cost, uses readily available materials, no maintenance required.		with 74.69% CH_4				
Digester assisted with microaerobic desulfurization of biogas	Incorporates microaerobic conditions for H_2S removal.	.2	NS	Pilot scale (with potential for scaling to various sizes and conditions)	Sewage sludge	NS	Donoso-Bravo et al. (2018)
MGA	Integrates dry and wet AD processes in a single modular system enhanced with a real-time command and control system powered by solar energy.	NS	.409 of biogas per day	Micro-agricultural farm	Ground miscanthus with silage corn, and manure with organic waste from food residues	NS	Voicea et al. (2017)

NS, Not specified; *AD*, anaerobic digestion.

The techno-economic analysis for large-scale biogas digester projects can be conducted by focusing on three key aspects: material flow analysis, based on the physical quantification of process inputs and outputs; energy flow analysis, which identifies the actual processes and amounts of energy produced; and financial flow analysis, which tracks the precise amounts of capital inflow and outflow to assess financial performance (Wang et al., 2021). In addition, the environmental benefits of the biogas sector can be analyzed using the life cycle assessment methodology. This approach quantifies and identifies potential environmental impacts throughout the entire process, from beginning to end, including significant contributors such as transport, electrical and thermal energy consumption, and the handling of by-products. Furthermore, a vital component of feasibility studies is the social study. In situations when the facility uses the proper biogas technology to handle wastewater and human waste, it can lessen the spread of disease, enhance local sanitation, and generate job opportunities. In a rural context, the social benefits of biogas installation include significant reductions in energy expenditure, time spent collecting wood, particularly for women, and improved general community health through reduced air pollution.

Feasibility studies examine every aspect of converting biomass into bioenergy in detail, incorporating qualitative elements such as feedstock quality assessment, energy conversion efficiency, and supply stability (Kalinichenko & Havrysh, 2019; Meng et al., 2020). Fig. 3.2 provides an overview of the essential elements for assessing the feasibility of a large-scale digestion plant.

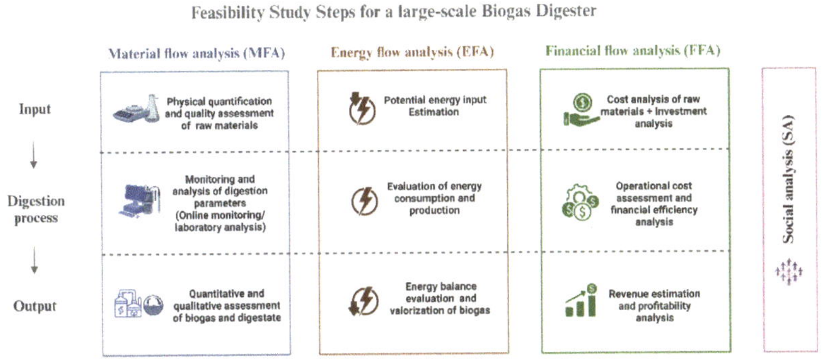

FIGURE 3.2 Feasibility study for a large-scale biogas digester.

3.5.3 Economic analysis and cost–benefit assessment

The economic benefits of using anaerobic digesters in wastewater treatment and industrial applications are highly desired by companies because they consume less energy than aerobic methods, due to the elimination of active aeration and produce fewer bio-solids, thus reducing postprocessing costs. Additionally, the biogas generated can help offset electricity and natural gas expenses at treatment facilities, which reveals that it is a robust method for waste management and sustainable energy production (Lisowyj and Wright, 2018). The cost of transforming feedstocks into biofuels varies depending on the technology used and the efficiency of the production process (Rocha-Meneses et al., 2023). As these technologies continue to evolve, researchers are developing new methods to enhance the value of raw biogas.

Among these innovations, the integration of AI stands out as an effective method for boosting the efficiency of biogas digesters. This technology helps in reducing operating costs and improving yield through optimized process control by predicting the process for better management, making biogas an attractive option for large- and small-scale applications (Ling et al., 2024; Swami et al., 2023). Regarding the economic analysis of hybrid systems integrating biogas with other renewable energy sources such as solar energy and wind, production costs are influenced by various factors, including component prices, which vary by region, and the lifespan and maintenance requirements of these components. Although initial costs may be high, long-term savings in energy costs can offset these expenses. This approach to hybrid energy systems supports sustainable development objectives and promotes a transition to a greener, more sustainable future. Advanced microbial optimization techniques are considered more cost-effective and advantageous for large-scale applications than various chemical treatments, due to the high amounts of chemicals and detoxification requirements involved (Nwokolo and Enebe, 2022). Additionally, sustainability in terms of economic, energy, and environmental aspects can be achieved through the use of decentralized digesters (Ajay et al., 2021). Cost analysis also indicates that fixed biogas plants are approximately 1.27 times more expensive than portable biogas plants. The integration of nutrient recovery technologies can enhance both the sustainability and economic viability of various processes, although many of these technologies are still under development (Wadchasit et al., 2023). A significant challenge remains in marketing digestates as organic fertilizers. Consumer hesitancy and competition with conventional fertilizers underscore the urgent need to boost digestates' marketability and secure greater governmental support (Romero-Güiza et al., 2016). Moreover, a real sustainable system must take into account factors other than nutrient recovery, such as the associated transport costs, which are crucial to the transition to a circular economy solution.

In conclusion, continuous and sustainable government support through measures such as renewable energy credits and carbon pricing is essential for creating a favorable market environment for biogas production. This involvement is crucial for encouraging the development of biogas technologies, promoting investment, and facilitating market growth.

3.6 Technological advantages and limitations

3.6.1 Evaluation of artificial intelligence monitoring system

AI and ML are increasingly combined with condition monitoring techniques to allow machines and equipment to learn from historical data and identify patterns that may predict a malfunction. There are numerous strategies available to address the limits of condition monitoring systems, including deep learning, image processing, support vector machines, natural language processing, and others (Naik et al., 2023). Several ML techniques have been employed to predict AD; however, research on deep learning (DL) is limited. DL architectures employ many network layers to extract representations for the detection or classification of higher-level features from raw input. Among the various DL approaches, recurrent neural networks (RNNs) are best suited for sequence prediction issues, such as time series data (Jeong et al., 2021). Support ML, or SVM, is a method for data analysis and pattern recognition in input/output data that was first created by Cortes in 1995. SVM-based algorithms are reliable for modeling and engineering applications and have simple theoretical underpinnings (Zarei et al., 2023). As one of the most advanced nonparametric supervised classifiers available today, SVM may be configured in a variety of ways using the kernel function, which generates the transformation for transforming the input space to the output space. Several functions, including linear, polynomial, radial basis function, and multilayer perceptron, are frequently employed as the kernel function in SVM. Patent text data can be a valuable source of information on technological advancements in certain industries (biogas and AD), but interpreting such text requires a substantial investment of time and resources. Because of this, text analysis and natural language processing (NLP) methods have received a lot of praise for their capacity to glean insights from patent databases. Sentiment analysis, relationship extraction, machine translation, and topic segmentation and recognition are a few NLP applications (De Clercq et al., 2019). By thoroughly evaluating these aspects, organizations can select an AI monitoring system that aligns with their specific requirements and objectives while ensuring effective risk management and operational excellence. When evaluating an AI monitoring system, several key aspects should be considered to determine its effectiveness and suitability for the intended purpose. Fig. 3.3 presents a comprehensive evaluation framework.

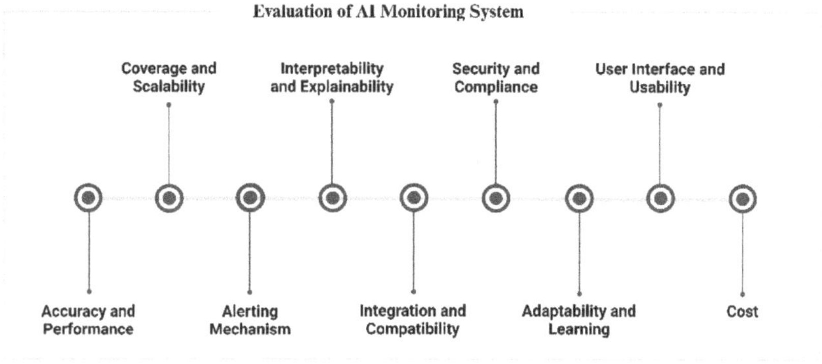

FIGURE 3.3 Evaluation of AI monitoring system. *AI*, Artificial intelligence.

3.6.2 Comparative analysis of hybrid systems

Renewable energy technologies are inherently more complex than conventional fossil fuel-based energy systems. They are usually characterized by intermittent energy resources, nonconventional energy conversion mechanisms, and less developed energy carriers and storage methods. Consequently, it is often difficult to compare the performance and economics of these renewable energy technologies to the conventional methods. Such comparison is an essential step in determining the role of a renewable energy technology in a market primarily driven by cost and in identifying areas for further research and development.

Simulation is a tool that has been employed to compare renewable energy systems with conventional methods. The flexibility of simulation makes it possible to simulate complex energy systems and to vary the conditions under which they operate. Simulation has been employed in identifying and resolving energy system operation issues, in predicting economic and environmental impacts, and in optimizing the design of a new or existing energy system (Bazmi & Zahedi, 2011). A study analyzed 37 computer tools used to assess how renewable energy can be integrated into different energy systems. There is no single "perfect" tool, as the best choice depends on the specific goals of the analysis (e.g., analyzing a single building vs. a national grid). The study considered factors like the type of energy system, included technologies, timeframes, and tool availability to help decision-makers pick the right tool.

For example, a study used HOMER software to design affordable hybrid renewable energy systems. The study results revealed that the solar–biomass system is significantly cheaper than the wind–biomass option based on total cost, energy cost, and operational costs (Mishra et al., 2016).

A new hybrid system combining solar panels and biomass is designed to provide clean heating and cooking fuel. It uses solar energy to heat biogas reactors and buildings, while leftover biogas is used for cooking fuel. This system boasts impressive efficiency (25% annual), saves energy (95%), and cuts carbon emissions by almost 3000 tons a year compared with traditional systems. With a payback period of 5 years, this cost-effective solution is especially attractive for rural areas with abundant sunshine, biomass, and limited economic development. In a relevant study, a pilot solar–biogas system designed to treat waste from Morocco's recycled paper industry shows promise for clean energy production. Combining waste processing, solar panels for heat, and thermal storage, the system achieved high efficiency (70% thermal) and biogas output (1100L per month with 74% methane). This self-sufficient system can significantly reduce reliance on external energy (2 MWh per month savings) and greenhouse gas emissions (69% CO_2 reduction). Scaling up this system could generate enough biogas to meet a large portion (64%) of a recycled paper industry company's fuel need (Ouhammou et al., 2022). As electric vehicles (EVs) gain popularity, charging them strains the power grid. This study proposes a solution: solar–biogas-powered EV charging stations. The system combines solar panels, biogas generators, battery storage, and a special converter for EVs. This design reduces reliance on the main grid and is economically viable. Researchers found a 4.5 kW solar panel setup to be most cost-effective, with the system paying for itself in 12 years. Beyond cost benefits, this method also significantly cuts greenhouse gas emissions compared with traditional EV charging methods. Overall, solar–biogas-powered charging stations show promise for a clean and sustainable future of EVs (Kah et al., 2022). An article reviewed hybrid renewable energy systems that combine multiple renewable energy sources like solar, wind, and biomass. The authors found that the hybrid system can be more economical and environmentally friendly than traditional energy sources, and that combining biomass with concentrated solar energy (biomass-CSP) is a promising option for both energy production and waste management (Alhijazi et al., 2023). In the same line, research addressed a key challenge in renewable energy: ensuring consistent power despite resource fluctuations. The study proposed a new, comprehensive multigenerator system using wind, concentrated solar, and biogas. This system showed that it can produce electricity, hot air, hydrogen, fresh water, and cooling. They analyzed the system's efficiency under four scenarios and incorporated thermal storage to manage variations in solar radiation. The results are promising: The system achieved high efficiency (64.91% to 71.06% energetic and 31.80% to 53.81% exergetic) and significantly reduced CO_2 emissions compared with fossil fuels. This suggests this design has the potential to be a reliable and eco-friendly renewable energy source (Bamisile et al., 2020). A study investigated a promising solution for providing clean and reliable electricity to remote healthcare facilities. They analyzed a system combining wind turbines and biogas generators, with the option for a backup diesel generator. This hybrid system showed a significant advantage: It could be cost-effective, generating electricity as low as 7.34 cents per kilowatt-hour. Even more importantly, it drastically reduced greenhouse gas emissions compared with traditional diesel-powered generators. This suggests the technology has the potential to be a clean and affordable way to power healthcare facilities in remote areas (Alrbai et al., 2023).

Research proposes a wind–biogas–hydrogen hybrid system that can generate both electricity and hydrogen fuel. The biogas acts as a reliable source when wind is low and excess wind can be used to create clean-burning hydrogen through a special wind-powered pyrolysis process. This system can be adjusted up or down depending on energy needs and even has the benefit of removing carbon dioxide from the atmosphere. Overall, the study provides a strong theoretical framework for this type of hybrid system and suggests future research areas like improving the efficiency of the market participation model and finding the optimal sizing for real-world applications (Liu, 2022).

Although there is a lot of research on solar–wind, solar–biogas, and solar–wind–biogas hybrids, little is known about wind–biogas and other combinations, such as hydropower and geothermal. Investigating these fields is essential to maximizing their effectiveness and reaping the rewards of improved system output, improved grid integration, and heightened energy security for a future powered by renewable energy sources. It can be difficult to assess an energy system's economic viability, particularly when comparing small-scale systems to larger utilities. Large systems only consider the cost of electricity.

3.6.3 Microbial optimization strategies

Microbial optimization strategies in biogas production have different advantages. However, they also face several challenges. Table 3.4 summarizes the advantages and disadvantages of microbial optimization techniques. These approaches can greatly enhance biogas and methane production by the optimization of microbial communities implicated in the degradation of organic matter and AD. Additionally, microbial optimization helps to stabilize the digestion operation, by minimizing the possibility of imbalances and increasing the efficiency of the process (Shahzad et al., 2024). Another advantage is the use of a broader array of feedstocks, including those with high-fiber content or lower degradation rates to monitor microbiota. Additionally, the progress concerning genetic engineering and synthetic biology offers great advancement for designing microbial consortia that are efficient in biogas production. However, several limitations hinder microbial optimization strategies, among them the complex interactions between various microorganisms within a biogas reactor. Indeed, small changes can have unpredictable effects on the overall process, which makes it intricate to control the microbial consortia with precision. For example, acidification due to organic overloading led to low methane yields, which extended recovery

TABLE 3.4 Advantages and disadvantages of microbial optimization techniques.

Technique	Advantages	Disadvantages	References
Bacterial	• Sustainable and environmentally friendly. • Degradation of cellulose and hemicellulose. • Compared with single strains, bacterial consortium through synergetic interactions, reduce of the need for pretreatment, broaden the types of usable feedstocks, and enhance the overall metabolic efficiency and stability of the digestion process.	• Slow digestion. • pH sensitivity and temperature dependence. • Substrate inhibition by toxicity of some feedstocks and rapid production of acids that reduce pH. • Operational complexity in monitoring pH, temperature, nutrient balance, and sludge handling. • Dependency on microbial health due to sudden changes in environmental conditions or feedstock composition, and a long recovery times after a disruption.	Ali et al. (2017)
Fungal	• Better lignocellulosic degradation due to a reduced pretreatment needs and an efficient substrate breakdown. • High tolerance to temperature and pH fluctuations and potential for sequential batch processing. • Allowance of the utilization of nonconventional and low-quality feedstock due to the synergistic interactions between fungi and bacteria.	• Slow growth rate that could increase operational costs. • High cost of media and nutrients and sensitivity to pH changes. • Risk of growth of inhibitory or competitive contaminant fungi. • Handling and mixing problems due to clumping and channeling of mycelia in digesters. • Growth and enzyme activity inhibition by the production of some organic acids and metabolic by-products.	Mustafa et al. (2016) Zhao et al. (2014) Tišma et al. (2018); Yadav and Vivekanand (2020); Mustafa et al. (2016)

(Continued)

TABLE 3.4 (Continued)

Technique	Advantages	Disadvantages	References
Enzymes	• Improvement in substrate accessibility. • Decrease in the digestion period through acceleration of hydrolysis. • Increase in process stability by breaking down substances that could inhibit microbial activity and improving buffering capacity. • Lowering chemical and thermal pretreatments, which reduces energy costs. • Specific enzyme mixtures can be selected to target wide range of feedstocks.	• High cost • Susceptibility to inactivation due to process conditions. • Dependency on specific conditions, which may require different enzymes or dosing strategies, complicating operations. • Regulatory and environmental concerns especially for genetically modified enzymes.	Kendir Çakmak and Ugurlu, (2020), Abbas et al. (2022)
Rumen fluid	• Reduction in pretreatment needs through natural enzyme use. • Rapid process initiation especially in digesters recovering from operational disruption.	• Variability in microbial composition depending on livestock management practices (diet, health, and species of the donor animals). • Logistical challenges related to collection, storage, and transportation requirements. • Scalability and cost issues. • Health and safety problems due to pathogen risk. • Process stability and efficiency due to nonadaptation to handling nonplant materials. • Ethical and environmental concerns.	Takizawa et al. (2018)

times, increased operational expenses for wastewater treatment facilities, and negatively impacted the environment. Another limitation is the possibility of long-term ecological concerns. In addition to ethical and regulatory rules to consider when dealing with genetically engineered microbes, their introduction into biogas systems raises serious concerns about the probability of altering natural ecosystems in the case of uncontrolled escape. Moreover, advanced metagenomic analysis and genetic engineering require specialized, expensive equipment and expertise, and the continuous monitoring and adjustment of microbial communities in a biogas reactor can also increase operational costs.

3.7 Conclusion

Biogas technology plays a pivotal role in the circular economy, as it enables the conversion of biomass into renewable energy and valuable by-products. To enhance the overall efficiency and reliability of biogas production, it is essential to employ advanced approaches, such as the design of efficient digesters, the optimization of AD, the integration of biogas production with renewable energy systems, and the integration of data monitoring and control systems using AI. These advanced approaches can improve the overall efficiency and reliability of biogas digester technology, thereby enhancing energy resilience and optimizing resource use. By capitalizing on these insights, innovative biogas digester technology can assume a pivotal role in advancing the circular economy agenda, facilitating the transition to renewable energy, and mitigating the environmental challenges associated with organic waste management.

AI disclosure

During the preparation of this work, the author(s) used quillbot and deepl in order to For paraphrasing!. After using this tool/service, the author(s) reviewed and edited the content as needed and take(s) full responsibility for the content of the publication.

References

Abbas, I., Liu, J., Noor, R. S., Faheem, M., Farhan, M., Ameen, M., & Shaikh, S. A. (2022). Development and performance evaluation of small size household portable biogas plant for domestic use. *Biomass Conv. Bioref, 12*, 3107–3119. https://doi.org/10.1007/s13399-020-00956-y.

Ajay, C. M., Mohan, S., & Dinesha, P. (2021). Decentralized energy from portable biogas digesters using domestic kitchen waste: A review. *Waste Management, 125*, 10–26. https://doi.org/10.1016/j.wasman.2021.02.031.

Alexandropoulou, M., Antonopoulou, G., Fragkou, E., Ntaikou, I., & Lyberatos, G. (2017). Fungal pretreatment of willow sawdust and its combination with alkaline treatment for enhancing biogas production. *Journal of Environmental Management, 203*, 704–713. https://doi.org/10.1016/j.jenvman.2016.04.006.

Alhijazi, A. A. K., Almasri, R. A., & Alloush, A. F. (2023). A hybrid renewable energy (solar/wind/biomass) and multi-use system principles, types, and applications: A review. *Sustainability, 15*(24), 16803. https://doi.org/10.3390/su152416803.

Ali, S. S., & Sun, J. (2015). Physico-chemical pretreatment and fungal biotreatment for park wastes and cattle dung for biogas production. *SpringerPlus, 4*(1), 712. https://doi.org/10.1186/s40064-015-1466-9.

Ali, S. S., Abomohra, A. E.-F., & Sun, J. (2017). Effective bio-pretreatment of sawdust waste with a novel microbial consortium for enhanced biomethanation. *Bioresource Technology, 238*, 425–432. https://doi.org/10.1016/j.biortech.2017.03.187.

Ali, S. S., Mustafa, A. M., Kornaros, M., Manni, A., Sun, J., & Khalil, M. A. (2020). Construction of novel microbial consortia CS-5 and BC-4 valued for the degradation of catalpa sawdust and chlorophenols simultaneously with enhancing methane production. *Bioresource Technology, 301*, 122720. https://doi.org/10.1016/j.biortech.2019.122720.

Alrbai, M., Al-Dahidi, S., Al-Ghussain, L., Hayajneh, H., & Alahmer, A. (2023). A sustainable wind–biogas hybrid system for remote areas in Jordan: A case study of mobile hospital for a Zaatari Syrian refugee camp. *Sustainability, 15*(20), 14935. https://doi.org/10.3390/su152014935.

Andlar, M., Belskaya, H., Morzak, G., Šantek, M. I., Rezić, T., Tominac, V. P., & Šantek, B. (2021). Biogas production systems and upgrading technologies: A review. *Food Technology and Biotechnology, 59*(4), 387–412. https://doi.org/10.17113/ftb.59.04.21.7300.

Avila, R., Carrero, E., Vicent, T., & Blánquez, P. (2021). Integration of enzymatic pretreatment and sludge co-digestion in biogas production from microalgae. *Waste Management, 124*, 254–263. https://doi.org/10.1016/j.wasman.2021.01.042.

Baba, Y., Matsuki, Y., Mori, Y., Suyama, Y., Tada, C., Fukuda, Y., Saito, M., & Nakai, Y. (2017). Pretreatment of lignocellulosic biomass by cattle rumen fluid for methane production: Bacterial flora and enzyme activity analysis. *Journal of Bioscience and Bioengineering, 123*(4), 489–496. https://doi.org/10.1016/j.jbiosc.2016.11.008.

Bamisile, O., Huang, Q., Li, J., Dagbasi, M., Kemena, A. D., Abid, M., & Hu, W. (2020). Modelling and performance analysis of an innovative CPVT, wind and biogas integrated comprehensive energy system: An energy and exergy approach. *Energy Conversion and Management, 209*(2020), 11261. https://doi.org/10.1016/j.enconman.2020.112611.

Bartha, S., Carvalheiro, F., Moniz, P., Duarte, L., Berardino, S., Reis, A., & Vajda, B. (2019). *One Integrated Biorefinery Model For Use The Straw Base Horse Manure Resulted In Rural Truism Activity*.

Baruah, A., Basu, M., & Amuley, D. (2021). Modeling of an autonomous hybrid renewable energy system for electrification of a township: A case study for Sikkim, India. *Renewable and Sustainable Energy Reviews, 135*, 110158. https://doi.org/10.1016/j.rser.2020.110158.

Bazmi, A. A., & Zahedi, G. (2011). Sustainable energy systems: Role of optimization modeling techniques in power generation and supply—A review. *Renewable and Sustainable Energy Reviews, 15*(8), 3480–3500. https://doi.org/10.1016/j.rser.2011.05.003.

Boukrouh, S., Noutfia, A., Moula, N., Avril, C., Louvieaux, J., Hornick, J. L., Chentouf, M., & Cabaraux, J. F. (2023b). Ecological, morpho-agronomical, and nutritional characteristics of Sulla flexuosa (L.) Medik. ecotypes. *Scientific Reports, 13*, 13300. https://doi.org/10.1038/s41598-023-40148-y.

Boukrouh, S., Noutfia, A., Moula, N., Avril, C., Louvieaux, J., Hornick, J. L., Cabaraux, J. F., & Chentouf, M. (2023a). Ecological, morpho-agronomical, and bromatological assessment of sorghum ecotypes in Northern Morocco. *Scientific Reports, 13*, 15548. https://doi.org/10.1038/s41598-023-41565-9.

Chamazkoti, S. S., Hajinezhad, A., & Moosavian, S. F. (2024). Feasibility study of a flexible hybrid energy model with power-hydrogen production in wastewater treatment plant. *Energy Reports, 11*(2024), 3189–3201. https://doi.org/10.1016/j.egyr.2024.02.062.

De Clercq, D., Wen, Z., & Song, Q. (2019). Innovation hotspots in food waste treatment, biogas, and anaerobic digestion technology: A natural language processing approach. *The Science of the Total Environment, 673*, 402–413. https://doi.org/10.1016/j.scitotenv.2019.04.051.

Domingues, R. F., Sanches, T., Silva, G. S., Bueno, B. E., Ribeiro, R., Kamimura, E. S., Neto, R. F., & Tommaso, G. (2015). Effect of enzymatic pretreatment on the anaerobic digestion of milk fat for biogas production. *Food Research International, 73*, 26–30. https://doi.org/10.1016/j.foodres.2015.03.027.

Donoso-Bravo, A., Sadino-Riquelme, M. C., Díaz, I., & Muñoz, R. (2018). Mathematical modelling of in-situ microaerobic desulfurization of biogas from sewage sludge digestion. *Biotechnology Reports, 20*, e00293. https://doi.org/10.1016/j.btre.2018.e00293.

Dzobo, O. (2022). Design of an off-grid hydro-solar-biogas energy system for sustainable energy supply in rural communities. *IEEE International Conference on Power Systems Technology (POWERCON), 2022*, 1–4. https://doi.org/10.1109/POWERCON53406.2022.9930065 Kuala Lumpur, Malaysia.

Fajobi, M. O., Lasode, O. A., Adeleke, A. A., Ikubanni, P. P., & Balogun, A. O. (2022). Effect of biomass co-digestion and application of artificial intelligence in biogas production: A review. *Energy Sources, Part A: Recovery, Utilization, and Environmental Effects, 44*, 5314–5339.

Fang, W., Zhang, X., Zhang, P., Morera, X. C., van Lier, J. B., & Spanjers, H. (2020). Evaluation of white rot fungi pretreatment of mushroom residues for volatile fatty acid production by anaerobic fermentation: Feedstock applicability and fungal function. *Bioresource Technology, 297*, 122447. https://doi.org/10.1016/j.biortech.2019.122447.

Flavio, O.-Y., Abaase, S., Zebilila, M., & Atepor, L. (2022). Feasibility analysis of solar PV/biogas hybrid energy system for rural electrification in Ghana. *Cogent Engineering, 9*(1), 2034376. https://doi.org/10.1080/23311916.2022.2034376.

Forgács, G., Alinezhad, S., Mirabdollah, A., Feuk-Lagerstedt, E., & Horváth, I. S. (2011). Biological treatment of chicken feather waste for improved biogas production. *Journal of Environmental Sciences, 23*(10), 1747–1753. https://doi.org/10.1016/S1001-0742(10)60648-1.

Franca, L., Ferreira, B. O., Pereira, C., & Bassin, J. (2022). Performance of a percolation reactor integrated to solid-state anaerobic garage-type digesters with leachate recirculation for OFMSW treatment. *SSRN Electronic Journal*. https://doi.org/10.2139/ssrn.4090179.

Guo, H., Oosterkamp, M. J., Tonin, F., Hendriks, A., Nair, R., van Lier, J. B., & de Kreuk, M. (2021). Reconsidering hydrolysis kinetics for anaerobic digestion of waste activated sludge applying cascade reactors with ultra-short residence times. *Water Research, 202*, 117398. https://doi.org/10.1016/j.watres.2021.117398.

Habchi, S., Lahboubi, N., Karouach, F., Naim, I., Lahlou, Y., Bakraoui, M., Sallek, B., & El Bari, H. (2022). Effect of thermal pretreatment on the kinetic parameters of anaerobic digestion from recycled pulp and paper sludge. *Ecological Engineering & Environmental Technology 2022, 23*(1), 192–201. https://doi.org/10.12912/27197050/143568 ISSN 2719-7050.

Helal, A. A., Ghoneim, W. A., & Halaby, A. S. (2013). Feasibility study for self-sustained wastewater treatment plants—Using biogas CHP fuel cell, micro-turbine, PV and wind turbine systems. *Smart Grid and Renewable Energy, 2013*, 227–235.

Hernandez-Shek, M. A., Coutu, A., Fayolle, J., Peultier, P., Pauss, A., & Ribeiro, T. (2023). Determination residence time distribution of the solid phase in a dry anaerobic semi-continuous horizontal reactor of 0.5 m3 using innovative 3D printed tracers. *Bioresource Technology Reports, 22*, 101438. https://doi.org/10.1016/j.biteb.2023.101438.

Hua, B., Dai, J., Liu, B., Zhang, H., Yuan, X., Wang, X., & Cui, Z. (2016). Pretreatment of non-sterile, rotted silage maize straw by the microbial community MC1 increases biogas production. *Bioresource Technology, 216*, 699–705. https://doi.org/10.1016/j.biortech.2016.06.001.

Igielski, M. (2023). Energy security management system in small ports. *Współczesna Gospodarka, 17*(2), 15–27.

Ilangovan, P., Sharmila Begum, M., & Srividhya, P. K. (2023). Development of online monitoring device and performance evaluation of biogas plants using enhanced methane prediction algorithm (EMPA). *Sustainable Energy Technologies and Assessments, 56*, 103041. https://doi.org/10.1016/j.seta.2023.103041.

Imran, M., Ali, U., & Hasnain, A. (2021). Impact of blends of aqueous amines on absorber intercooling for post combustion CO_2 capture system. *Energy & Environment, 32*, 921–944.

Ivanovs, K., & Blumberga, D. (2023). Psychrophilic plug-flow digester with assisted solar heat—small-scale system feasibility. *Agronomy Research*, 21. https://doi.org/10.15159/AR.23.040.

Jean-Jacques, D., & Alexander, M. (2012). Infrastructure in developing countries: An overview of some economic issues, ZEF- Discussion Papers on Development Policy No. 165. *Center for Development Research,* 51 ISSN: 1436-9931.

Jin, W., Xu, X., Yang, F., Li, C., & Zhou, M. (2018). Performance enhancement by rumen cultures in anaerobic co-digestion of corn straw with pig manure. *Biomass and Bioenergy, 115*, 120–129.

Karouach, F., Bakraoui, M., Zguani, A., Hammadi, A., & El Bari, H. (2021). Co-digestion of industrial recycled pulp and paper sludge with vinasse wastewater: experimental and theoretical study. *International Journal of Environmental.*

Karouach, F., BenBakrim, W., Ezzariai, A., Mnaouer, I., Ibourki, M., Kibret, M., Sobeh, M., Hafidi, M., & Kouisni, L. (2024). Valorization of water hyacinth to biomethane and biofertilizer through anaerobic digestion technology. *Fuel, 358*(Part A), 130008. https://doi.org/10.1016/j.fuel.2023.130008 15 February 2024.

Kong, X., Du, J., Ye, X., Xi, Y., Jin, H., Zhang, M., & Guo, D. (2018). Enhanced methane production from wheat straw with the assistance of lignocellulolytic microbial consortium TC-5. *Bioresource Technology, 263*, 33–39. https://doi.org/10.1016/j.biortech.2018.04.079.

Lin, Y., Liang, J., Zeng, C., Wang, D., & Lin, H. (2017). Anaerobic digestion of pulp and paper mill sludge pretreated by microbial consortium OEM1 with simultaneous degradation of lignocellulose and chlorophenols. *Renewable Energy, 108*, 108–115. https://doi.org/10.1016/j.renene.2017.02.049.

Nagler, M., Kozjek, K., Etemadi, M., Insam, H., & Podmirseg, S. M. (2019). Simple yet effective: Microbial and biotechnological benefits of rumen liquid addition to lignocellulose-degrading biogas plants. *Journal of Biotechnology, 300*, 1–10. https://doi.org/10.1016/j.jbiotec.2019.05.004.

Passos, F., Hom-Diaz, A., Blanquez, P., Vicent, T., & Ferrer, I. (2016). Improving biogas production from microalgae by enzymatic pretreatment. *Bioresource Technology, 199*, 347–351. https://doi.org/10.1016/j.biortech.2015.08.084.

Poszytek, K., Ciezkowska, M., Sklodowska, A., & Drewniak, L. (2016). Microbial consortium with high cellulolytic activity (MCHCA) for enhanced biogas production. *Frontiers in*

Microbiology, 7. https://doi.org/10.3389/fmicb.2016.00324, https://www.frontiersin.org/journals/microbiology/articles/.

Regattieri, A., Bortolini, M., Ferrari, E., Gamberi, M., & Piana, F. (2018). Biogas Micro-Production from Human Organic Waste—A Research Proposal. *Sustainability, 10*(2), 330. https://doi.org/10.3390/su10020330.

Sandhya Kiran, J. K., Abhishek, P., Manikanta, P., Pavan, K. S., Chaitra, K., & Kiran, M. S. (2023). Waste to wealth: Vegetable market waste to generate electricity, a design proposal suitable for any medium scale vegetable market in Indian cities. *Materials Today: Proceedings*. https://doi.org/10.1016/j.matpr.2023.04.032.

Schroyen, M., Vervaeren, H., Vandepitte, H., Van Hulle, S. W. H., & Raes, K. (2015). Effect of enzymatic pretreatment of various lignocellulosic substrates on production of phenolic compounds and biomethane potential. *Bioresource Technology, 192*, 696–702. https://doi.org/10.1016/j.biortech.2015.06.051.

Suksong, W., Kongjan, P., Prasertsan, P., & O-Thong, S. (2019). Thermotolerant cellulolytic Clostridiaceae and Lachnospiraceae rich consortium enhanced biogas production from oil palm empty fruit bunches by solid-state anaerobic digestion. *Bioresource Technology, 291*, 121851. https://doi.org/10.1016/j.biortech.2019.121851.

Ta, T., Lin, C.-Y., Ta, N., & Chu, C. (2019). Biohythane production via single-stage anaerobic fermentation using entrapped hydrogenic and methanogenic bacteria. *Bioresource Technology, 300*, 122702. https://doi.org/10.1016/j.biortech.2019.122702.

Takizawa, S., Baba, Y., Tada, C., Fukuda, Y., & Nakai, Y. (2018). Pretreatment with rumen fluid improves methane production in the anaerobic digestion of paper sludge. *Waste Management, 78*, 379–384.

Voicea, I., Gageanu, I., Matache, M., & Vladut, V. *Innovative technology for obtaining bioenergy through the process of advanced anaerobic digestion*. https://doi.org/10.22616/ERDev2017.16.N035.

Weide, T., Baquero, C. D., Schomaker, M., Brügging, E., & Wetter, C. (2020). Effects of enzyme addition on biogas and methane yields in the batch anaerobic digestion of agricultural waste (silage, straw, and animal manure). *Biomass and Bioenergy, 132*, 105442. https://doi.org/10.1016/j.biombioe.2019.105442.

Yuan, X., Ma, L., Wen, B., Zhou, D., Kuang, M., Yang, W., & Cui, Z. (2016). Enhancing anaerobic digestion of cotton stalk by pretreatment with a microbial consortium (MC1). *Bioresource Technology, 207*, 293–301.

Further reading

Çakmak, E. K., & Ugurlu, A. (2020). Enhanced biogas production of red microalgae via enzymatic pretreatment and preliminary economic assessment. *Algal Research, 50*, 101979. https://doi.org/10.1016/j.algal.2020.101979.

Cruz, I. A., Chuenchart, W., Long, F., Surendra, K. C., Andrade, R. S. L., Bilal, M., Liu, H., Figueiredo, R. T., Khanal, S. K., & Ferreira, L. F. R. (2022). Application of machine learning in anaerobic digestion: Perspectives and challenges. *Bioresource Technology, 345*, 126433. https://doi.org/10.1016/j.biortech.2021.126433.

Jeong, K., Abbas, A., Shin, J., Son, M., Kim, Y. M., & Cho, K. H. (2021). Prediction of biogas production in anaerobic co-digestion of organic wastes using deep learning models. *Water Research, 205*, 117697. https://doi.org/10.1016/j.watres.2021.117697.

Kabeyi, M. J. B., & Olanrewaju, O. A. (2022). Biogas production and applications in the sustainable energy transition. *Journal of Energy, 2022*, 8750221. https://doi.org/10.1155/2022/8750221.

Kah, Y. Y., Chin, H. H., & Klemes, J. J. (2022). Solar energy-powered battery electric vehicle charging stations: Current development and future prospect review. *Renewable and Sustainable Energy Reviews, 169*(2022), 112862. https://doi.org/10.1016/j.rser.2022.112862.

Kalinichenko, A., & Havrysh, V. (2019). Feasibility study of biogas project development: Technology maturity, feedstock, and utilization pathway. *Archives of Environmental Protection, 45*, 68–83. https://doi.org/10.24425/aep.2019.126423.

Kong, X., Du, J., Ye, X., Xi, Y., Jin, H., Zhang, M., & Guo, D. (2018). Enhanced methane production from wheat straw with the assistance of lignocellulolytic microbial consortium TC-5. *Bioresource Technology, 263*, 33–39. https://doi.org/10.1016/j.biortech.2018.04.079.

Kumar, A., Mandal, B., & Sharma, A. (2015). Advancement in biogas digester. *Green Energy and Technology,* 351–382. https://doi.org/10.1007/978-81-322-2337-5_14.

Kwang, C., Hidaka, T., Nomura, Y., Fujiwara, T., Akimoto, S., & Tsubota, J. (2023). Insight into the mechanism of biogas upgrading with hydrogen addition in thermophilic anaerobic digestion of sewage sludge. *Journal of Water and Environment Technology, 21*, 323–341. https://doi.org/10.2965/jwet.23-080.

Lahboubi, N., Karouach, F., Bakraoui, M., El Gnaoui, Y., Essamri, A., & El Bari, H. (2022). Effect of alkali-NaOH pretreatment on methane production from anaerobic digestion of date palm waste. *Ecological Engineering & Environmental Technology, 23*(2), 78–89. https://doi.org/10.12912/27197050/144846 ISSN 2719-7050, License CC-BY 4.0.

Lamas, W. D., & Giacaglia, G. E. (2020). Fuzzy logic application in multi-period performance analysis of a SWTP. *IEEE Latin America Transactions, 19*, 500–506.

Ling, J., Chan, Y., Chen, J., Chong, D. J. S., Tan, A., Arumugasamy, S. K., & Lau, P.-L. (2024). Machine learning methods for the modelling and optimisation of biogas production from anaerobic digestion: A review. *Environmental Science and Pollution Research, 31*, 1–20. https://doi.org/10.1007/s11356-024-32435-6.

Lisowyj, M., & Wright, M. (2018). A review of biogas and an assessment of its economic impact and future role as a renewable energy source. *Reviews in Chemical Engineering, 36*(3), 401–421. https://doi.org/10.1515/revce-2017-0103.

Liu, X. (2022). *The optimal bidding strategy of a wind-biogas-hydrogen hybrid system (Dissertation).* https://urn.kb.se/resolve?urn=urn:nbn:se:uu:diva-466359.

Machado, A., da, S., & Ferraz, A. (2017). Biological pretreatment of sugarcane bagasse with basidiomycetes producing varied patterns of biodegradation. *Bioresource Technology, 225*, 17–22. https://doi.org/10.1016/j.biortech.2016.11.053.

Mackuľak, T., Prousek, J., Švorc, Ľ., & Drtil, M. (2012). Increase of biogas production from pretreated hay and leaves using wood-rotting fungi. *Chemical Papers, 66*(7), 649–653.

Meng, L., Alengebawy, A., Ai, P., Jin, K., Chen, M., & Pan, Y. (2020). Techno-economic assessment of three modes of large-scale crop residue utilization projects in China. *Energies, 13*, 3729. https://doi.org/10.3390/en13143729.

Mishra, S., Panigrahi, C. K., & Kothari, D. P. (2016). Design and simulation of a solar–wind–biogas hybrid system architecture using HOMER in India. *International Journal of Ambient Energy, 37*(2), 184–191. https://doi.org/10.1080/01430750.2014.915886.

Monnerie, N., et al. (2014). Coupling of wind energy and biogas with a high temperature steam electrolyser for hydrogen and methane production. *Green and Sustainable Chemistry, 4*, 60–69. https://doi.org/10.4236/gsc.2014.42010.

Mustafa, A. M., Poulsen, T. G., & Sheng, K. (2016). Fungal pretreatment of rice straw with *Pleurotus ostreatus* and *Trichoderma reesei* to enhance methane production under solid-state anaerobic digestion. *Applied Energy, 180*, 661–671. https://doi.org/10.1016/j.apenergy.2016.07.135.

Naik, J., Acharya, A., & Thaker, J. (2023). Revolutionizing condition monitoring techniques with integration of artificial intelligence and machine learning. *Materials Today: Proceedings*. https://doi.org/10.1016/j.matpr.2023.08.262.

Nkemka, V. N., Gilroyed, B., Yanke, J., Gruninger, R., Vedres, D., McAllister, T., & Hao, X. (2015). Bioaugmentation with an anaerobic fungus in a two-stage process for biohydrogen and biogas production using corn silage and cattail. *Bioresource Technology, 185*, 79–88. https://doi.org/10.1016/j.biortech.2015.02.100.

Norouzi, N. (2021). Exergy and energy analysis of a CHP system with a gas turbine and a horizontal axis wind turbine. *Trends in Renewable Energy, 7*(1).

Nwokolo, N., & Enebe, M. (2022). An insight on the contributions of microbial communities and process parameters in enhancing biogas production. *Biomass Conversion and Biorefinery*, 14. https://doi.org/10.1007/s13399-022-02580-4.

Okolie, J. A., Epelle, E. I., Tabat, M. E., Orivri, U., Amenaghawon, A. N., Okoye, P. U., & Gunes, B. (2022). Waste biomass valorization for the production of biofuels and value-added products: A comprehensive review of thermochemical, biological and integrated processes. *Process Safety and Environmental Protection, 159*, 323–344. https://doi.org/10.1016/j.psep.2021.12.049.

Olabi, A. G., Obaideen, K., Abdelkareem, M. A., AlMallahi, M. N., Shehata, N., Alami, A. H., Mdallal, A., Hassan, A. A. M., & Sayed, E. T. (2023). Wind energy contribution to the sustainable development goals: Case study on London array. *Sustainability, 15*(5), 4641. https://doi.org/10.3390/su15054641.

Oluwaseun, A. T., Mgbachi, C. A., Okelola, M. O., & Ajenikoko, G. A. (2018). A comparative analysis of renewable energy using biogas and solar photovoltaic systems: A case study of Ajaba. *In Osun State Control Theory and Informatics, 7*, 7–13.

Ouhammou, B., Mohammed, A., Sliman, S., Jamil, A., Mohammed, B., Karouach, F., Hassan, E. B., & Kousksou, T. (2022). Experimental conception and thermo-energetic analysis of a solar biogas production system. *Case Studies in Thermal Engineering, 30*(2022), 101740. https://doi.org/10.1016/j.csite.2021.101740.

Pérez-Rodríguez, N., García-Bernet, D., & Domínguez, J. M. (2017). Extrusion and enzymatic hydrolysis as pretreatments on corn cob for biogas production. *Renewable Energy, 107*, 597–603. https://doi.org/10.1016/j.renene.2017.02.030.

Prabhu, A. V., Antony Raja, S., Avinash, A., & Pugazhendhi, A. (2021). Parametric optimization of biogas potential in anaerobic co-digestion of biomass wastes. *Fuel, 288*, 119574. https://doi.org/10.1016/j.fuel.2020.119574.

Priyanka, Y., Sudeep, Y., Singh, D., Shekher Giri, B., & Mishra, P. K. (2022). Barriers in biogas production from the organic fraction of municipal solid waste: A circular bioeconomy perspective. *Bioresource Technology, 362*, 127671. https://doi.org/10.1016/j.biortech.2022.127671 ISSN 0960-8524.

Raheem, A., Hassan, M. Y., Shakoor, R., & Rasheed, N. (2014). Economic feasibility of stand-alone wind energy hybrid with bioenergy from anaerobic digestion for electrification of remote area of Pakistan. *International Journal of Integrated Engineering, 6*.

Reuchlin, S., Joshi, R., & Schmehl, R. (2023). Sizing of hybrid power systems for off-grid applications using airborne wind energy. *Energies, 16*(10), 4036.

Rocha-Meneses, L., Luna-delRisco, M., González, C. A., Moncada, S. V., Moreno, A., Sierra-Del Rio, J., & Castillo-Meza, L. E. (2023). An overview of the socio-economic, technological, and environmental opportunities and challenges for renewable energy generation from residual biomass: A case study of biogas production in Colombia. *Energies, 16*, 5901. https://doi.org/10.3390/en16165901.

Romero-Güiza, M., Mata-Álvarez, J., Chimenos, J., & Astals, S. (2016). Nutrient recovery technologies for anaerobic digestion systems: An overview. *Revista Investigación, Optimización y Nuevos procesos en Ingeniería, 29*, 7–26. https://doi.org/10.18273/revion.v29n1-2016001.

Sadoune, H., Rihani, R., & Marra, F. S. (2023). DNN model development of biogas production from an anaerobic wastewater treatment plant using Bayesian hyperparameter optimization. *Chemical Engineering Journal, 471*, 144671. https://doi.org/10.1016/j.cej.2023.144671.

Shahsavar, M. M., Akrami, M., Gheibi, M., Kavianpour, B., Fathollahi-Fard, A. M., & Behzadian, K. (2021). Constructing a smart framework for supplying the biogas energy in green buildings using an integration of response surface methodology, artificial intelligence and petri net modelling. *Energy Conversion and Management, 248*, 114794. https://doi.org/10.1016/j.enconman.2021.114794.

Shahzad, H. M. A., Almomani, F., Shahzad, A., Mahmoud, K. A., & Rasool, K. (2024). Challenges and opportunities in biogas conversion to microbial protein: A pathway for sustainable resource recovery from organic waste. *Process Safety and Environmental Protection, 185*, 644–659. https://doi.org/10.1016/j.psep.2024.03.055.

Sharma, V., Sharma, D., Tsai, M.-L., Ortizo, R. G. G., Yadav, A., Nargotra, P., Chen, C.-W., Sun, P.-P., & Dong, C.-D. (2023). Insights into the recent advances of agro-industrial waste valorization for sustainable biogas production. *Bioresource Technology, 390*, 129829. https://doi.org/10.1016/j.biortech.2023.129829.

Solomou, N., Malamis, D., Barampouti, E., Mai, S., & Loizidou, M. (2023). The design and performance prediction model of an integrated scheme of a membrane bioreactor and anaerobic digester for the treatment of domestic wastewater and biowaste. *Sustainability, 15*, 11455. https://doi.org/10.3390/su151411455.

Swami, S., Suthar, S., Singh, R., Thakur, A. K., Gupta, L. R., & Sikarwar, V. S. (2023). Integration of anaerobic digestion with artificial intelligence to optimise biogas plant operation. *Environment Development and Sustainability*. https://doi.org/10.1007/s10668-023-04326-2.

Tamoor, M., Bhatti, A. R., Hussain, M. I., Miran, S., Kiren, T., Ali, A., & Lee, G. H. (2023). Optimal sizing and technical assessment of a hybrid renewable energy solution for offgrid community center power. *Fronteries in Energy Research, 11*, 1283586. https://doi.org/10.3389/fenrg.2023.1283586.

Tantayotai, P., Pornwongthong, P., Muenmuang, C., Phusantisampan, T., & Sriariyanun, M. (2017). Effect of cellulase-producing microbial consortium on biogas production from lignocellulosic biomass. *Energy Procedia, 141*, 180–183. https://doi.org/10.1016/j.egypro.2017.11.034.

Thirunavukkarasu, M., Sawle, Y., & Lala, H. (2023). A comprehensive review on optimization of hybrid renewable energy systems using various optimization techniques. *Renewable and Sustainable Energy Reviews, 176*(2023), 113192. https://doi.org/10.1016/j.rser.2023.113192 Received 15 May 2021.

Tufaner, F., & Demirci, Y. (2020). Prediction of biogas production rate from anaerobic hybrid reactor by artificial neural network and nonlinear regressions models. *Clean Technologies and Environmental Policy, 22*, 713–724. https://doi.org/10.1007/s10098-020-01816-z.

Tumusiime, E., Kirabira, J. B., & Musinguzi, W. B. (2022). Optimization of substrate mixing ratios for wet anaerobic digestion of selected organic waste streams for productive biogas systems. *Energy Reports, 8*, 10409–10417. https://doi.org/10.1016/j.egyr.2022.08.189.

Wadchasit, P., Rakmak, N., O-Thong, S., Rattanasak, U., Imai, T., Jitpinit, S., & Nuithitikul, K. (2023). Improvement of biogas production and quality by addition of struvite precipitates derived from liquid anaerobic digestion effluents of palm oil wastes. *Journal of Environmental Chemical Engineering, 11*, 109081. https://doi.org/10.1016/j.jece.2022.109081.

Wang, J., Chai, Y., Shao, Y., & Qian, X. (2021). Techno-economic assessment of biogas project: A longitudinal case study from Japan. *Resources, Conservation and Recycling, 164*, 105174. https://doi.org/10.1016/j.resconrec.2020.105174.

Wen, B., Yuan, X., Li, Q. X., Liu, J., Ren, J., Wang, X., & Cui, Z. (2015). Comparison and evaluation of concurrent saccharification and anaerobic digestion of Napier grass after pretreatment by three microbial consortia. *Bioresource Technology, 175*, 102–111. https://doi.org/10.1016/j.biortech.2014.10.043.

Yadav, D. S., & Paul, B. (2022). Conversion of the domestic LPG cook stove to use biogas fuel: An experimental approach. *Transactions of the Indian National Academy of Engineering, 7*, 1213–1222. https://doi.org/10.1007/s41403-022-00355-x.

Yıldırım, E., Ince, O., Aydin, S., & Ince, B. (2017). Improvement of biogas potential of anaerobic digesters using rumen fungi. *Renewable Energy, 109*, 346–353. https://doi.org/10.1016/j.renene.2017.03.021.

Zarei, M., Bayati, M. R., Ebrahimi-Nik, M., Rohani, A., & Hejazi, B. (2023). Modelling the removal efficiency of hydrogen sulfide from biogas in a biofilter using multiple linear regression and support vector machines. *Journal of Cleaner Production, 404*, 136965. https://doi.org/10.1016/j.jclepro.2023.136965.

Zhang, H., Wu, J., Zhao, X., Yan, P., Yang, R., Yan, J., Yuan, X., & Cui, Z. (2023). Improving aerobic stability and methane production of maize stover silage with lactic acid bacteria inoculants: Focus on pentose-fermentation. *Industrial Crops and Products, 201*, 116861. https://doi.org/10.1016/j.indcrop.2023.116861.

Zhao, J., Zheng, Y., & Li, Y. (2014). Fungal pretreatment of yard trimmings for enhancement of methane yield from solid-state anaerobic digestion. *Bioresource Technology, 156*, 176–181. https://doi.org/10.1016/j.biortech.2014.01.011.

Zheng, Y., Wang, X., & Yang, F. (2019). Improving the anaerobic digestion of switchgrass via cofermentation of rumen microorganisms (rumen bacteria, protozoa, and fungi) and a biogas slurry. *Energy & Fuels, 33*(2), 1185–1195. https://doi.org/10.1021/acs.energyfuels.8b03496.

Zhong, W., Zhang, Z., Luo, Y., Sun, S., Qiao, W., & Xiao, M. (2011). Effect of biological pretreatments in enhancing corn straw biogas production. *Bioresource Technology, 102*(24), 11177–11182. https://doi.org/10.1016/j.biortech.2011.09.077.

Zhu, L., & Zhang, C. (2020). Feasibility. *Processes, 8*, 180. https://doi.org/10.3390/pr8020180.

Zieliński, M., Dębowski, M., & Kazimierowicz, J. (2021). Microwave radiation influence on dairy waste anaerobic digestion in a multi-section hybrid anaerobic reactor (M-SHAR). *Processes, 9*(10), 1772. https://doi.org/10.3390/pr9101772.

Chapter 4

Applications and future trends of in situ technologies for biogas purification and upgrading

Margarida Gonçalves[1,2,3], Catarina Viegas[2,4], Bruna Petry[1,2], Ana Paula Ramos[1,2], Nuno Lapa[1,5], and José Paulo Mota[1,5]
[1]*Chemistry Department, NOVA School of Science and Technology, Universidade NOVA de Lisboa, Caparica, Portugal,* [2]*MEtRICs – Mechanical Engineering and Resource Sustainability Center, Universidade NOVA de Lisboa, Caparica, Portugal,* [3]*VALORIZA – Research Centre for Endogenous Resource Valorization, Polytechnic Institute of Portalegre, Portalegre, Portugal,* [4]*LEAF– Linking Landscape, Environment, Agriculture and Food, Associate Laboratory TERRA, School of Agriculture, University of Lisbon, Lisbon, Portugal,* [5]*LAQV-REQUIMTE – Associated Laboratory for Green Chemistry, Universidade NOVA de Lisboa, Campus de Caparica, Caparica, Portugal*

Abbreviation

AMP	2-amino-2-methyl-1-propanol
AMs	Acetoclastic methanogens
AC	Activated carbon
AD	Anaerobic digestion
Fe–AnGS	Anaerobic granular sludge and iron oxides complexes
AnMBR	Anaerobic membrane bioreactor
UASB	Anaerobic sludge blanket reactor
AP	Applied potential
bioLNG	Biological liquid natural gas
BTF	Biotrickling filter
CACOI	Carbon-coated zero-valent iron
CC	Closed circuit
CZVI	Coarse zero-valent iron particles
DESs	Deep eutectic solvents
MEA	Diethanolamine
DGA	Diglycolamine
DET	Direct electron transfer
ESA	Electrical swing adsorption
EC-AnMBR	Electrochemical anaerobic membrane bioreactor

Biogas in the Circular Economy. DOI: https://doi.org/10.1016/B978-0-443-29230-9.00005-X

GAC	Granular activated carbon
HPAD	High-pressure anaerobic digestion
HRAP	High-rated algal pond
HBA	Hydrogen bond acceptor
HBD	Hydrogen bond donor
HOB	Hydrogen-oxidizing bacteria
HMs	Hydrogenotrophic methanogens
IL	Ionic liquid
AHA-ILs	Ionic liquids with aprotic heterocyclic anions
MOFs	Metal-organic frameworks
MMEA	Methyl monoethanolamine
MDEA	Methyldiethanolamine
MEC	Microbial electrolysis cells
MEA	Monoethanolamine
NZVI	Nano zero-valent iron
NPs	Nanoparticles
MDEA	N-methyldiethanolamine
OTC	Open trickling column
PBR	Photobioreactor
PZ	Piperazine
PSA	Pressure swing adsorption
SCP	Single-cell protein
TSA	Temperature swing adsorption
TAN	Total ammonia nitrogen
TEA	Triethanolamine

4.1 Introduction

Anaerobic digestion (AD) is a fundamental technology needed to achieve the decarbonization of energy production activities and contribute to the circular economy in the waste management sector. All over the world, anaerobic decomposition of bio-wastes with CH_4 recovery and upgrading has the potential to significantly reduce natural gas consumption while decreasing the carbon footprint associated with waste transportation and storage. Biogas production and upgrading avoids environmental pollution and generates value, job opportunities, and marketable products for local populations, thus contributing to the implementation of United Nations Sustainable Development Goals (Piadeh, 2024).

Nonetheless, the full potential of AD has been limited by several constraints of technological, socio-economic, and political nature. The recalcitrance of some feedstocks that limits biogas productivity and the need for robust, sustainable, and cost-effective technologies for biogas upgrading are the main technological challenges to the deployment of a successful biogas production industry (Archana, 2024).

Political instruments are also fundamental tools to stimulate the adoption of AD as a widespread technology of waste treatment and valorization, emission reduction strategy, and a source of renewable methane (Bangalore et al., 2016; Zglobisz, 2010).

The revised European Renewable Energy Directive (EU, 2023) includes a subtarget of 3.5% for advanced biofuels and biogas that must complement electric mobility in the decarbonization of the transport sector. Selective collection of bio-wastes and their valorization as feedstocks for composting or biogas production has been defined as a priority in EU state members (European Parliament and of the Council, 2018).

To reach these strategic targets, it is essential to develop adequate and sustainable technologies to convert biogas to pure biomethane, thus producing a high heating value renewable fuel with quality for use in internal combustion engines or injection in the natural gas grid (Angelidaki, 2018).

Biogas upgrading is an important component of AD technology that involves separating all biogas components including methane (CH_4), carbon dioxide (CO_2), water (H_2O), hydrogen (H_2), hydrogen sulfide (H_2S), nitrogen (N_2), ammonia (NH_3), siloxanes (Si-O-Si), droplets, and particulates. In addition to the main constituent, methane, secondary products, and contaminants are also present in biogas in different concentrations. These constituents can be separated and/or recovered for use in other industrial processes. (Fig. 4.1)

Methane, carbon dioxide, hydrogen, hydrogen sulfide, ammonia, and carbon particulates are the products of the aerobic and anaerobic decomposition of the organic components of the feedstocks. Nitrogen can be present if atmospheric air intrusion in the anaerobic digester occurs; oxygen from atmospheric air is generally not detected because is rapidly consumed by

FIGURE 4.1 Main components of biogas and average composition.

aerobic processes. Siloxanes are often present when the organic fraction of municipal solid waste is used in the AD process because these are common components of hygiene products and cosmetics that reach significant concentrations in urban wastewaters (Calbry-Muzyka, 2022). Biogas upgrading allows to increase the heating value, eliminate components related with corrosion, and deposit formation in engines or turbines, producing a gas fuel that fulfills the specifications for injection to the natural gas grid or use as a vehicle fuel (Petersson & Wellinger, 2009).

4.2 Technologies for biogas upgrading

Biogas upgrading technologies involve physical, chemical, biochemical, and electrochemical processes (Fig. 4.2) that allow to obtain a methane-rich gas fuel by separation and eventual recovery or conversion of other biogas components including impurities present at low concentrations (Aghel et al., 2022; Angelidaki et al., 2019; Bauer et al., 2013; Khan et al., 2021) (Fig. 4.2.).

Absorption, adsorption, and membrane technologies for biogas upgrading are traditionally used in AD industry, but new approaches are recently proposed aiming to contribute to a higher sustainability of the overall process or as a strategy for biorefinery concepts. In the last decades, several authors addressed the comparative analysis of biogas upgrading processes with focus in their performance, energy, and material requirements and potential for further development and optimization (Bauer et al., 2013; Golmakani et al., 2022; Karne et al., 2023; Khan et al., 2021; López & Rodríguez, 2024; Mwacharo, 2020; Patil et al., 2019; Petersson & Wellinger, 2009; Sun et al., 2015; Upadhyay et al., 2023).

Aghel et al. (2022) compared different biogas upgrading technologies with emphasis on their carbon capture characteristics and efficiency. Some authors focused their analysis on in situ upgrading technologies, namely those focused

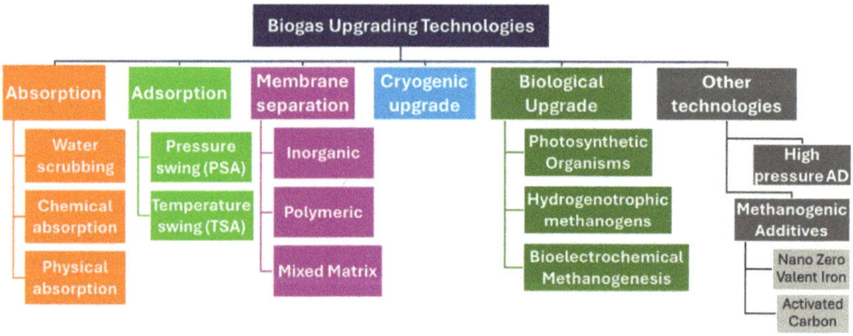

FIGURE 4.2 Traditional and advanced biogas upgrading technologies.
Adapted from Khan et al., 2021.

on CO_2 consumption and conversion by different microorganisms (Comesaña-Gándara, 2022; Sun, 2022b; Zhang et al., 2020; Zhao et al., 2021).

4.2.1 Absorption technologies

Absorption technologies are based on the capacity to remove CO_2 from raw biogas by reaction and conversion to water-soluble species or transient complexes with organic or inorganic molecules.

4.2.1.1 Water scrubbing

Physical absorption of biogas components in aqueous media is the most commonly used method for upgrading biogas. Alkaline species favor the reactions of carbonate and bicarbonate formation; therefore, high pH values are generally used:

$$CO_2 + H_2O \leftrightarrow HCO_3^- + H^+ \tag{4.1}$$

$$HCO_3^- + OH^- \leftrightarrow CO_3^{2-} + H_2O \tag{4.2}$$

This technology allows the efficient separation of the biogas major by-products (H_2S and CO_2) and can also contribute to a reduction of ammonia, siloxanes, and carbon particulates (Fig. 4.3).

The raw biogas is compressed and admitted to the absorption column that may be operated at pressures up to 10 bar (Bauer, 2013). The higher solubilities of H_2S and CO_2 in water, at room temperature, favor their

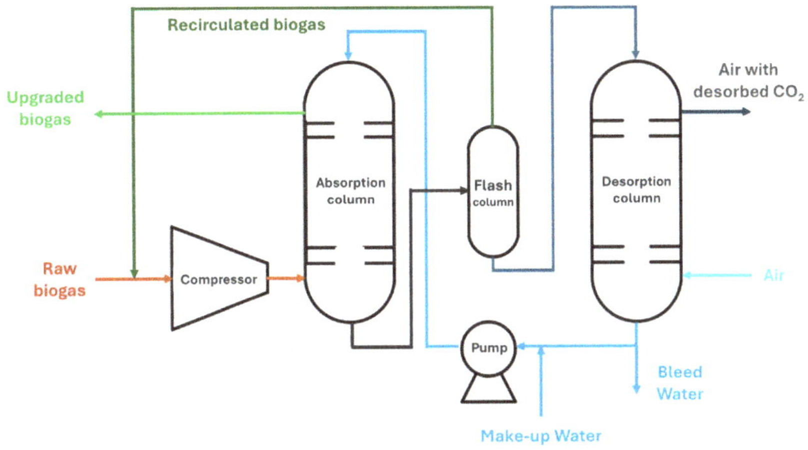

FIGURE 4.3 Biogas upgrading using water scrubbers.
Adapted from Angelidaki, 2019; Bauer, 2013.

solubilization in the aqueous medium while CH_4 is much less retained. The upgraded biogas is recovered in the top of the absorption column, while the water phase with dissolved biogas components exits in the bottom being transferred to the flash column. This column is operated at lower pressures (< 3 bar) to reduce the retention of CH_4 in the water phase. The released gases are recirculated to the raw biogas feed, while the water phase is transferred to wastewater facilities or to a desorption column, for CO_2 removal by air stripping. Regenerating CO_2-loaded water allows its reuse in the absorption column, reducing overall water consumption and wastewater generation (Khan, 2021).

To promote optimal mass transfer between the raw biogas and the water phase in the absorption column, some kind of packing material is generally used, thus improving the column efficiency and reducing investment costs (Tayar, 2019). Examples of packaging materials are granite stones, candlenut marble, plastic balls, and rings, calcite marble, and steel wire mesh (Walozi et al., 2016).

An innovative packing material with a high specific surface area (800 m^2/m^3) achieved CO_2 and H_2S removal efficiencies of 84%, and 96%, showing a better performance than other classical packing materials (Herdem et al., 2024).

Seawater supplemented with 0.1 wt% of Iranian-modified clinoptilolite zeolite and precipitated phosphogypsum was tested for physical adsorption of CO_2 using a countercurrent U-shaped micro-channel reactor. A CO_2 capture efficiency of 96.85% was reached for a liquid flow rate of 150 mL/h and a biogas flow rate of 50 mL/min. The zeolite and mineral additives contribute to the liquid medium alkalinity, and the reactor microchannels contribute to a high mass transfer coefficient with low energy consumption. Nonetheless, the low reactor volume will require multiple parallel units to design a plant with proper upgrading capacity (Aghel et al., 2022).

Another alkaline additive proposed to improve biogas upgrading in water scrubbing systems was oil palm ash dissolved in the maturation pond effluent. A solution with pH 9.3 and alkalinity of 7.5 g of $CaCO_3L^{-1}$ was obtained and used to remove CO_2 from biogas with an efficiency of 53% at a flow rate of 210 L. h^{-1} (Rattanaya, 2022).

Wantz et al. (2022)propose a model for optimization of water scrubbing design and operational parameters to maximize CH_4 purity and recovery while improving water regeneration performance and energy efficiency. The same authors developed and tested a high-pressure water scrubbing prototype, adequate for small-scale anaerobic digesters, operating on farms or other installations with biogas production rates in the range of 20 to 40 $Nm^3.h^{-1}$. The proposed system uses anisotropic packing on the absorption column to improve mass transfer, reaching a CH_4 purity higher than 97%. Water regeneration is performed under vacuum (0.8 to 0.2 bar) to improve CH_4

recovery by 10%, and the system energy efficiency was 91% for an electrical consumption of 0.38 kWh Nm^{-3} (Wantz, 2023).

Biogas recirculation on the anaerobic digestor is an approach studied by different authors for in situ biogas upgrading, by maximization of CO_2 retention in the liquid phase. The gas flow rate must be optimized to ensure maximum performance, and this option should not be applied; the reactor start-up is not recommended to avoid excessive hydrolysis and acidification. Although the physicochemical reactions of carbonate and bicarbonate formation are probable mechanisms of CO_2 removal, the recirculation of biogas also has a positive effect on the growth of methanogenic microorganisms leading to a higher conversion rate by the biochemical pathway (Yuan, 2022).

A countercurrent column for CO_2 removal by the anaerobic digestor liquid effluent was also proposed as a variation of water scrubbing systems. This system was able to produce a gas stream with 90.3% CH_4, for a biogas flow of 0.36 $L.h^{-1}$, at room temperature and atmospheric pressure without consumption of additional water resources (Zhao et al., 2021).

4.2.1.2 High-pressure anaerobic digestion

A variant of biogas recirculation for in situ upgrading is the operation of the anaerobic digestor at pressures 3 to 10 times higher than atmospheric pressure, designated as high-pressure anaerobic digestion (HPAD). HPAD involves the pressurization of the anaerobic digestor with raw biogas reaching an equilibrium pressure above atmospheric pressure. Since CO_2 dissolves more readily in the liquid phase than CH_4, when the pressure increases, the gas phase becomes richer in CH_4, resulting in an upgraded biogas (Lindeboom et al., 2012).

However, although several studies showed the potential of HPAD to achieve higher CH_4 content, the unconventional reactor environments relating to the pressurized gas and a pH drop could impact microbial activity and hamper its feasibility (Bär et al., 2018; Chen, 2014; Lindeboom et al., 2013; Lindeboom et al., 2012; Zhao et al., 2020).

Operation at high-pressure favored the growth of some bacterial phyla like *Bacteroidetes*, *Thermotogae*, and *Firmicutes* but decreased some methanogenic species such as *Thiopseudomonas*, causing a delay on the methanogenic phase (Thapa, 2023).

A two-stage HPAD integrating biogas production and upgrading operating at a working pressure of 9 bar allowed to reach CH_4 concentrations up to 77%. However, the high-pressure condition also increases ammonia concentration in the digestate leading to a lower specific methane yield (Lemmer, 2015).

Regardless of different studies showing the potential of HPAD for increasing CH_4 concentration in the biogas, there are some limitations to the efficiency of this technology, namely the impact of CO_2 solubilization in the pH of the liquid phase and the effect on microbial populations composition and activity, that can negatively influence biogas productivity (Bär et al., 2018; Chen et al., 2014; Lindeboom et al., 2013; Lindeboom et al., 2012; Zhao et al., 2020).

4.2.1.3 Chemical absorption

Chemical scrubbing technology promotes biogas upgrading using aqueous chemical agents that react with CO_2 in the scrubbing column, retaining it in the liquid phase and releasing the upgraded biogas from the top. The aqueous solution with the captured CO_2 is then transferred to the regeneration column where CO_2 is released by heating, thus regenerating the absorption solution that will be recirculated to the scrubbing column (Fig. 4.4).

Typical agents for chemical scrubbing are amines such as mono-, di-, and tri-ethanolamine (MEA, MDA, and TEA), diglycolamine (DGA), methyl-diethanolamine (MDEA), 2-amino-2-methyl-1-propanol (AMP), and piperazine (PZ), given their capacity to form complexes with CO_2 at room temperature and pressures of 1 to 2 bar (Kadam & Panwar, 2017; Kapoor, 2019). These reactions involve simultaneous participation of more than one amine molecule and require the catalytic assistance of a Brønsted base such as water as shown in Eqs. 4.3 to 4.5 (Said, 2020).

$$R{-}NH_2 + CO_2 \rightarrow R{-}NH{-}COO^- + RNH_3^+ \quad (4.3)$$

$$R{-}NH_2 + CO_2 + H_2O \rightarrow HCO_3^- + RNH_3^+ \quad (4.4)$$

$$[R{-}NH{-}COO^-\ RNH_3^+] + H_2O \rightarrow R{-}NH_2 + [HCO_3^-\ RNH_3^+] \quad (4.5)$$

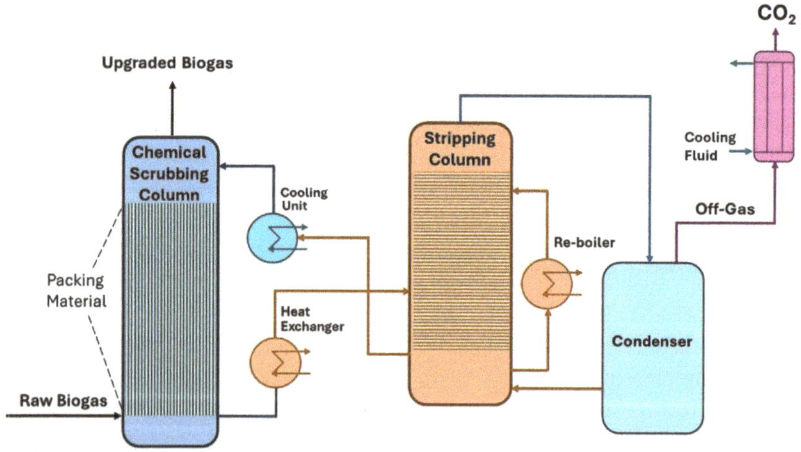

FIGURE 4.4 Chemical scrubbing technology.
Adapted from Kapoor, 2019.

The balance between CH_4 purity, amine price, system configuration and energy requirements has been addressed by several authors for optimization of the process efficiency and economic viability (Abd et al., 2022; Kim et al., 2024; Meng et al., 2022).

Mixed amines produced by blending methyl monoethanolamine (MMEA), and N-methyldiethanolamine (MDEA) showed a higher biogas upgrading performance than aqueous monoethanolamine (MEA). In particular, lower regeneration energy and higher CO_2/CH_4 selectivity were reported, suggesting that these mixed solutions could be interesting options chemical scrubbing (Meng, 2022).

Energy consumption requirements, amine degradation, foaming, precipitate formation, and corrosion problems are some of the drawbacks of this technology along with purification efficiency (Kapoor, 2019). The search for low-cost, renewable, and high-performance adsorbents is, therefore, a growing interest of the scientific community.

Kim et al. evaluated the use of hydrolyzed urine as a source of ammonium ions to sequester CO_2 in chemical scrubbing systems. A mole ratio of 0.52 mol $CO_2.mol^{-1}$ TAN (total ammonia nitrogen) was achieved demonstrating the potential of valorization of the nitrogen components as low-cost CO_2 sequestering agents (Kim et al., 2024).

CO_2 chemical absorption using alkaline solutions in an electrochemical cell was proposed by Mohammadpour et al. to improve the regeneration of the absorbent solution and achieve high CO_2 recovery. Carbonate ions formed by reaction of the alkaline solution with CO_2 moved from the cathode to the anode, through the anion-exchange membrane. The reverse reaction of CO_2 production from carbonate ions takes place at the anode compartment, releasing gaseous CO_2, that may be recovered for further use. The energy input was between 0.25 and 0.92 kWh/kg of CO_2 at optimum conditions (pH = 9), which represents a potential 50% reduction of energy requirements when comparing with commercially available biogas upgrading technologies (Mohammadpour et al., 2022).

Minimization of the energy requirements of a biogas upgrading system combining chemical absorption and CH_4 liquefaction was achieved by modeling and parameter optimization based on an energy and exergy analysis. This approach enabled the reduction of specific energy consumption by 20.4% and 5.8%, for the biogas upgrading and the CH_4 liquefaction, respectively, what corresponded to a 15.4% decrease of specific energy consumption for the overall process (Khan et al., 2022).

4.2.1.4 Organic solvent scrubbing

Solvent absorption systems also designated as physical absorption systems rely on the solvation of CO_2 through weaker forces such namely Lewis's acid-base interactions between the solvent and the biogas components especially CO_2

TABLE 4.1 Solvents used in organic scrubbing and corresponding commercial products (adapted from Carranza-Abaid, 2021).

Solvent used	Commercial product
Methanol	Rectisol™/Ifpexol™
n-formyl-morpholine	Morphysorb™
N-methyl-2-pyrrolidone	Purisol™
Propylene carbonate	Fluor™
Poly(ethylene glycol) dimethyl ether	Selexol™/Genesorb™
Tributyl phosphate	Estasolvan™
Tetramethylene sulfone	Sulfinol™
Glycerol	Glycerin

and H_2S. Given the weaker nature of these interactions, solvent regeneration has lower energy requirements than those necessary for chemical scrubbing (Carranza-Abaid, 2021). The commercial products used for organic scrubbing and the solvents they contain are presented in Table 4.1. They share a common trend, the presence of nucleophilic oxygenated groups able to interact with positive dipoles.

More recently, alternative solvents with polar or charged groups, also able to absorb CO_2 by electrostatic interaction, have been proposed for biogas upgrading. That is the case of ionic liquids that are solvents composed of two salts with opposite charge whose mixture remains liquid at room temperature and has a range of adjustable properties justifying their application in extraction or reaction processes (Lei, 2017).

The application of ionic liquids in biogas upgrading has been reviewed by Haider et al. (2022) that pointed out their advantages such as high CO_2 uptake, thermal stability, and negligible vapor pressure, but also their limitations related to production costs and environmental impacts.

Ionic liquids with aprotic heterocyclic anions (AHA-ILs) have been proposed for CO_2 capture through chemical absorption. Triethyl(octyl)phosphonium 2-cyanopyrrole ([P2228][CNPyr]) is an example of an AHA-IL studied as chemical adsorbent for biogas upgrading, due to its lower solvent and energy consumption requirements. A CH_4 stream with 97% purity was produced from a raw biogas with 40% (mol) of CH_4 and 60% of CO_2, operating the absorption and stripping columns at the temperatures ranges of 50°C and 95°C, respectively, using a pressure of 6 bar in the absorption step.

In those conditions, it was possible to reach a specific energy consumption of 0.211 kWh/Nm^3 of CH_4, lower than the values reported for other technologies (Moya, 2022).

The price, recyclability, and environmental impacts of the IL solvent are still critical parameters to optimize, but the capacity to also remove other biogas impurities such as H_2S, ammonia, water, or siloxanes can influence the overall costs and energy consumption by eliminating the need of using pretreatments or further upgrading steps (Moya, 2022; Seo, 2014).

Deep eutectic solvents (DESs) are another type of synthetic solvents also proposed as alternative to conventional solvents for biogas upgrading, due to their non-volatility, nonflammability, high absorption capacity, nontoxic character, and high thermal stability (Słupek et al., 2020). DESs are obtained by direct mixing of a hydrogen bond acceptor (HBA) with a hydrogen bond donor (HBD) that form a stable homogeneous mixture with a lower melting point than its individual components (Abbott, 2003).

Two DESs composed of choline chloride + urea and choline chloride +oxalic acid have been proposed for efficient removal of CO_2 and H_2S from biogas, and the theoretical calculation of the system performance estimated an upgrading cost of 0.35–0.37 €/Nm^3 of pure CH_4, that is in the range of the costs of current upgrading technologies (Słupek et al., 2020).

Hydrophobic DESs and conventional solvents (propylene carbonate, propylene glycol, and glycerol) have been studied for the absorption of typical biogas impurities (toluene, limonene, and siloxane D4). The hydrophobic DESs showed higher absorption capacity and were not affected by water co-absorption that negatively impacted the upgrading capacity of the polar conventional solvents (Villarim, 2022).

4.2.2 Adsorption technologies

Adsorption of CO_2 and other biogas components and impurities in solid porous materials enables their selective removal and recovery of high purity methane. Adsorption technologies take advantage of the different adsorption capacities of porous materials toward CO_2 and CH_4 to separate these two main components of biogas. After reaching the maximum retention capacity, the adsorbent must be regenerated by decreasing pressure, increasing temperature, and eventually flushing with an appropriate gas (Hosseini & Denayer, 2022; Selim, 2024).

4.2.2.1 Adsorbent materials

High surface area, high micropore relative volume, and high CO_2/CH_4 selectivity are the main properties sought after in materials used for CO_2 adsorption, in the context of biogas upgrading. Those properties can be found in activated chars, carbon molecular sieves, zeolite, activated carbons (ACs), and metal-organic frameworks (MOFs) (Abd, 2024).

Pérez-Huertas et al. reviewed the works focused on the production of activated chars from polymeric wastes, an approach that allows the transformation of nonbiodegradable wastes in materials with industrial relevance. The production process involves pyrolysis at temperatures typically above 400°C, followed by activation also at high temperatures by treatment with concentrated hydroxides (KOH and NaOH), concentrated acids (H_3PO_4 and H_2SO_4), or other chemical reagents such as zinc chloride ($ZnCl_2$), or potassium carbonate (K_2CO_3). The role of these chemicals is to react with the carbon structure generating volatile products and creating micropores. Physical activation by heating under a gas flow (nitrogen, argon, carbon dioxide, or steam) does not require caustic reagents and also improves the char porosity by removing carbonization products absorbed in the pore structure and by moderate oxidation of the carbon structure (Pérez-Huertas et al., 2023).

The selectivity toward CO_2 is a critical property for biogas upgrading performance because it is decisive for the CO_2/CH_4 separation and for CH_4 recovery efficiency. Rainone et al. (2021) demonstrated that two commercial carbon molecular sieves (produced by Carbotech and Xintao) showed better performances for biogas upgrading than an AC (Desotec) due to their higher selectivity toward CO_2 that ensured lower CH_4 losses, regardless of the higher CO_2 adsorption capacity of the AC.

Porous zeolites and MOFs have shown high performances in the upgrading of dry biogas, leading to high CH_4 purity and recovery. Nonetheless, these adsorbents are also highly hydrophilic, thus requiring complete drying of the biogas before CO_2 removal to prevent deterioration of the upgrading efficiency by competitive co-adsorption of water molecules (Abd & Othman, 2022; Chidambaram, 2021).

The tunability of MOF properties can be used to overcome these limitations. Porous MOFs containing aluminum metal ions (Al(III)) and carboxylate ligands with two hydrophobic binding pockets with two aromatic rings each were shown to have a strong interaction with CO_2 without competitive co-adsorption of water, therefore able to upgrade biogas with water contents up to 85% (Chidambaram, 2021).

4.2.2.2 Pressure swing adsorption (PSA)

Pressure swing adsorption involves a sequence of four operations: (1) column pressurization—filling the adsorption column with raw biogas until reaching the operational pressure (4–10 bar); (2) CO2 adsorption—continuously feed the column with raw biogas, at constant pressure, to retain CO_2 and release purified methane; (3) column blowdown—when the adsorbent is saturated with CO_2, column regeneration is initiated by closing the feed valve, decreasing column pressure to 1 bar or less, and collecting the desorbed CO_2; and (4)

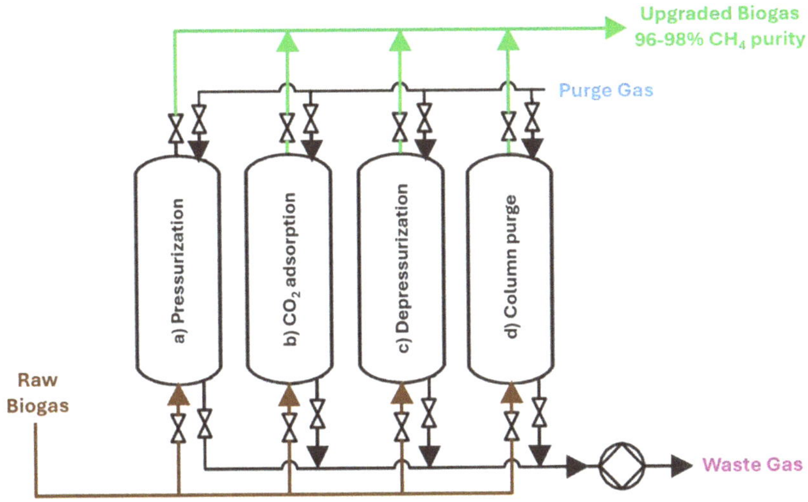

FIGURE 4.5 Pressure swing adsorption.
Adapted from Bauer, 2013; Iglesias, 2021.

purge phase—in which residual CO_2 present in the adsorbent is removed by purging with some inert gas or with upgraded biogas that may be recycled to the feed (Fig. 4.5). Since each adsorption column will undergo these four steps sequentially, a continuous PSA system must include several parallel columns, typically 2 or 4 (Bauer, 2013; Iglesias, 2021).

The main advantages of this process are the efficiency of CH_4 purification and the capacity of recovering purified CO_2 for additional valorization pathways. Nonetheless, the techno-economic viability of PSA technology is strongly influenced by the price, availability, and selectivity of the packing material and by the need of a pretreatment for H_2S removal before adsorption.

A commercial carbon molecular sieve (MSC CT-350) has been used as packing material for a fixed-bed column employed to evaluate CO_2/CH_4 separation, by PSA. The adsorbent showed selectivity for CO_2, in the pressure range of 1 to 8 bar. The adsorbent had the same adsorption capacity over six consecutive adsorption/desorption cycles, therefore could be fully regenerated under low pressure conditions (Pancione, 2024).

Commercial molecular sieves (Honeywell 13X, 4 A, and 5 A) were tested for a single-step treatment of biogas without requiring multiple-stage recirculation. The adsorption columns were packed with a first layer of zeolite 4 A (to retain all biogas moisture) followed by additional layers of zeolites 13X or 5 A that were able to remove CO_2 from the dried biogas. The system showed a CO_2 retention rate of 42.4 g CO_2 kg^{-1}, but optimization of the molecular sieve regeneration procedure is required before application in an industrial scale (Bonaccorsi, 2024).

In a recent study, an Al-dicarboxylate microporous MOF has been evaluated for dynamic fixed-bed adsorption of carbon dioxide and methane. The proposed system allowed the production of 99% purity CH_4 from a feed of CO_2:CH_4 with 40%–60% vol. and had a CO_2 recovery efficiency of 97% (Karimi, 2024).

Heat and mass balance models of biogas upgrading using PSA technology were developed to simulate the process and compare the performance of six adsorbents (silica gel, CMS-3 k, zeolite-NAUSY, Zeolite KZ10-04, porous polymeric beads, and zeolite 5 A). Results elucidate the effects of operational parameters such as feed flow rate, adsorption step duration, axial dispersion coefficient, and bed dimensions on methane purity and recovery efficiency. CMS-3 K was the adsorbent with better performance allowing to reach CH_4 purity and recovery values of 97.2% and 86.2%, respectively, in agreement with biogas quality standards (Abd & Othman, 2022).

4.2.2.3 Temperature swing adsorption (TSA)

Temperature swing adsorption is another process for desorbing CO_2 from porous adsorbents, typically involving inorganic materials or biochars that can withstand high temperature conditions.

The use of waste materials locally available for CO_2 adsorption is way of promoting sustainability of the biogas upgrading process by preserving raw materials and reducing waste production. Calcinated eggshells (850°C, particle size 280 μm) were evaluated as adsorbents for CO_2 using a biogas flow rate of 0.03 $m^3\ h^{-1}$. A sorption capacity of 5.0 g of CO_2/100 g of adsorbent was achieved corresponding to a removal efficiency and 82.5 % (Mrosso et al., 2024).

Natural and modified clay (activated with 2 M NaOH) were used for biogas upgrading. The highest CO_2 adsorption capacity (5.72 mmol g^{-1}) was obtained for modified clay, reaching a CO_2 removal efficiency of 93.8% at 7 g l^{-1} ratio. Complete removal of hydrogen sulfide was also achieved with this adsorbent (Mulu, 2021).

4.2.2.4 Electrical swing adsorption (ESA)

The regeneration of the adsorbent in systems operating under the thermal swing adsorption principle can be co-adjuvated by the supply of electrical current that will cause the direct, resistive, or Joule heating of the adsorbent, in a short period of time. This option designated as electrical swing adsorption typically reduces the time of adsorption/desorption cycles and reduces heat losses for the overall process. An electrically conductive AC monolith containing a potassium silicate binder was heated to 150°C in less than 30 s at a potential of 8 V. After using the AC monolith for upgrading a synthetic biogas mixture (30% CO_2 + 70% CH_4), the adsorbent was regenerated by increasing the applied voltage from 4 to 8 V, under a purge gas flow of 100 NmL min^{-1}, for 15 s. In these conditions, it was possible to obtain CH_4 with 84.5% purity, reaching a recovery yield of 87.3% (Verougstraete, 2022).

4.3 Membrane separation technology

Membrane separation technology exploits differences in the molecular size and properties of CH_4 and CO_2, to design materials with different permeabilities for these gases. Typically, water, ammonia, and hydrogen sulfide are previously removed with adequate pretreatments to obtain a CH_4 stream that fulfills the requirements for grid injection or engine applications. Membrane separation also allows the recovery of a CO_2 stream sufficiently concentrated to be converted into added-value products, such as methane, other fuels, or other chemicals with industrial application (Gkotsis, 2023; Roozitalab et al., 2023).

Membrane permeability is often insufficient to yield high-purity CH_4 with a single-stage system; therefore, multistage arrangements are typically used to achieve the appropriate concentrations of CH_4 in the upgraded gas stream (Fig. 4.6).

Membranes used for biogas upgrading can be designated as inorganic when they are produced from ceramic, metallic, or glass materials and organic materials, while organic membranes are composed of polymers or macromolecules. Mixed matrix membranes combine polymers with different inorganic fillers that modulate their properties (Fajrina, 2023). Mixed matrix composite membranes add to this configuration a porous support layer with the purpose of improving mechanical properties and membrane durability and performance (Molavi & Shojaei, 2019). Thin film composite (TFC) membranes integrate a nano-thin selective polymeric layer added to the porous

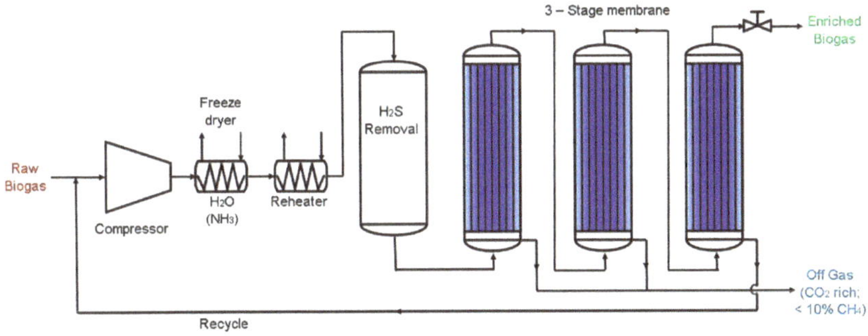

FIGURE 4.6 Membrane technology for biogas upgrading.
Adapted from Katariya & Patolia, 2023.

support by dip-coating or interfacial polymerization reactions. These TFC membranes are characterized by low energy requirements, simplicity of operation, and high specificity (Khulbe & Matsuura, 2018). Thin film nanocomposite membranes integrate porous nanoparticles (NPs) into the selective polymer layer, thus increasing selectivity and modulating other properties to maximize their performance in gas or liquid separations (Guo, 2017).

Taherizadeh et al. (2024) synthetisized SSZ-13 membranes (high-silica aluminosilicate zeolites) with a pore size of 200 nm, a CO_2/CH_4 selectivity of 173, and a CO_2 permeance of 6.5×10^{-7} [mol/(m^2 s Pa)] for equimolar CO_2/CH_4 mixtures, thus adequate for CO_2 separation from natural gas or biogas.

Koutsiantzi et al. demonstrated the feasibility of biogas upgrading at pilot scale, using commercial polyimide (PI) membranes. The system was operated for 6 months, achieving an average CH_4 recovery of 95.7% for an average CH_4 inlet concentration of 55% (Koutsiantzi et al., 2024).

Cellulose-based carbon hollow fiber membranes were tested for the separation of simulated biogas mixtures, in 183 days of continuous operation tests. The membranes showed selectivity between CO_2 and CH_4 both in sorption (2.98 and 2.00 mmol.g^{-1} respectively, at 298 K and 10 bar) and in permeation properties (120.9 and 2.3 barrer, respectively, at 308 K). The results indicate that the permeation mechanism is mainly controlled by diffusion and that the presence of H_2S and moisture, at typical biogas concentrations, decreased selectivity by 23% due to the reduction of CO_2 and CH_4 permeability by 43% and 25%, respectively (Brunetti, 2022).

Nano-engineered grazyne membranes were proposed as suitable materials for biogas upgrading based on density functional theory and molecular dynamics computational simulations. Grazynes are composed by graphene stripes with sp^2 hybridization, linked by acetylenic linkages, with sp hybridization. The calculations suggest that these membranes could upgrade biogas yielding a CH_4 stream with 97 % purity (Viñes et al., 2023).

Roozitalab et al. reviewed the available membrane materials for biogas or natural gas upgrading and concluded that 6FDA-based polyimides, and PIMs are the two suitable polymers for biogas upgrading membranes due to their high gas solubility and permeability that allows operation at moderate pressures. Furthermore, the incorporation of fillers such as zeolite or MOFs enables modulation of mechanical properties and selectivity to reach high upgrading performances (Roozitalab et al., 2023).

Zito et al. performed a comparative techno-economic analysis of the performance zeolite and polymeric membranes for biogas upgrading in a multistage system. The feed contained 60% methane, and the system was operated at 25°C and 30 bar. Both types of membranes achieved CH_4 recoveries higher than 90% but zeolite membranes also enabled high CO_2 recovery levels (≥93%) and produced CH_4 and CO_2 streams with higher purity. Regardless of the higher cost of zeolite membrane modules, their good separation performance is responsible by their slightly higher economic potential. Increasing biogas flow rate from 100 to 1000 $Nm^3 h^{-1}$ was beneficial for the process economic viability (revenue/year increase of about 12 times), suggesting that centralized upgrading units maybe a more profitable option for regions with high density of anaerobic digesters (Zito et al., 2024).

4.4 Cryogenic upgrading

Cryogenic distillation is a biogas upgrading technique that exploits the differences of boiling points between CO_2 and CH_4 (o (−78.46°C and −161.5°C, respectively), operating at pressure and temperature conditions where CO_2 is liquified while CH_4 remains in the gaseous state. Using multistage compression with intercalary cooling, it is possible to compress and cool raw biogas until CO_2 condenses, being separated from gaseous CH_4 in a distillation column (Fig. 4.7).

The process must consider the CO_2 freezing temperature and pressure to avoid formation of solid CO_2 at the lower operating temperatures. This process enabled the recovery of liquified CO_2 and a CH_4 gas stream with 94.4% purity by pressurizing raw biogas to 4.9 MPa and cooling it to −61°C. The system presented a specific energy consumption of 0.31 kWh Nm^{-3} raw biogas (Yousef, 2018).

Ghorbani et al. proposed a biogas upgrading process based on the liquefaction of both biomethane and carbon dioxide, coupled with amine

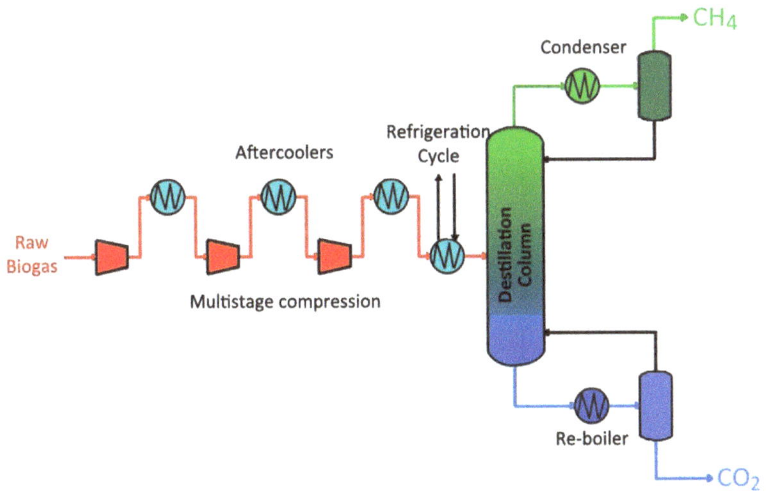

FIGURE 4.7 Cryogenic upgrading of biogas.
Adapted from Naquash, 2022.

scrubbing. The system integrates a mixed refrigerant cascade to recover waste heat from exothermal operations and use it in the endothermal steps. This integrated system enables processing of 2.393 kg s^{-1} of raw biogas producing 0.5387 kg s^{-1} of bioLNG, 0.1280 kg s^{-1} of CH_4, and 0.2102 kg s^{-1} of liquid CO_2 and 1.625 kg s^{-1} of hot water. The specific power consumption and coefficient of performance are calculated to be 0.4761 kW h kg^{-1} bioLNG and 3.675, respectively (Ghorbani et al., 2021).

Nevertheless, as pointed out by Naquash et al. (2022), cryogenic distillation remains the more energy- and cost intensive technology for biogas upgrading; therefore, its viability depends on the integration of methane liquefaction as a critical step, in particular, for long-distance transportation.

Alternative cryogenic technologies based in CO_2 solidification and melting are the anti-sublimation process (Clodic & Younes, 2006) and the controlled freeze zone technology created by ExxonMobil in 1983 for removal of CO_2 from natural gas (Parker, 2011).

4.5 Heterogeneous catalysis

The biogas catalytic methanation with hydrogen (Sabatier process) can be achieved using fixed-bed reactors with nickel (Ni) or ruthenium (Ru) based catalysts, at high-temperature and under moderate pressures, according to Eqs. 4.6 to 4.8 (Mercader et al., 2024).

$$CO_2 + 4H_2 \rightleftharpoons CH_4 + 2H_2O \Delta H_r^{\circ} = -165.1\,[kJ{\cdot}mol^{-1}] \tag{4.6}$$

$$CO_2 + H_2 \rightleftharpoons CO + H_2O \Delta H_r^{\circ} = +41.2\,[kJ{\cdot}mol^{-1}] \tag{4.7}$$

$$CO + 3H_2 \rightleftharpoons CH_4 + H_2O \Delta H_r^{\circ} = -206.3\,[kJ{\cdot}mol^{-1}] \tag{4.8}$$

Mercader et al. evaluated the effect of water removal in the efficiency of CO_2 conversion by the Sabatier reaction using Ni-Fe/Al_2O_3 catalyst. Coupling the catalytic conversion with dehydration of the reaction medium using an LTA 5 A zeolite (Sorption Enhanced Sabatier Reaction), the CO_2 hydrogenation reaction was increased more than 20% for temperatures over 350°C. This option requires zeolite periodic heating at 500°C, to desorb the water and recover activity; therefore, the catalytic reaction and the adsorption process should occur in separate reactors (Mercader et al., 2024).

Nanostructured mixed oxide catalysts with different Ni/Ce molar ratios were evaluated for the catalytic hydrogenation of CO_2 using a model biogas sample containing CO_2 (12.9 mol%), CH_4 (25.9 mol%), H_2 (55.2 mol%), and N_2 (6.0 mol%), corresponding to a biogas composition of CH_4, 67 mol% + CO_2, 33 mol%, supplemented with enough hydrogen to support the methanation reaction. After pretreatment with H_2, at 400°C for 1 h, the catalysts were able to achieve CO_2 conversion efficiencies of 83 to 89 mol% at 275°C and atmospheric pressure (Atzori, 2023).

Durán et al. proposed a design with distributed biogas feed over the reactor height instead of a single biogas inlet as an option to reach higher conversion efficiency. The distributed biogas feed enables a more homogeneous temperature profile over the reactor and a lower production of carbon monoxide and carbonaceous residue, unwanted by-products of CO_2 methanation (Durán et al., 2024).

Minardi et al. made a techno-economic analysis of use of methanation units and hydrogen produced by electrolysis to convert CO_2 captured from anaerobic digestors and concluded that the total production of CH_4 could be increased by 50% by adding 165 kNm^3/y of CH_4, produced by methanation of the recovered CO_2, to the 330 $kNm^3\ y^{-1}$ produced in the digestion unit. This study identified as limiting factor to the economic viability of the system the cost of electricity consumed by the electrolyzer units and determined that a competitive cost of 1.39 €/Nm^3 of synthetic natural gas could be reached when electricity costs reached a value of about 50 €/MWh (Minardi et al., 2023).

4.6 Biological upgrading

Biological upgrading can be achieved using microorganisms like microalgae or homoacetogens that are able of consuming CO_2 to produce biomass or other chemicals que remain dissolved in the liquid medium, thus increasing CH_4 concentration in the biogas.

Microalgae can also be used to treat and convert the anaerobic digestate, normally very rich in nitrogen and minerals. This strategy may be performed in parallel reactors with biogas and digestate recirculation or following an integrated in situ approach, with co-cultures of microalgae with fungi or bacteria. The in situ approach exploits the synergies between microalgae and anaerobic bacteria but can have cause some microalgae growth limitations related to the reactor operational conditions and light availability (Yang, 2022).

Another possibility of biogas biological upgrading is the use of anaerobic bacteria able to reduce CO_2 to CH_4 in the presence of hydrogen. These microorganisms designated as hydrogenotrophic methanogens can operate within the anaerobic digestor (in situ) or in parallel upgrading systems (Fu et al., 2021; Hoyos, 2024; Zhang et al., 2020).

4.6.1 Photosynthetic organisms

Biogas upgrading with microalgae cultures explores the high capacity of these photosynthetic microorganisms for CO_2 capture and conversion as well as their ability to metabolize nitrogen and minerals. These characteristics led to the proposal of integrated systems of CO_2 capture and digestate treatment using bubble for countercurrent contact between the microalgae culture and biogas

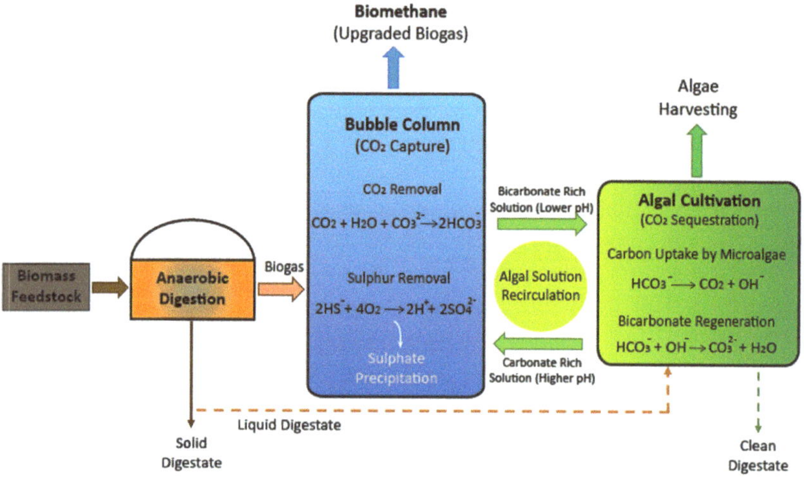

FIGURE 4.8 Biogas upgrading with microalgae.
Adapted from Bose, 2019.

(Fig. 4.8), while feeding the microalgae culture thanks with the nutrient-rich anaerobic digestate (Bose, 2019).

Microalgae, bacteria, and fungi consortia have been proposed as integrated biotreatment system for biogas upgrading. There are several advantages in these synergistic associations: microalgae and bacteria exchange nutrients, genetic material, and growth stimulants (Sun, 2022a), while filamentous fungi provide a surface for microalgae fixation and also exchange critical nutrients, avoiding accumulation or deficiency conditions (Wang, 2022). These microbial associations benefit the growth and physiological status of the consortia microorganisms, therefore generally improves their bioremediation efficiency.

A consortium of *Chlorella vulgaris* + endophytic bacteria (S395–2) + *Clonostachys rosea* symbiont showed higher performance of biogas upgrading and nutrient removal from the culture medium than *C. vulgaris* monocultures or binary combinations of *C. vulgaris* with endophytic bacteria (S395–2) or with *C. rosea* (Xu, 2024).

Raw biogas with 60% CH_4 and 39% CO_2 obtained from AD of food waste in a 100 L reactor was upgraded by CO_2 removal using a 2.5 L absorption column interconnected with a 1.2 m^2 high-rate algal pond (HRAP). CO_2 was captured with an efficiency of 90%, and an upgraded biogas with 93.9% CH_4 was obtained. The dominant species in the algal pond were the cyanobacteria *Pseudoanabaena* sp. and the microalgae *Chlorella vulgaris* that grew in synthetic wastewater and centrifuged liquid digestate, showing the capacity to completely remove nitrogen and phosphorus from these liquid effluents (Marín, 2022).

Removal of both the CO_2 and H_2S from biogas and nutrients from digestate was achieved by an algal-bacterial consortia including the species *Chlorella* sp. (15%), *Tetradesmus obliquus* (33%), *Lindavia* sp. (24%), and *Synechococcus* sp. (28%). Biogas upgrading was optimized by sequential supplementation of the HRAP with $NaHCO_3$ and Na_2CO_3 that allowed a biomass productivity of 22 $g.m^{-2}.d^{-1}$, CO_2 removal efficiencies higher than 95.7% reaching CH_4 concentrations higher than 90% in the treated biogas and complete removal of N and P from the liquid media (Méndez, 2022).

Various additives can be used to improve the biogas upgrading efficiency by photosynthetic organisms. Their action may involve the supplementation with critical nutrients, chemicals with metabolic effects, or any other additives that benefit the microalgae growth and physiological state, indirectly improving their CO_2 capture capacity.

NPs may enhance CO_2 capture in biogas scrubbing systems by three potential mechanisms: (1) bubble breaking effect, that decreases bubble size and increases overall contact area between gas bubbles and the liquid phase; (2) shuttle effect, that involves adsorption of gas bubbles in the NPs surface followed by bubble release into the liquid medium; (3) hydrodynamic effect, due to the NP fast movements and collisions that induce turbulence at the liquid–gas boundary layer, thus improving mass transfer (Choi et al., 2015; Vargas-Estrada, 2023).

Hoyos et al. studied biogas upgrading in a pilot scale biogas plant interconnected to an external biogas absorption column where a microalgae culture contacts with raw biogas. The authors evaluated the influence of carbon-coated zero-valent iron nanoparticles (NPs) in the overall upgrading process and concluded that this supplementation improves microalgae productivity. This higher microalga density had a positive impact on biogas upgrading efficiency achieving CO_2 concentrations $< 1\%$ and CH_4 concentrations $> 94\%$ in the treated biogas (Hoyos et al., 2024).

Three different NPs (Fe_2O_3, carbon-coated zero-valent iron (CACOI) and SiO_2) were added to a mixed microalgae culture consisting of *Chlorella* sp. (91.3%), *Nitzschia* sp. (7.6%), and *Pseudanabaena* sp. (1.2%) to improve photosynthetic biogas upgrading. The NP addition at concentrations from 10 to 70 $mg\ L^{-1}$ had positive effects in CO_2 consumption but also in biomass productivity as well as lipid and carbohydrate synthesis. Nevertheless, a shorter lag phase and delayed exponential growth were observed for higher NP concentrations, and the beneficial effects of NPs were reduced by exposure to UV light (Vargas-Estrada, 2022).

Separation between the CO_2 adsorption column and the HRAP has the advantage of avoiding the contamination of upgrade biogas with oxygen produced by the microalgae during the photosynthetic metabolism (Bose, 2019). Nonetheless, oxygen concentrations in the upgraded biogas can reach values higher than 3% (v/v); thus, some strategies for O_2 stripping from the liquid medium have been proposed. Circulation of the culture broth by an (OTC), packed with Raschig rings with a specific area of 640 m^2 m^{-3} promoted the transfer of dissolved O_2 to atmospheric air due to the high gas–liquid contact area of this column. The increase in turbulence regime of the HRAP also promoted O_2 desorption from the aqueous microalgae culture. Both effects enabled to reach a final O_2 concentration of 1.5% in the upgraded biogas (Franco-Morgado, 2021).

Intermittent supply of biogas, at flow rate of 14.4 L d^{-1} only during the day period, to an microalgae photobioreactor (PBR) connect with a biogas scrubbing column enabled a good upgrading efficiency producing a biogas stream with 95.9 ± 1.3% v/v of CH_4, 2.5 ± 0.8% v/v of CO_2, 1.4 ± 0.6% v/v of N_2, 0.1 ± 0.1% v/v of O_2, and negligible H_2S concentrations. The photobioreactor contained *Chlorella vulgaris, Chlorella ellipsoidea, Tetracoccus* sp., and the cyanobacterium *Pseudanabaena* sp., whose biomass productivity was not affected by the biogas supply. On the contrary, when biogas was continuously supplied during the day and the night periods, reaching a flow rate of 28.8 L d^{-1}, the CH_4 and CO_2 concentrations were 88.8 ± 1.6% v/v and 9.3 ± 1.1% v/v, respectively, showing a lower upgrading efficiency (Ángeles, 2021).

Strigolactones are phytohormones that can promote metabolism in microalgae and arbuscular mycorrhizal fungi and therefore can improve the photosynthetic capture of CO_2 by microbial consortia (H. Janeeshma, 2024; Wang, 2023a).

A consortium of *B. licheniformis* + *C. vulgaris* + *G. lucidum* showed stronger metabolic viability and higher photosynthetic and biogas upgrading efficiencies when the mixed culture was supplemented with 10^{-11} M of 5-Deoxystrigol, a natural strigolactone that enhances the metabolism and photosynthesis of the algal co-culture systems (Wang, 2023b).

A strigolactone analog (GR24) was used to stimulate the metabolic processes in three types of microalga-fungi/bacteria symbiotic systems to improve nutrient removal from piggery wastewaters and biogas upgrading by the microalgae. The co-cultivation system with *Chlorella vulgaris* + *Ganoderma lucidum* achieved the best photosynthetic performance and microalgae growth. Supplementation with 10^{-9} M GR24 increased CO_2 conversion and COD removal efficiencies, allowing to reach CH_4 concentrations in the biogas higher than 80% (Wang, 2023a).

The biogas upgrading using photosynthetic organisms does not require the use of specialty adsorbents or energy-intensive processes but does not allow the recovery of CO_2 to produce CH_4 or other fuels or chemicals. On the contrary, the captured CO_2 will be converted to algal biomass that can be valorized in various applications. The removal of nitrogen and phosphorus from the algal culture medium also contributes for the techno-economical balance of the overall process, by coupling biogas upgrading and digestate processing and valorization (Bose, 2019; Yang, 2022).

4.6.2 Fermentative CO_2 reduction

Some nonphotosynthetic microorganisms can also consume CO_2 from biogas, converting it into added-value chemicals, by fermentative processes (Angelidaki, 2018).

Hydrogen can be used by homoacetogens to convert CO_2 from biogas to acetic acid, enabling a CH_4 concentration to values higher than 95% (Omar, 2018).

Zhao et al. (2020) developed a membrane biofilm reactor with external hydrogen supply through a gas-permeable hollow fiber for continuous biogas upgrading and simultaneous production of acetate and ethanol. Conversion efficiencies of 93.8% and 98.1% were reached for CO_2 and H_2, respectively, leading to a biogas purity of 96.7%. At the same time, acetate and ethanol production rates reached the values of 37.8 and 13.5 mmol $L^{-1}d^{-1}$, respectively, for a hydraulic retention time of 2 days (Zhao et al., 2020).

Using a homoacetogens consortium enriched from peatland soil, a 95% CO_2 consumption efficiency could be achieved, and acetic acid was simultaneously produced reaching a concentration of 120 mg L^{-1}. An upgraded biogas with 98% CH_4 was obtained, at 30°C, pH = 8 and hydrogen supplementation enough to reach a H_2 to CO_2 ratio of 2:1 in the raw biogas. *Clostridium* sp., *Proteiniclasticum* sp., and *Petrimonas* sp. were the main homoacetogens responsible for the CO_2 reduction and acetic acid production, using H_2 as an energy source (Chaikitkaew, 2021).

Gas fermentation for biogas upgrading was demonstrated using *Clostridium autoethanogenum* to convert synthetic biogas supplemented with H_2 into an upgraded biogas with a CH_4/CO_2 molar ratio of 2.2 and reaching ethanol and acetate concentrations of 2.8 ± 0.1 and 3.9 ± 0.1 g L^{-1}, respectively, in the liquid phase (Heffernan, 2023).

López et al. proposed an integrated system for biogas upgrading and succinic acid production using bioreactors *Actinobacillus succinogenes* cultures fed with sugar-rich confectionery wastewaters. Succinic acid concentrations of 77.99 ± 3.90 g L^{-1} could be achieved in the fermentation broth, while upgraded biogas with CH_4 contents higher than 98% could be obtained when using high gas residence times and biogas pressurization (López et al., 2023).

4.6.3 Biological methanation

The two main methanogen groups in the anaerobic digester are acetoclastic methanogens (AMs) (that use acetic acid as a substrate), and hydrogenotrophic methanogens (HMs) (that use hydrogen and carbon dioxide as substrates), in both cases producing methane and therefore increasing its concentration in the biogas (Antukh, 2022).

Hydrogenotrophic methanogens are anaerobic bacteria able to convert CO_2 from raw biogas to CH_4 according to Eq. 4.9, using an *in situ* process, inside the anaerobic digester or through an ex situ approach, in a separate bioreactor (Comesaña-Gándara, 2022; Lóránt & Tardy, 2022).

$$4H_2 + CO_2 \rightarrow CH_4 + 2H_2O(\Delta G = -131 \text{ kJ. mol}^{-1}) \tag{4.9}$$

Carbon dioxide bioconversion to CH_4 constitutes an innovative solution for biogas upgrading. In situ bioconversion captures endogenous CO_2 and promotes its reduction using the hydrogen available in the biogas as well as additional hydrogen supplied by an external source, ideally using renewable electricity. However, the biological methanation approach faces several challenges for its widespread industrial use, including a need for better understanding of conversion mechanisms, improved hydrogen supply, and the relationship between operating conditions and microbial population distribution. (Fu et al., 2021; Rao, 2023).

Zhang et al. reviewed the different approaches to promote biogas upgrading by hydrogenotrophic methanogenesis (Zhang et al., 2020). Three main pathways for increasing CO_2 reduction were identified as follows:

- Hydrogen injection in the anaerobic digestor
- Bioelectrochemical technologies
- Additives (e.g., ash, zero-valent iron, and biochar)

Despite technical simplicity and convenience of in situ upgrading, this approach has some associated difficulties namely how to supply additional H_2 in conditions that increase its contact with the liquid phase and reach the H_2/CO_2 ratio adequate for the methanation reaction to proceed. Other problems are related to the influence of H_2 addition in microbial populations stability and metabolic behavior and to the changes of liquid phase pH due to CO_2 depletion. The availability and price of external H_2 supply from water electrolysis using renewable electricity are also a key factor to the sustainability and economic viability of in situ biological upgrading technologies (Antukh, 2022; Lai, 2021).

4.6.3.1 Direct hydrogen injection in the anaerobic digestor

This approach favors the biological reduction of CO_2 to CH_4, by increasing hydrogen availability in the liquid phase and by favoring the development of hydrogenotrophic methanogens populations (Angelidaki, 2018; Xu, 2021).

Hydrogen may be obtained from a sustainable process such as water electrolysis powered with renewable electricity and water recovered from the anaerobic digestor could be used for that purpose.

Nevertheless, since the process is occurring *in situ*, the higher consumption of CO_2 leads to an alkalinization of the digestion medium that must be compensated to prevent inhibition of methanogenesis (Ravi, 2019).

Controlling hydrogen mass transfer to the liquid phase is an essential factor for optimizing the process and avoid the use of elevated hydrogen concentrations that may lead to acetate accumulation and inhibition of the digestion process (Agneessens, 2017). Reactor design and innovation in the hydrogen injection and dispersion systems can contribute to enhance hydrogen availability to the hydrogenotrophic methanogens populations, thus increasing the CO_2 conversion efficiency (Alfaro, 2019; Bassani et al., 2016).

Supplying hydrogen to a three-phase up-flow anaerobic reactor through a permeable membrane increased effluent CH_4 purity from 74.1% to 95.2% and favored the prevalence of *Methanobacteriaceae* on the biofilm (Rao, 2024).

The biological in situ biogas upgrading via hydrogenotrophic methanogenesis with an external H_2 addition was evaluated using an anaerobic membrane bioreactor (AnMBR) and operated in semi-batch mode using waste-activated sludge as substrate. Hydrogen was supplied by pulsed additions in other to achieve different molar equivalents relatively to the CO_2 concentration. Supplying H_2 at a H_2:CO_2 molar equivalent ratio of 11, increased acetate concentration and led to a CH_4 yield and CH_4 production rate of 0.31 L/g of VS_{input} and 0.086 L/L/d, respectively, with the CH_4 concentration in biogas reaching 92%, as result of an efficient CO_2 reduction (Hafuka, 2022).

Membrane technology can also be integrated in the anaerobic digestor for in situ biogas upgrading and recovery of valuable chemicals. A gas permeable membrane was used to supply hydrogen to the liquid phase of an anaerobic digestor, stimulating the growth of a biofilm of *Clostridium ragsdalei P11* a microorganism able to promote CO_2 hydrogenation. Methane concentrations of 96.7% were achieved when using a hydrogen partial pressure of 1.55 atm (Zhao et al., 2020).

The use of hydrogen oxidizing bacteria (HOB) *Cupriavidus necator* H16 allowed the upgrading of biogas CH_4 concentration from 57.1% to 94.2%, within 96 h, while producing single-cell protein (SCP), with a protein content from 44% to 71%, a valuable product that contributes to the overall economic feasibility of the upgrading process. A comparative techno-economic analysis showed that the HOB upgrading process had a higher net revenue (547.22 €/1000 m^3 of biogas) than biological upgrading with hydrogenotrophic methanogens (157.21€/1000 m^3 of biogas), and even of water scrubbing upgrading (368.52€/1000 m^3 of biogas), due to the diversity of products obtained (Li, 2024).

Besides feeding H_2 to the anaerobic reactor, the additional supply of CO_2 from an external source has been proposed to improve the CH_4 production rate, by avoiding H_2 accumulation and consequent reactor acidification as endogenous CO_2 is consumed. Biogas production rates of 4954–6605 NmL/L.d were obtained using a 3 L batch reactor, upon supplementation with an H_2/CO_2

gas mixture (4:1 molar ratio), at flow rates of 960 to 2400 NmL/L.d. Maximum CH_4 concentrations of 92–97 vol % were achieved in these conditions (Sekoai, 2020).

Sun et al. developed a machine learning-aided inverse design for biogas upgrading using data collected from literature and a multi-objective optimization algorithm to define optimal operating conditions for biological CO_2 conversion. Higher efficiency of biogas upgrading (CH_4 concentration of 98.33% and H_2 conversion rate of 97.05%) was obtained at 37.12°C with an H_2/CO_2 ratio of 3.73 using an up-flow anaerobic sludge blanket reactor in continuous mode. Hydrogen should be supplied dispersed in microbubbles, and no gas recirculation step is needed because in these conditions H_2 is almost fully consumed (Sun et al., 2024).

In situ approach for CO_2 conversion in an anaerobic reactor with a continuous supply of exogenous H_2 allowed to produce biogas with 80–85% CH_4 and increase acetate concentration in the digestate to 5–6 g/L due to homoacetogenic activity. In a second approach, a trickling bed anaerobic biofilter was used to upgrade raw biogas exiting the anaerobic digestor also using an exogenous supply of H_2; in this configuration, a CH_4 content of 85% to 90% was reached in the biogas, showing that the ex situ parallel biological upgrading may be more efficient due to a better gas–liquid mass transfer and by avoiding negative impacts on the CH_4 production yield of the anaerobic digestor (Tartakovsky, 2021).

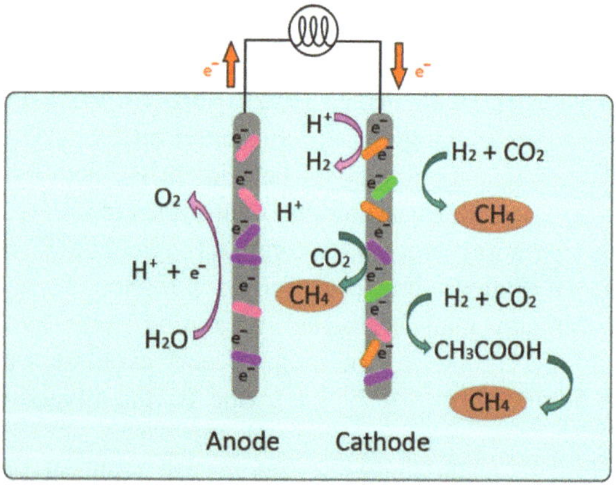

FIGURE 4.9 Bioelectrochemical pathways for biogas upgrading.
Adapted from Chu, 2024.

4.6.3.2 Bioelectrochemical technologies

Bioelectrochemical technologies combine the presence of hydrogenotrophic methanogen populations and *in situ* hydrogen generation by water electrolysis to achieve high efficiencies of CO_2 reduction by electroactive biofilms deposited in the surface of an electrode (Fig. 4.9).

Protons and electrons are generated in the anode and transferred to the cathode where CO_2 is converted to CH_4 by electroactive bacteria (Fig. 4.9), able to use H_2 as an electron donor (hydrogenotrophic methanogenesis) or to receive electrons directly from the electrode (direct electrical methanogenesis) (Chu, 2024; Zhang et al., 2020).

Carbon-based materials such as graphite rods, carbon cloth, carbon brushes, or carbon nanotubes are preferred electrode materials due to their surface properties that benefit biofilm growth and electron-transfer mechanisms, but some metallic components are needed to improve the electrical conductivity and catalytic activity of the electrodes (Zhen et al., 2018; Erbay et al., 2015). Platinum (Pt), molybdenum (Mo), nickel (Ni), and copper (Cu) have been used for that purpose but nonprecious metal materials such as iron (Fe) and stainless steel have also shown good methanogenic performance (Chu, 2024).

An integrated system with a two-chamber microbial electrochemical cell and anaerobic reactor was used to study the effect of the supplementation with complexes of anaerobic granular sludge and iron oxides (Fe–AnGS), in biogas productivity and in situ upgrading. Supplementation with Fe–AnGS resulted in a 23%–29% increase in the energy recovery rate of external circuit current and a 26%–31% decrease in volatile fatty acids. The microbial populations were selectively enriched in methanogens and electroactive bacteria what resulted in hydrogenotrophic and acetoclastic methanogenesis in the cathode and anode chambers, respectively (Wei, 2024).

Comparing the performance of an electrochemical anaerobic membrane bioreactor (EC-AnMBR) with a traditional anaerobic membrane bioreactor (AnMBR) it was observed that the bioelectrochemical processes occurring in the electroactive biofilms at the anode and cathode enabled higher methane productivity but also contributed to mitigate membrane fouling by improving the oxidation of organic metabolites. Genomic and metabolomic analysis showed that EC-AnMBR promoted the up-regulation of genes involved in carbohydrate metabolism and methane metabolism thus increasing the degradation of extracellular metabolites at the anode and enhancing the methanogenic activity on the cathode (Han, 2024).

A bioelectrochemical system operated at applied cathode potential of only −0.5 V was used to promote biogas upgrading via direct electron transfer (DET). The CH_4 content of the treated biogas reached 97% and CO_2 was consumed at a rate of 318.5 mol/d/m^2. The cathode surface biofilm was dominated by AMs from the genus *Methanothrix* (92.7%), and proton consumption led to a higher pH thus improving CO_2 dissolution in the liquid phase (Liu, 2021).

Optimization of biological methanation systems has been proposed following opposite strategies: either integration of AD and microbial electrolysis cells (MEC) as a strategy to enhance biogas production and upgrading in situ (Ao, 2024) or two-stage AD systems integrating various possible in situ upgrading techniques to form a hybrid system that is more suitable to overcome constraints such as pH fluctuations and gas–liquid mass transfer limitations (Yellezuome, 2023).

The production of edible single-cell protein can also be coupled with biogas upgrading thus contributing to the economic viability of the process. For this purpose, a bioinorganic electrosynthesis process was tested using raw biogas (70% CH_4 + 30% CO_2), at a flow rate of 50 mL d^{-1} and an applied voltage of 3.0 V. The CO_2 bioconversion efficiency reached $92.97 \pm 5.61\%$ coupled with a protein production of 472.04 ± 22.05 mg L^{-1}. The methanotrophic biomass produced in the system showed a total amino acid content of 62.8% (Xu, 2021).

High initial investment, scale-up difficulties, and low technology maturity are the reasons why no commercial AD-MEC systems are available today, but the good performance results explain why there is a growing interest in this technology.

The different operating conditions, in particular in situ, ex situ, batch mode, continuous mode, reactor configuration, and electrode materials, are critical parameters requiring further optimization and innovative solutions to improve the viability of microbial electrochemical systems for biogas upgrading. Choice of sustainable nutrient sources and recovery of valuable by-products are also challenges that must be addressed (Aryal, 2022).

Electricity cost and the initial high investment required are the present limiting factor to the application of biological upgrading at a commercial scale. Biological upgrading can become a sustainable alternative to water scrubbing by using surplus electricity purchased at 50% of the regular electricity price and if the initial investment could be decreased to approximately one-third of 2020 investment costs. A sensitivity analysis demonstrated that economic profitability and the production cost of biomethane would be highly associated with the price of surplus electricity and the subsidy granted for biomethane production (Elyasi, 2021).

4.6.3.3 *Additives (biochars, activated carbon, and zero-valent iron)*

Singh et al. compared different additives and configurations to promote hydrogenotrophic methanogenesis and improve *in situ* biological reduction of CO_2. The conditions evaluated were as follows: (1) addition of granular activated arbon (GAC); (2) addition of nano zero-valent iron (NZVI); (3)

electro methanogenic system in closed circuit (CC); and (4) electro methanogenic system with applied potential (AP). The better performance of CH_4 production was obtained for the NZVI system (5.1 L), followed by AP (2.9 L), CC (2.74 L), GAC (2.32 L), and control, (1.75 L), indicating that electron-transfer mechanisms are crucial to achieve high CO_2 reduction efficiencies (Singh et al., 2024).

Biochars and Activated Carbon—The addition of biochars or ACs to the digestion medium has been shown to improve gas–liquid mass transfer and to provide a support for biofilm immobilization. These carbon additives may also adsorb some of the organic molecules that could act as process inhibitors and participate in electron-transfer processes thus contributing to a higher methane yield (Wang, 2021).

The packing of a bubble tower and biotrickling filter (BTF) with AC allowed to achieve the best microbial immobilization level for hydrogenotrophic methanogens and led to the production of an upgraded gas with 91.9% CH_4, from a H_2/CO_2 feed with a ratio of 2.5:1 (Huang, 2024).

GAC was evaluated as an additive for improving the conversion of CO_2 to acetic acid by *Clostridium thailandense* bacteria. Hydrogen was supplied to the reactor at a H_2 to CO_2 ratio of 2:1 (v/v). Addition of 7.5 g/L GAC improved CH_4 purity in the upgraded biogas and increased CO_2 and H_2 consumption efficiencies to 88.3% and 98.7%, respectively. Moreover, the production of acetic acid by conversion of CO_2 in biogas constitutes an additional source of revenue contributing to the process economic potential (Chaikitkaew, 2024).

NZVI is able to react with water to produce hydrogen (Eq 4.10 to 4.12), thus improving the efficiency of hydrogenotrophic methanogenesis (Hao, 2017).

$$Fe^{\circ} + 2H_2O \rightarrow Fe^{2+} + H_2 + 2OH^{-}\Delta G^{\circ} = -5.02\ kJ.\ mol^{-1}Fe^{\circ} \quad (4.10)$$

$$CO_2 + 4H_2 \rightarrow CH_4 + 2H_2O\Delta G^{\circ} = -131\ kJ.\ mol^{-1} \quad (4.11)$$

$$8H^{+} + Fe^{\circ} + CO_2 \rightarrow CH_4 + Fe^{2+} + 2H_2O\Delta G^{\circ} = -150.5\ kJ.\ mol^{-1}CH_4 \quad (4.12)$$

The divalent iron ions (Fe^{2+}) influence the metabolic status of microorganisms by stimulating the oxidoreductase activity. NZVI has been shown to promote acidogenesis, increasing concentrations of formic and acetic acids, but also favored the development of *Methanobacterium* populations, thus having a positive influence in biological and bioelectrochemical technologies for biogas upgrading (Wang, 2021).

The loss of activity of NZVI NPs can occur due to agglomeration and passivation phenomena. This problem could be minimized by modification of NZVI with an organic surfactant (sodium dodecyl sulfate), what allowed the production of CH_4 with 93.32% purity, and increased conversion efficiency by 10.30% relatively to the system with nonmodified NZVI NPs (Yang, 2024).

Supplementation with NZVI at levels from 0 mg.L^{-1} to 2000 mg.L^{-1} for H_2-assisted ex situ biogas upgrading showed positive effects for all concentrations tested but achieved the highest biomethane content (94.1 %, v/v), CO_2 utilization ratio (95.9%), and acetate yield (19.4 mmol.L^{-1}) for the concentration of 500 mg L^{-1}. In these conditions, an increase in the production of extracellular polymeric substances was observed and the microbiome characterization showed predominance of *Petrimonas*, *Romboutsia*, *Acidaminococcus*, and *Clostridium* species (Wu, 2023).

Coarse zero-valent iron particles (CZVI) have also been tested as additives for biogas upgrading, in batch tests, at a concentration of 10 g.L^{-1}. Methane yield and methane concentration in the biogas reached the values of 286.4 mL.g^{-1}-volatile solids and 95.5%, respectively, corresponding to a 2.2-fold and 1.5-fold increase (Sun, 2023).

Ammonia nitrogen—The relative abundance of hydrogenotrophic methanogens may be favored by the concentration of ammonia nitrogen concentrations up to 5.5 g.L^{-1}. In those conditions, these bacteria became dominant (73.1%) leading to a CH_4 concentration of 94.1% and a 57.5% increase in CH_4 yield (Li, 2020).

The biological methanation of CO_2 in a trickle bed reactor with a reaction volume of 0.8 m^3 allowed to obtain CH_4 with a purity > 98% at a flow rate of 6.1 $m^3 \cdot (m^3_{RV} \cdot d)^{-1}$. Keeping ammonium concentration above 400 mg.L^{-1} enabled control of the reactor pH and H_2S concentration in the biogas was reduced to around 100 ppm (Fenske, 2023).

4.7 Centralized biogas upgrading systems

In regions with a high density of AD units, the transportation of raw biogas produced in different locations to a centralized upgrading unit using an appropriate pipeline grid may be an alternative option to in situ upgrading. Application of this model to a real-world scenario of 633 biogas plants, in Northern Germany, demonstrated a higher total profit for the optimized network compared to the scenario of decentralized biogas upgrading. The centralized model allows economy of scale in some upgrading components such as maintenance operations or biogas injection in the natural gas grid (Schröer et al., 2024).

4.8 Conclusions

Biogas upgrading transforms raw biogas into biomethane suitable for use as a renewable fuel. Multiple technologies exist, ranging from commercial solutions to emerging methods, each with specific advantages and challenges concerning efficiency, costs, and infrastructure requirements.

Absorption technologies, like water scrubbing, are widely used but require significant water volumes and treatment for reuse. Chemical absorption offers high CO_2 removal efficiency but involves costly absorbent regeneration and corrosion issues. Physical absorption uses reusable absorbents but requires significant energy for regeneration and may struggle with varying biogas compositions.

Adsorption techniques, such as pressure swing adsorption (PSA) and temperature swing adsorption (TSA), recover CO_2 for reuse. PSA is flexible and scalable but costly and dependent on periodic adsorbent replacement. TSA is efficient for diverse impurities but slower due to heat transfer limitations, requiring innovative heating methods to improve speed.

Membrane separation uses varying component permeabilities to separate CO_2 and CH_4. Inorganic membranes are durable but costly and prone to fouling, while polymeric membranes are more affordable but less stable. Mixed matrix membranes offer a balance of efficiency and durability but remain expensive and complex to produce.

Cryogenic upgrading achieves high biomethane purity but is energy-intensive and expensive. Coupling it with biomethane liquefaction may offset costs by facilitating distribution.

Catalytic hydrogenation of CO_2 yields nearly complete conversion to CH_4, but catalysts are costly and sensitive to impurities. Sustainable use requires recovery and regeneration of rare elements used in catalysts.

HPAD directly produces high-pressure biogas but raises safety concerns and operational costs.

Biological upgrading leverages microbial processes for CO_2 conversion. Photosynthetic organisms convert CO_2 using sunlight but need countercurrent columns for improved gas-medium contact. Nonphotosynthetic microbes can ferment CO_2 into valuable chemicals like ethanol, while hydrogenotrophic methanogens convert CO_2 to CH_4 using renewable hydrogen. Bioelectrochemical methanogenesis offers high efficiency but involves high initial costs and complex operations.

Emerging methods, such as methanogenic additives like NZVI and AC, boost methane production and stability but present environmental risks. Careful monitoring and NP recovery solutions are needed.

In summary, biogas upgrading technology selection depends on biogas composition, desired purity, costs, and environmental impact. While established methods like absorption and adsorption dominate, membrane technology balances cost and efficiency but faces scalability challenges. Cryogenic and biological methods are effective but limited by energy and robustness issues. Emerging innovations show promise but require further validation. Ongoing research aiming to optimize CH_4 production and CO_2 valorization is essential for decarbonizing AD technology.

References

Abbott, A. P., Capper, G., Davies, D. L., Rasheed, R. K., & Tambyrajah, V. (2003). Novel solvent properties of choline chloride/urea mixtures. *Chemical Communications, 1*, 70–71. https://doi.org/10.1039/b210714g, http://www.rsc.org/suppdata/cc/b2/b210714g/.

Abd, A., & Othman, M. (2022). Biogas upgrading to fuel grade methane using pressure swing adsorption: Parametric sensitivity analysis on an industrial scale. *Fuel, 308*, 121986. https://doi.org/10.1016/j.fuel.2021.121986.

Abd, A. A., Othman, M., Helwani, Z., & Kim, J. (2024). An overview of biogas upgrading via pressure swing adsorption: Navigating through bibliometric insights towards a conceptual framework and future research pathways. *Energy Conversion and Management, 306*, 118268. https://doi.org/10.1016/j.enconman.2024.118268.

Abd, A. A., Shabbani, H. J. K., Helwani, Z., & Othman, M. R. (2022). Experimental study and static numerical optimization of scalable design of non-adiabatic and non-isothermal pressure swing adsorption for biogas upgrading. *Energy, 257*, 124781. https://doi.org/10.1016/j.energy.2022.124781.

Aghel, B., Gouran, A., Behaien, S., & Vaferi, B. (2022). Experimental and modeling analyzing the biogas upgrading in the microchannel: Carbon dioxide capture by seawater enriched with low-cost waste materials. *Environmental Technology and Innovation, 27*, 102770. https://doi.org/10.1016/j.eti.2022.102770.

Agneessens, L. M., Ottosen, L. D. M., Voigt, N. V., Nielsen, J. L., de Jonge, N., Fischer, C. H., & Kofoed, M. V. W. (2017). In-situ biogas upgrading with pulse H2 additions: The relevance of methanogen adaption and inorganic carbon level. *Bioresource Technology, 233*, 256–263. https://doi.org/10.1016/j.biortech.2017.02.016.

Alfaro, N., Fdz-Polanco, M., Fdz-Polanco, F., & Díaz, I. (2019). H2 addition through a submerged membrane for in-situ biogas upgrading in the anaerobic digestion of sewage sludge. *Bioresource Technology, 280*, 1–8. https://doi.org/10.1016/j.biortech.2019.01.135.

Angelidaki, I., Xie, L., Luo, G., & Zhang, Y. (2019). Biogas upgrading: Current and emerging technologies. In A. Pandey, C. Larroche, C.-G. Dussap, E. Gnansounou, S. K. Khanal, & S. Ricke (Eds.), Biomass, biofuels, biochemicals: Biofuels: Alternative feedstocks and conversion processes for the production of liquid and gaseous biofuels (2nd ed., pp. 817–843). Academic Press. https://doi.org/10.1016/B978-0-12-816856-1.00033-6

Angelidaki, I., Treu, L., Tsapekos, P., Luo, G., Campanaro, S., Wenzel, H., & Kougias, P. G. (2018). Biogas upgrading and utilization: Current status and perspectives. *Biotechnology Advances, 36*(2), 452–466. https://doi.org/10.1016/j.biotechadv.2018.01.011.

Antukh, T., Lee, I., Joo, S., & Kim, H. (2022). Hydrogenotrophs-based biological biogas upgrading technologies. *Frontiers in Bioengineering and Biotechnology, 10*, 833482. https://doi.org/10.3389/fbioe.2022.833482.

Ao, T. J., Liu, C.-G., Sun, Z.-Y., Zhao, X.-Q., Tang, Y.-Q., & Bai, F.-B. (2024). Anaerobic digestion integrated with microbial electrolysis cell to enhance biogas production and upgrading in situ. *Biotechnology Advances, 73*, 108372. https://doi.org/10.1016/j.biotechadv.2024.108372.

Archana, K., Liu, C. G., Sun, Z. Y., Zhao, X. Q., Tang, Y. Q., & Bai, F. W. (2024). A review on recent technological breakthroughs in anaerobic digestion of organic biowaste for biogas generation: Challenges towards sustainable development goals. *Fuel, 358*, 130298. https://doi.org/10.1016/j.fuel.2023.130298.

Aryal, N., Zhang, Y., Bajracharya, S., Pant, D., & Chen, X. (2022). Microbial electrochemical approaches of carbon dioxide utilization for biogas upgrading. *Chemosphere, 291*, 132843. https://doi.org/10.1016/j.chemosphere.2021.132843.

Atzori, L., Cutrufello, M. G., Meloni, D., & Secci, F. (2023). Soft-templated NiO–CeO2 mixed oxides for biogas upgrading by direct CO_2 methanation. *International Journal of Hydrogen Energy, 48*(64), 25031–25043. https://doi.org/10.1016/j.ijhydene.2023.03.420.

Bangalore, M., Hochman, G., & Zilberman, D. (2016). Policy incentives and adoption of agricultural anaerobic digestion: A survey of Europe and the United States. *Renewable Energy, 97*, 559–571. https://doi.org/10.1016/j.renene.2016.05.062.

Bär, K., Merkle, W., Tuczinski, M., Saravia, F., Horn, Harald, Ortloff, F., Lemmer, A., & Kolb, T. (2018). Development of an innovative two-stage fermentation process for high-calorific biogas at elevated pressure. *Biomass and Bioenergy, 115*(May), 186–194. https://doi.org/10.1016/j.biombioe.2018.04.009.

Bassani, I., Kougias, P. G., & Angelidaki, I. (2016). In-situ biogas upgrading in thermophilic granular UASB reactor: key factors affecting the hydrogen mass transfer rate. *Bioresource Technology, 221*, 485–491. https://doi.org/10.1016/j.biortech.2016.09.083.

Bauer, F., Persson, T., Hulteberg, C. P., & Tamm, D. (2013). Biogas upgrading - technology overview, comparison and perspectives for the future. *Biofuels, Bioproducts and Biorefining, 7*(5), 499–511. https://doi.org/10.1002/bbb.1423.

Bonaccorsi, L., Fazzino, F., Fotia, A., & Malara, A. (2024). Comparison at laboratory scale of different types and configurations of commercial molecular sieves for one-step direct upgrading of biogas produced by anaerobic digestion. *Fuel, 365*, 131292. https://doi.org/10.1016/j.fuel.2024.131292.

Bose, A., Lin, R., Rajendran, K., O'Shea, R., Xia, A., & Murphy, J. D. (2019). How to optimise photosynthetic biogas upgrading: a perspective on system design and microalgae selection. *Biotechnology Advances,* 107444. https://doi.org/10.1016/j.biotechadv.2019.107444 Elsevier Inc.

Brunetti, A., Lei, L., Avruscio, E., & Karousos, D. (2022). Long-term performance of highly selective carbon hollow fiber membranes for biogas upgrading in the presence of H2S and water vapor. *Chemical Engineering Journal, 448*, 137615. https://doi.org/10.1016/j.cej.2022.137615.

Calbry-Muzyka, A., Madi, H., Rüsch, F., & Gandiglio, M. (2022). Biogas composition from agricultural sources and organic fraction of municipal solid waste. *Renewable Energy, 181*, 1000–1007. https://doi.org/10.1016/j.renene.2021.09.100.

Carranza Abaid, A., Wanderley, R., Knuutila, H. K., & Jakobsen, J. P. (2021). Analysis and selection of optimal solvent-based technologies for biogas upgrading. *Fuel, 303*, 121327. https://doi.org/10.1016/j.fuel.2021.121327.

Chaikitkaew, S., Seengenyoung, J., Mamimin, C., & Birkeland, N. K. (2021). Simultaneous biogas upgrading and acetic acid production by homoacetogens consortium enriched from peatland soil. *Bioresource Technology Reports, 15*, 100701. https://doi.org/10.1016/j.biteb.2021.100701.

Chaikitkaew, S., Wongfaed, N., Mamimin, C., & O-Thong, S. (2024). Biogas upgrading towards acetic acid production using Clostridium thailandense supplemented with granular activated carbon (GAC) and L-arginine: A genomic analysis approach. *Carbon Resources Conversion, 7*(4), 100236. https://doi.org/10.1016/j.crcon.2024.100236.

Chen, J. L., Ortiz, R., Steele, T. W., & Stuckey, D. C. (2014). Toxicants inhibiting anaerobic digestion: A review. *Biotechnology Advances, 32*(8), 1523–1534. https://doi.org/10.1016/j.biotechadv.2014.10.005.

Chidambaram, A., Le, D. H., Navarro, J. A. R., & Stylianou, K. (2021). Robust metal-organic frameworks for dry and wet biogas upgrading. *Applied Materials Today, 22*, 100933. https://doi.org/10.1016/j.apmt.2020.100933.

Choi, I. D., Lee, J. W., & Kang, Y. T. (2015). CO_2 capture/separation control by SiO_2 nanoparticles and surfactants. *Separation Science and Technology (Philadelphia), 50*(5), 772–780. https://doi.org/10.1080/01496395.2014.965257.

Chu, X., Ren, Y., Qu, G., & Ren, N. (2024). Current status of research on microbial electrocatalytic CH_4 production for biogas upgrading and challenges. *Journal of Environmental Chemical Engineering, 12*(2), 112088. https://doi.org/10.1016/j.jece.2024.112088.

Clodic, D. and Younes, M. (2006) 'METHOD AND SYSTEM FOR EXTRACTING CARBON DOXDE BY ANTI-SUBLMATION FOR STORAGE THEREOF'.

Comesaña-Gándara, B., García-Depraect, O., Santos-Benei, F., & Bordel, S. (2022). Recent trends and advances in biogas upgrading and methanotrophs-based valorization. *Chemical Engineering Journal Advances, 11*, 0–6. https://doi.org/10.1016/j.ceja.2022.100325.

Durán, P., Aragues-Aldea, P., Gonzalez-Pizarro, R., Mercader, V. D., Cazana, F., Frances, E., Pena, J. A., & Herguido, J. (2024). Biogas upgrading through CO_2 methanation in a polytropic - distributed feed fixed bed reactor. *Catalysis Today, 440*, 114849. https://doi.org/10.1016/j.cattod.2024.114849.

Elyasi, S. N., He, L., Tsapekos, P., & Rafiee, S. (2021). Could biological biogas upgrading be a sustainable substitution for water scrubbing technology? A case study in Denmark. *Energy Conversion and Management, 245*, 114550. https://doi.org/10.1016/j.enconman.2021.114550.

EU (2023) *Renewable Energy Directive - Revised Directive EU/2023/2413, European Commission*. Available at: https://energy.ec.europa.eu/topics/renewable-energy/renewable-energy-directive-targets-and-rules/renewable-energy-directive_en.

European Parliament and of the Council (2018) *DIRECTIVE (EU) 2018/851, Amending Directive 2008/98/EC on waste, 14.6.2018; Official Journal of the European Union*.

Fajrina, N., Yusof, N., Ismail, A. F., & Aziz, F. (2023). A crucial review on the challenges and recent gas membrane development for biogas upgrading. *Journal of Environmental Chemical Engineering, 11*(3), 110235. https://doi.org/10.1016/j.jece.2023.110235.

Fenske, C. F., Kirzeder, F., Strübing, D., & Koch, K. (2023). Biogas upgrading in a pilot-scale trickle bed reactor – Long-term biological methanation under real application conditions. *Bioresource Technology, 376*, 128868. https://doi.org/10.1016/j.biortech.2023.128868.

Franco-Morgado, M., Tabaco-Angoa, T., Ramírez-García, M. A., & González-Sánchez, A. (2021). Strategies for decreasing the O2 content in the upgraded biogas purified via microalgae-based technology. *Journal of Environmental Management, 279*, 111813. https://doi.org/10.1016/j.jenvman.2020.111813.

Fu, S., Angelidaki, I., & Zhang, Y. (2021). In situ Biogas Upgrading by CO_2-to-CH_4 Bioconversion. *Trends in Biotechnology, 39*(4), 336–347. https://doi.org/10.1016/j.tibtech.2020.08.006.

Ghorbani, B., Ebrahimi, A., & Ziabasharhagh, M. (2021). Thermodynamic and economic evaluation of biomethane and carbon dioxide liquefaction process in a hybridized system of biogas upgrading process and mixed fluid cascade liquefaction cycle. *Process Safety and Environmental Protection, 151*, 222–243. https://doi.org/10.1016/j.psep.2021.05.014.

Gkotsis, P., Kougias, P. G., Mitrakas, M., & Zouboulis, A. (2023). Biogas upgrading technologies – Recent advances in membrane-based processes. *International Journal of Hydrogen Energy, 48*(10), 3965–3993. https://doi.org/10.1016/j.ijhydene.2022.10.228.

Golmakani, A., Nabavi, A., Wadi, B., & Manovic, V. (2022). Advances, challenges, and perspectives of biogas cleaning, upgrading, and utilisation. *Fuel, 317*, 123085. https://doi.org/10.1016/j.fuel.2021.123085.

Guo, X., Liu, D., Han, T., & Huang, H. (2017). Preparation of thin film nanocomposite membranes with surface modified MOF for high flux organic solvent nanofiltration. *AIChE Journal, 63*(4), 1303–1312. https://doi.org/10.1002/aic.15508.

Hafuka, A., Fujino, S., Kimura, K., Oshita, K., Konakahara, N., & Takahashi, S. (2022). In-situ biogas upgrading with H2 addition in an anaerobic membrane bioreactor (AnMBR) digesting waste activated sludge. *Science of the Total Environment, 828*, 154573. https://doi.org/10.1016/j.scitotenv.2022.154573.

Haider, J., Qyyum, M. A, Riaz, A, Naquash, A., Kazmi, B., Yasin, M., Nizami, A., Byun, M., Lee, M., & Lim, H. (2022). State-of-the-art process simulations and techno-economic assessments of ionic liquid-based biogas upgrading techniques: Challenges and prospects. *Fuel, 314*, 123064. https://doi.org/10.1016/j.fuel.2021.123064.

Han, Y., Li, W., Gao, Y., Teng, C., Wang, J., Liu, Z., Yin, J., Lu, X., & Zhen, G. (2024). Biogas upgrading and membrane anti-fouling mechanisms in electrochemical anaerobic membrane bioreactor (EC-AnMBR): Focusing on spatio-temporal distribution of metabolic functionality of microorganisms. *Water Research, 256*, 121557. https://doi.org/10.1016/j.watres.2024.121557.

Hao, X. (2017). Analysing the mechanisms of sludge digestion enhanced by iron. *Water Research, 117*, 58–67. https://doi.org/10.1016/j.watres.2017.03.048.

Heffernan, J. K., Lai, C. Y., Gonzalez-Garcia, A, & Nielsen, L. K. (2023). Biogas upgrading using Clostridium autoethanogenum for value-added products. *Chemical Engineering Journal, 452*, 138950. https://doi.org/10.1016/j.cej.2022.138950.

Herdem, H., Şahin, U., & Ülgen, A. (2024). An innovative approach to a pressurized water scrubbing system with several packing materials for upgrading biogas to biomethane. *Process Safety and Environmental Protection, 188*, 538–548. https://doi.org/10.1016/j.psep.2024.05.053.

Hosseini, S. S., & Denayer, J. F. M. (2022). Biogas upgrading by adsorption processes: Mathematical modeling, simulation and optimization approach - A review. *Journal of Environmental Chemical Engineering, 10*(3), 107483. https://doi.org/10.1016/j.jece.2022.107483.

Hoyos, E. G., Kuri, R., Toda, T., & Muñoz, R. (2024). Innovative design and operational strategies to improve CO_2 mass transfer during photosynthetic biogas upgrading. *Bioresource Technology, 391*, 129955. https://doi.org/10.1016/j.biortech.2023.129955.

Huang, J. H., Fan, Z. L., Li, R., & Sun, M. T. (2024). Biogas upgrading by biotrickling filter: Effects of temperature and packing materials. *Chemical Engineering Journal, 481*, 148367. https://doi.org/10.1016/j.cej.2023.148367.

Janeeshma, E., Habeeb, H., Am, S., & Sinisha, A. K. (2024). Strigolactone and analogues: A new generation of plant hormones with multifactorial benefits in environmental sustainability. *Environmental and Experimental Botany, 223*, 105775. https://doi.org/10.1016/j.envexpbot.2024.105775.

Kadam, R., & Panwar, N. L. (2017). Recent advancement in biogas enrichment and its applications. *Renewable and Sustainable Energy Reviews, 73*(September 2016), 892–903. https://doi.org/10.1016/j.rser.2017.01.167.

Kapoor, R., Ghosh, P., Kumar, M., & Vijay, V. K. (2019). Evaluation of biogas upgrading technologies and future perspectives: A review. *Environmental Science and Pollution Research, 26*(12), 11631–11661. https://doi.org/10.1007/s11356-019-04767-1.

Karimi, M. (2024). Biogas upgrading using shaped MOF MIL-160(Al) by pressure swing adsorption process: Experimental and dynamic modelling assessment. *Separation and Purification Technology, 344*, 127260. https://doi.org/10.1016/j.seppur.2024.127260.

Karne, H., Mahajan, U., Ketkar, U., & Kohade, A. (2023). A review on biogas upgradation systems. *Materials Today: Proceedings, 72*, 775–786. https://doi.org/10.1016/j.matpr.2022.09.015.

Katariya, H. G., & Patolia, H. P. (2023). Advances in biogas cleaning, enrichment, and utilization technologies: a way forward. *Biomass Conversion and Biorefinery, 13*, 9565–9581. https://doi.org/10.1007/s13399-021-01750-0.

Khan, J., & Saif-Ul-Allah, M. W. (2022). Reduction in Specific Energy Consumption of Overall Biogas Upgrading and Biomethane Liquefaction Process: Energy and Exergy Analysis. *Energy Conversion and Management, 271*, 116269. https://doi.org/10.1016/j.enconman.2022.116269.

Khan, M. U., Lee, J. T. E., Bashir, M. A., Dissanayake, P. D., Ok, Y. S., Tong, Y. W., Shariati, M. A, Wu, S., & Ahring, B. K. (2021). Current status of biogas upgrading for direct biomethane use: A review. *Renewable and Sustainable Energy Reviews, 149*, 111343. https://doi.org/10.1016/j.rser.2021.111343.

Khulbe, K. C., & Matsuura, T. (2018). Thin film composite and/or thin film nanocomposite hollow fibermembrane forwater treatment, pervaporation, and gas/vapor separation. *Polymers, 10*(10), 1–22. https://doi.org/10.3390/polym10101051.

Kim, H., Park, H., Kim, K., & Lee, C. (2024). Electrochemical recovery of nutrients from human urine coupled with biogas upgrading. *Bioresource Technology, 394*, 130298. https://doi.org/10.1016/j.biortech.2023.130298.

Koutsiantzi, C., Koukovinos, K., Liatsou, A., & Gkotsis, P. (2024). Anaerobic digestion biogas upgrading using a two-stage membrane system under pilot-scale conditions. *Environmental Research, 245*, 118080. https://doi.org/10.1016/j.envres.2023.118080.

Lai, C. Y., Zhou, L., Yuan, Z., & Guo, J. (2021). Hydrogen-driven microbial biogas upgrading: Advances, challenges andsolutions. *Water Research, 197*, 117120. https://doi.org/10.1016/j.watres.2021.117120.

Lei, Z., Chen, B., Koo, Y. M., & MacFarlane, D. R. (2017). Introduction: Ionic liquids. *Chemical Reviews, 117*(10), 6633–6635. https://doi.org/10.1021/acs.chemrev.7b00246.

Lemmer, A., Chen, Y., Wonneberger, A. M., & Frank, G. (2015). Integration of a water scrubbing technique and two-stage pressurized anaerobic digestion in one process. *Energies, 8*(3), 2048–2065. https://doi.org/10.3390/en8032048.

Li, R., Jiang, Y., Huang, J., & Luo, K. (2024). Simultaneous biogas upgrading and single cell protein production using hydrogen oxidizing bacteria. *Chemical Engineering Journal, 490*, 151576. https://doi.org/10.1016/j.cej.2024.151576.

Li, Z. (2020). High levels of ammonia nitrogen for biological biogas upgrading. *International Journal of Hydrogen Energy, 45*(53), 28488–28498. https://doi.org/10.1016/j.ijhydene.2020.07.247.

Lindeboom, R., Ferrer, I., & Weijma, J. (2013). Silicate minerals for CO_2 scavenging from biogas in Autogenerative High Pressure Digestion. *Water Research, 47*(11), 3742–3751. https://doi.org/10.1016/j.watres.2013.04.028.

Lindeboom, R. E., Weijma, J., & Lier, J. B (2012). High-calorific biogas production by selective CO_2 retention at autogenerated biogas pressures up to 20 bar. *Environmental Science & Technology, 46*(3), 1895–1902.

Liu, C., Xiao, J., Li, H., Chen, Q., Sun, D., Cheng, X., Li, P., Dang, Y., Smith, J. A., & Holmes, D. E. (2021). High efficiency in-situ biogas upgrading in a bioelectrochemical system with low energy input. *Water Research, 197*, 117055. https://doi.org/10.1016/j.watres.2021.117055.

López, A., & Rodríguez, T. L. F. (2024). From Biogas to Biomethane: An In-Depth Review of Upgrading Technologies That Enhance Sustainability and Reduce Greenhouse Gas Emissions. *Applied Sciences (Switzerland), 14*(6), 2342. https://doi.org/10.3390/app14062342.

López, J. C., Monsonis, R., de Los Mozos, E. L., Heredia, F., & Gómez-Pérez, P. (2023). Simultaneous biosuccinic production and biogas upgrading: Exploring the potential of sugar-based confectionery waste within a biorefinery concept. *Bioresource Technology, 384*, 129362. https://doi.org/10.1016/j.biortech.2023.129362.

Lóránt, B., & Tardy, G. M. (2022). Current status of biological biogas upgrading technologies. periodica. *Polytechnica Chemical Engineering, 66*(3), 465–481. https://doi.org/10.3311/PPch.19537.

De Jesus, D. F. M., Mendez, L., Suero, I., & Diaz, I. (2022). Anaerobic digestion of food waste coupled with biogas upgrading in an outdoors algal-bacterial photobioreactor at pilot scale. *Fuel, 324*, 124554. https://doi.org/10.1016/j.fuel.2022.124554.

Méndez, L. (2022). Photosynthetic upgrading of biogas from anaerobic digestion of mixed sludge in an outdoors algal-bacterial photobioreactor at pilot scale. *Journal of Water Process Engineering, 48*, 102891. https://doi.org/10.1016/j.jwpe.2022.102891.

Meng, F., Han, S., Meng, Y., & Ju, T. (2022). Effects of hydroxyethyl group on monoethanolamine (MEA) derivatives for biomethane from biogas upgrading. *Fuel, 325*, 124874. https://doi.org/10.1016/j.fuel.2022.124874.

Mercader, V. D., Durán, P., Aragüés-Aldea, P., & Francés, E. (2024). Biogas upgrading by intensified methanation (SESaR): Reaction plus water adsorption - desorption cycles with Ni-Fe/Al2O3 catalyst and LTA 5A zeolite. *Catalysis Today, 433*, 114667. https://doi.org/10.1016/j.cattod.2024.114667.

Minardi, M., Marocco, P., & Gandiglio, M. (2023). Carbon recovery from biogas through upgrading and methanation: A techno-economic and environmental assessment. *Journal of CO_2 Utilization, 78*, 102632. https://doi.org/10.1016/j.jcou.2023.102632.

Mohammadpour, H., Cord-Ruwisch, R., Pivrikas, A., & Ho, Goen (2022). Simple energy-efficient electrochemically-driven CO_2 scrubbing for biogas upgrading. *Renewable Energy, 195*, 274–282. https://doi.org/10.1016/j.renene.2022.05.155.

Molavi, H., & Shojaei, A. (2019). Mixed-matrix composite membranes based on UiO-66-derived MOFs for CO_2 separation. *ACS Appl. Mater. Interfaces, 11*(9), 9448–9461.

Moya, C., Santiago, R., Hospital-Benito, D., & Lemus, J. (2022). Design of biogas upgrading processes based on ionic liquids. *Chemical Engineering Journal, 428*, 132103. https://doi.org/10.1016/j.cej.2021.132103.

Mrosso, R., Mecha, A. C., & Kiplagat, J. (2024). Performance evaluation of calcined eggshell waste (Sorbent) for biogas upgrading: Adsorption isotherms, adsorption kinetics, and fixed bed studies. *Environmental Challenges, 16*, 100961. https://doi.org/10.1016/j.envc.2024.100961.

Mulu, E., M'arimi, M., Ramkat, R. C., & Kiprop, A. (2021). Biogas upgrade using modified natural clay. *Energy Conversion and Management: X, 12*, 100134. https://doi.org/10.1016/j.ecmx.2021.100134.

Mwacharo, F. (2020). Edited by Centria University of Applied Sciences. Centria University of Applied Sciences. *Biogas drying and purification methods, 28*(1), 135–148.

Naquash, A., Qyyum, M. A., Haider, J., & Bokhari, A. (2022). State-of-the-art assessment of cryogenic technologies for biogas upgrading: Energy, economic, and environmental perspectives. *Renewable and Sustainable Energy Reviews, 154*, 111826. https://doi.org/10.1016/j.rser.2021.111826.

Omar, B. (2018). Simultaneous biogas upgrading and biochemicals production using anaerobic bacterial mixed cultures. *Water Research, 142*, 86–95. https://doi.org/10.1016/j.watres.2018.05.049.

Pancione, E. (2024). Uncovering the potential of MSC CT-350 for CO_2/CH_4 separation toward the optimization of a Pressure Swing Adsorption process for biogas upgrading. In: *Fuel Processing Technology, 256*, 108065. https://doi.org/10.1016/j.fuproc.2024.108065.

Parker, P. E. M., Northrop, S., Valencia, J. A., & Foglesong, R. E. (2011). CO_2 management at ExxonMobil's LaBarge field, Wyoming, USA. *Energy Procedia, 4*, 5455–5470. https://doi.org/10.1016/j.egypro.2011.02.531.

Patil, D. S., Barjibhe, R. B., Chopade, V., & Patil, S. S. (2019). A retrospection on current trends in biogas purification. *International Journal of Innovative Technology and Exploring Engineering, 8*(12), 5147–5152. https://doi.org/10.35940/ijitee.L2768.1081219.

Pérez-Huertas, S., Calero, M., Ligero, A., Perez, A., Terpilowski, K, & Martin-Lara, M. A. (2023). On the use of plastic precursors for preparation of activated carbons and their evaluation in CO_2 capture for biogas upgrading: a review. *Waste Management, 161*, 116–141. https://doi.org/10.1016/j.wasman.2023.02.022.

Petersson, A. and Wellinger, A. (2009) *Biogas upgrading technologies-developments and innovations - Task 37-Energy from biogas and landfill gas.* Available at: www.iea-biogas.net.

Piadeh, F., Offie, I., Behzadian, K., Rizzuto, J. P., Bywater, A., Córdoba-Pachón, J. R., & Walker, M. (2024). A critical review for the impact of anaerobic digestion on the sustainable development goals. *Journal of Environmental Management, 349.* https://doi.org/10.1016/j.jenvman.2023.119458.

Rainone, F., D'Agostino, O., Erto, A., & Balsamo, M. (2021). Biogas upgrading by adsorption onto activated carbon and carbon molecular sieves: Experimental and modelling study in binary CO_2/CH_4 mixture. *Journal of Environmental Chemical Engineering, 9*(5), 106256. https://doi.org/10.1016/j.jece.2021.106256.

Rao, Y. (2023). Meta-analysis of biogas upgrading to renewable natural gas through biological CO_2 conversion. *Journal of Cleaner Production, 426*, 139128. https://doi.org/10.1016/j.jclepro.2023.139128.

Rao, Y., Lin, T. Y., Ling, F., & He, Z. (2024). Biological upgrading of biogas assisted with membrane supplied hydrogen gas in a three-phase upflow reactor. *Bioresource Technology, 394*, 130260. https://doi.org/10.1016/j.biortech.2023.130260.

Erbay, C., Yang, G., Figueiredo, P., Sadr, R., & Yu, C. (2015). Three-dimensional porous carbon nanotube sponges for high-performance anodes of microbial fuel cells. *Journal of Power Sources, 298*, 177–183. http://dx.doi.org/10.1016/j.jpowsour.2015.08.021.

Esteban, R. I., Muñoz, R., Fdz-Polanco, M., & Díaz, I. (2021). Biogas from anaerobic digestion as an energy vector: Current upgrading development. *Energies, 14*(10), 2742. https://doi.org/10.3390/en14102742.

Rattanaya, T., Kongjan, P., Cheewasedtham, C., Bunyakan, C., Yuso, P., Cheirsilp, B., & Jariyaboon, R. (2022). Application of palm oil mill waste to enhance biogas upgrading and hornwort cultivation. *Journal of Environmental Management, 309*, 114678. https://doi.org/10.1016/j.jenvman.2022.114678.

Ravi, P. P., Merkle, W., Tuczinski, M., Saravia, F., Horn, H., & Andreas, L. (2019). Integration of membrane filtration in two-stage anaerobic digestion system: Specific methane yield potentials of hydrolysate and permeate. *Bioresource Technology, 275*, 138–144. https://doi.org/10.1016/j.biortech.2018.12.043.

Roozitalab, A., Hamidavi, F., & Kargari, A. (2023). A review of membrane material for biogas and natural gas upgrading. *Gas Science and Engineering, 114*, 204969. https://doi.org/10.1016/j.jgsce.2023.204969.

Torres, R. A., Vega-Quiel, M. J., Batista, A., & Fernández-Ramos, O. (2021). Influence of biogas supply regime on photosynthetic biogas upgrading performance in an enclosed algal-bacterial photobioreactor. *Algal Research, 57*, 102350. https://doi.org/10.1016/j.algal.2021.102350.

Said, R. G. B., Kolle, J. M., Khaled, E., & Tangour, B. (2020). A Unified Approach to CO_2-Amine Reaction Mechanisms. *ACS Omega, 5*(40), 26125–26133. https://doi.org/10.1021/acsomega.0c03727.

Schröer, D., Herlicka, L., Heinold, A., Latacz-Lohmann, U., & Meisel, F. (2024). A network design problem for upgrading decentrally produced biogas into biomethane. *Journal of Cleaner Production, 452*, 142049–142064. In this issue. https://doi.org/10.1016/j.jclepro.2024.142049.

Sekoai, P. T., Engelbrecht, N., du Preez, S. P., & Bessarabob, D. (2020). Thermophilic Biogas Upgrading via ex Situ Addition of H2 and CO_2 Using Codigested Feedstocks of Cow Manure and the Organic Fraction of Solid Municipal Waste. *ACS Omega, 5*(28), 17367–17376. https://doi.org/10.1021/acsomega.0c01725.

Selim, M. M. (2024). Enhancing carbon capture efficiency in biogas upgrading: A comprehensive review on adsorbents and adsorption isotherms. *AIP Advances, 14*(4), 040703. https://doi.org/10.1063/5.0208686.

Seo, S., Quiroz-Guzman, M., De Silva, M. A., Lee, T. B., Huang, Y., Goodrich, B. F., Schneider, W. F., & Brennecke, J. F. (2014). Chemically tunable ionic liquids with aprotic heterocyclic anion (AHA) for CO_2 capture. *Journal of Physical Chemistry B, 118*(21), 5740–5751. https://doi.org/10.1021/jp502279w.

Singh, S., Mukherjee, T., & Venkata Mohan, S. (2024). Conductive materials and applied potential to augment methanogenesis for biogas upgradation: Acidogenic synergy, CO_2 dynamics and sustainability analysis. *Journal of Cleaner Production, 448*, 141419. https://doi.org/10.1016/j.jclepro.2024.141419.

Słupek, E., Makoś, P., & Gębicki, J. (2020). Theoretical and economic evaluation of low-cost deep eutectic solvents for effective biogas upgrading to bio-methane. *Energies, 13*(13), 3379. https://doi.org/10.3390/en13133379.

Sun, J., Rao, Y., & He, Z. (2024). Machine learning-aided inverse design for biogas upgrading through biological CO_2 conversion. *Bioresource Technology, 399*, 130549. https://doi.org/10.1016/j.biortech.2024.130549.

Sun, Q., Li, H., Yan, J., Liu, L., Yu, Z., & Yu, X. (2015). Selection of appropriate biogas upgrading technology—A review of biogas cleaning, upgrading and utilisation. *Renewable and Sustainable Energy Reviews, 51*, 521–532. https://doi.org/10.1016/j.rser.2015.06.029.

Sun, X. (2022a). A review on algal-bacterial symbiosis system for aquaculture tail water treatment. *Science of the Total Environment, 847*, 157620. https://doi.org/10.1016/j.scitotenv.2022.157620.

Sun, Z. F. (2022b). Overview of recent progress in exogenous hydrogen supply biogas upgrading and future perspective. *Science of the Total Environment, 848*, 157824. https://doi.org/10.1016/j.scitotenv.2022.157824.

Sun, Z.-F., Zhao, L., Wu, J. T., & Wang, Z. H. (2023). Exogenous hydrogen supply improves in-situ biogas upgrading of sewage sludge: Performance and mechanisms. *Chemical Engineering Journal, 477*, 147307. https://doi.org/10.1016/j.cej.2023.147307.

Taherizadeh, A., Simon, A., Richter, H., & Stelter, M. (2024). Characterization and synthesis of high permeance SSZ-13 membranes to separate CO_2 from CH_4 for biogas upgrading. *Journal of Membrane Science, 703*, 122845. https://doi.org/10.1016/j.memsci.2024.122845.

Tartakovsky, B. (2021). A comparison of microbial and bioelectrochemical approaches for biogas upgrade through carbon dioxide conversion to methane. *Sustainable Energy Technologies and Assessments, 45*, 101158. https://doi.org/10.1016/j.seta.2021.101158.

Tayar, S. P., Solcia, R., Hidalgo, L. F., & Bevilaqua, D. (2019). Evaluation of biogas biodesulfurization using different packing materials. *ChemEngineering, 3*(1), 1–12. https://doi.org/10.3390/chemengineering3010027.

Thapa, A, Park, J. H., Shin, S. G., Jo, H. M., Kim, M. S., Park, Y, Han, U, & Cho, S. K. (2023). Elucidation of microbial interactions, dynamics, and keystone microbes in high pressure anaerobic digestion. *Sci Total Environ, 858*(Pt 1), 159718.

Upadhyay, A., Kovalev, A. A., Zhuravleva, E. A., & Kovalev, D. A. (2023). Recent Development in Physical, Chemical, Biological and Hybrid Biogas Upgradation Techniques. *Sustainability, 15*(1), 476. https://doi.org/10.3390/su15010476.

Vargas-Estrada, L. (2022). Elucidating the role of nanoparticles on photosynthetic biogas upgrading: Influence of biogas type, nanoparticle concentration and light source. *Algal Research, 68*, 102899. https://doi.org/10.1016/j.algal.2022.102899.

Vargas-Estrada, L. (2023). Influence of mesoporous iron based nanoparticles on Chlorella sorokiniana metabolism during photosynthetic biogas upgrading. *Fuel, 333*, 126362. https://doi.org/10.1016/j.fuel.2022.126362.

Verougstraete, B. (2022). Electrical swing adsorption on 3D-printed activated carbon monoliths for CO capture from biogas. *Separation and Purification Technology, 299*, 121660. https://doi.org/10.1016/j.seppur.2022.121660.

Villarim, P. (2022). Deep eutectic solvents and conventional solvents as VOC absorbents for biogas upgrading: A comparative study. *Chemical Engineering Journal, 446*, 136875. https://doi.org/10.1016/j.cej.2022.136875.

Viñes, F., Calzada, A., & Gamallo, P. (2023). Thermodynamic, kinetic and dynamic aspects of biogas upgrading using nano-engineered grazynes. *Journal of CO_2 Utilization, 71*, 102459. https://doi.org/10.1016/j.jcou.2023.102459.

Walozi, R., Nabuuma, B., & Sebiti, A. (2016). Application of Low Pressure Water Scrubbing Technique for Increasing Methane Content in Biogas. *Universal Journal of Agricultural Research, 4*(2), 60–65. https://doi.org/10.13189/ujar.2016.040206.

Wang, H. (2023a). The effects of influent chemical oxygen demand and strigolactone analog concentration on integral biogas upgrading and pollutants removal from piggery wastewater by different microalgae-based technologies. *Bioresource Technology, 370*, 128483. https://doi.org/10.1016/j.biortech.2022.128483.

Wang, J., Wu, B., Jiang, N., Liu, J., Zhao, Y., Xu, J., & Wang, H. (2022). Synergism and mutualistic interactions between microalgae and fungi in fungi-microalgae symbiotic system. *Bioresource Technology, 361*, 127728. https://doi.org/10.1016/j.biortech.2022.127728.

Wang, R. (2021). Deeper insights into effect of activated carbon and nano-zero-valent iron addition on acidogenesis and whole anaerobic digestion. *Bioresource Technology, 324*, 124671. https://doi.org/10.1016/j.biortech.2021.124671.

Wang, Z. (2023b). Performance of different bacteria-microalgae-fungi consortium cultivation in nutrient removal and biogas upgrading by induction of GR24 and 5-Deoxystrigol. *Journal of Cleaner Production, 392*, 136292. https://doi.org/10.1016/j.jclepro.2023.136292.

Wantz, E. (2023). Innovative high-pressure water scrubber for biogas upgrading at farm-scale using vacuum for water regeneration. *Applied Energy, 350*, 121781. https://doi.org/10.1016/j.apenergy.2023.121781.

Wantz, E., Benizri, D., Dietrich, N., & Hebrard, G. (2022). Rate-based modeling approach for High Pressure Water Scrubbing with unsteady gas flowrate and multicomponent absorption applied to biogas upgrading. *Applied Energy, 312*, 118754. https://doi.org/10.1016/j.apenergy.2022.118754.

Wei, Y. (2024). Biogas upgrading performance and underlying mechanism in microbial electrolysis cell and anaerobic digestion integrated system. *Bioresource Technology, 400*, 130683. https://doi.org/10.1016/j.biortech.2024.130683.

Wu, K. K. (2023). Synergistic effect of hydrogen and nanoscale zero-valent iron on ex-situ biogas upgrading and acetate recovery. *Science of the Total Environment, 856*, 159100. https://doi.org/10.1016/j.scitotenv.2022.159100.

Xu, M. (2021). Biogas upgrading and valorization to single-cell protein in a bioinorganic electrosynthesis system. *Chemical Engineering Journal, 426*, 131837. https://doi.org/10.1016/j.cej.2021.131837.

Xu, W. (2024). Effect of different microalgae-bacterium-fungus symbiont technologies on nutrient removal from aquaculture wastewater and biogas upgrading under a variety of mixed light wavelengths. *Algal Research, 79*, 103486. https://doi.org/10.1016/j.algal.2024.103486.

Yang, J. (2024). Effective anaerobic digestion performance from kitchen wastewater by nanoscale zero-valent iron composite materials: Hydrogen corrosion capability and biological biogas upgrading. *Journal of Environmental Chemical Engineering, 12*(3), 113098. https://doi.org/10.1016/j.jece.2024.113098.

Yang, W. (2022). Microalgal cultivation for the upgraded biogas by removing CO_2, coupled with the treatment of slurry from anaerobic digestion: A review. *Bioresource Technology*. Elsevier Ltd https://doi.org/10.1016/j.biortech.2022.128118 128118 .

Yellezuome, D, Zhu, X, Liu, X, Liu, X, Liu, R, Wang, Z, Li, Y, Sun, C, Hemida Abd-Alla, M, Rasmey, AM, et al. (2023). Integration of two-stage anaerobic digestion process with in situ biogas upgrading. *Bioresour Technol, 369*, 128475.

Yousef, A. M. (2018). Low-temperature distillation process for CO_2/CH_4 CHseparation: a study for avoiding co2 freeze-out. *Journal of Heat and Mass Transfer, 140*(4), 042001. https://doi.org/10.1115/1.4038193.

Yuan, T., Zhang, Z., Lei, Z., Shimizu, K., & Lee, D. J. (2022). A review on biogas upgrading in anaerobic digestion systems treating organic solids and wastewaters via biogas recirculation. *Bioresource Technology, 344*, 126412. https://doi.org/10.1016/j.biortech.2021.126412.

Zglobisz, N. (2010). Influence of UK energy policy on the deployment of anaerobic digestion. *Energy Policy, 38*(10), 5988–5999. https://doi.org/10.1016/j.enpol.2010.05.054.

Zhang, L., Kuroki, A., & Tong, Y. W. (2020). A Mini-Review on In situ Biogas Upgrading Technologies via Enhanced Hydrogenotrophic Methanogenesis to Improve the Quality of Biogas From Anaerobic Digesters. *Frontiers in Energy Research, 8*(69). https://doi.org/10.3389/fenrg.2020.00069.

Zhao, J., Li, Y., & Dong, R. (2021). Recent progress towards in-situ biogas upgrading technologies. *Science of the Total Environment, 800*, 149667. https://doi.org/10.1016/j.scitotenv.2021.149667.

Zhao, L., Wang, Z., Ren, H. Y., Nan, J., Chen, C., & Ren, N. Q. (2020). Improving biogas upgrading and liquid chemicals production simultaneously by a membrane biofilm reactor. *Bioresource Technology, 313*(June), 123693. https://doi.org/10.1016/j.biortech.2020.123693.

Zhen, G., Zheng, S., Lu, X., Zhu, X., Mei, J., Kobayashi, T., Xu, K., Li, Y., & Zhao, Y. (2018). A comprehensive comparison of five different carbon-based cathode materials in CO2 electromethanogenesis: Long-term performance, cellelectrode contact behaviors and extracellular electron transfer pathways. *Bioresource Technology, 266*, 382–388. https://doi.org/10.1016/j.biortech.2018.06.101.

Zito, P. F., Barbieri, G., & Brunetti, A. (2024). Biogas upgrading to biomethane with zeolite membranes: Separation performance and economic analysis. *Chemical Engineering Research and Design, 207*, 121–132. https://doi.org/10.1016/j.cherd.2024.05.042.

Chapter 5

Carbon neutrality and challenges in biogas economy: bioeconomy case studies

Judith González-Arias[1,2], Tomás R. Reina[1], Soraya Zahedi[3], and Fernando G. Fermoso[3]

[1]*Inorganic Chemistry Department and Materials Sciences Institute, University of Seville-CSIC, Seville, Spain,* [2]*Chemical and Environmental Engineering Department, Technical School of Engineering, University of Seville, Seville, Spain,* [3]*Instituto de la Grasa, Spanish National Research Council (CSIC), Pablo de Olavide University Campus, Seville, Spain*

List of abbreviations

GHGs	greenhouse gases
UNFCCC	United Nations Framework Convention on Climate Change
AD	anaerobic digestion
CCS	carbon capture and storage
CCU	carbon capture and utilization
$CO_{2,\ eq}$	CO_2 equivalent
OECD	Organization for Economic Co-operation and Development
TEV	theoretical economic value
OLR	organic loading rates
BPR	biogas production rates
COD	chemical oxygen demand
EPR	energy production rates
OFMSW	organic fraction of municipal solid waste
NPV	net present value
LCA	life cycle assessment

5.1 Introduction

Global industrialization and nonrenewable energy use have significantly increased greenhouse gas (GHG) emissions, leading to rising global temperatures and environmental degradation (Wang et al., 2021). According to the measurements taken at Mauna Loa Observatory in Hawaii, the atmospheric

Biogas in the Circular Economy. DOI: https://doi.org/10.1016/B978-0-443-29230-9.00001-2

CO_2 concentration has increased from 316 to 422 ppm from 1960 to 2024 (NASA, 2024). Predictions suggest a continued CO_2 emissions increasing trend, with a 50% rise in global GHG emissions by 2050 (Rabaey & Ragauskas, 2014). This poses severe threats, including species extinction, biodiversity loss, and extreme weather events (Chen et al., 2022; Yang et al., 2022).

The Paris Agreement, adopted by 197 member parties of the United Nations Framework Convention On Climate Change (UNFCCC) in 2015, aims to limit global temperature increase to underneath 2°C and pursue efforts for 1.5°C (United Nations, 2015). As of February 2021, 124 countries committed to achieving carbon neutrality by 2050 or 2060 (Chen, 2021). This aligns with the Agreement's goals, emphasizing the need to not only reduce CO_2 emissions but also implement measures for net-zero or negative carbon emissions to support sustainable development (Berndes et al., 2016).

Following this line, carbon neutrality, also known as climate neutrality, refers to the state where an entity, such as an organization, project, or process, balances the amount of GHGs that it emits with an equivalent amount of GHGs removed or offset (Wei et al., 2022). Achieving carbon neutrality essentially means having no net impact on the overall concentration of GHG in the atmosphere.

The importance of reaching carbon neutrality relies on the consequences of this action. For instance, carbon neutrality is crucial for addressing climate change by reducing the net contribution of human activities to the accumulation of GHGs, which are major contributors to global warming (Wang et al., 2021). Similarly, it allows to meet environmental goals worldwide. Globally, governments, businesses, and individuals are setting carbon reduction goals to meet international agreements and contribute to sustainable development (Salvia et al., 2021). Carbon neutrality plays a vital role in meeting these targets. In this sense, carbon neutrality aligns with the principles of sustainable development by minimizing negative environmental impacts and promoting responsible resource use (Chen et al., 2022).

Pursuing carbon neutrality encourages innovation in cleaner technologies, renewable energy sources, and sustainable practices, fostering a transition to a low-carbon economy (Li & Chen, 2023). As governments worldwide implement stricter environmental regulations, achieving carbon neutrality can help organizations adapt to evolving legal frameworks and avoid potential penalties (Ark Resources, 2024). Similarly, carbon neutrality can lead to cost savings through energy efficiency measures, reduced waste, and the implementation of sustainable practices, contributing to long-term economic viability (Faster Capital, 2024).

In the context of biogas in the circular economy, carbon neutrality is pivotal for creating closed-loop systems, where the carbon emitted during biogas production is offset by sustainable practices within the circular economy.

Anaerobic digestion (AD) of biomass represents one among various waste-to-energy technologies capable of producing clean bioenergy, specifically in the form of biogas ($CH_4 + CO_2$). This method facilitates the conversion of diverse organic wastes, including agricultural residues, livestock manure, food waste, wastewater sludge, and macroalgae, into bioenergy (Yuhendra et al., 2021; Farghali et al., 2019; González-Arias et al., 2020). Biogas emerges as a highly sustainable energy source, offering benefits such as reducing dependence on fossil fuels, reducing carbon emissions, addressing the fossil fuel crisis, preventing deforestation, minimizing harmful waste, and improving soil fertility. Furthermore, the versatility of biogas applications and its efficient utilization make it a compelling choice for sustainable energy solutions (Farghali et al., 2022).

In this line, biogas stands as a key component of the bioeconomy, thus playing a crucial role in reducing GHG emissions. Traditional methods of organic waste disposal, such as landfilling, result in the release of methane. As previously stated, biogas production involves the AD of organic materials, mitigating methane emissions by capturing and utilizing the gas for energy purposes. This not only addresses the environmental impact of waste decomposition but also provides a cleaner alternative to fossil fuels. The reduction in methane emissions contributes significantly to overall GHG emission reduction, aligning bioeconomy practices with global efforts to combat climate change and achieve carbon neutrality.

Bioeconomy relies on renewable biological resources, utilizing organic waste and biomass as feedstock for biogas production (Feng et al., 2023). By harnessing these resources, bioeconomy reduces dependence on finite, non-renewable resources, that is, fossil fuels. This sustainable approach addresses the critical need to transition to energy sources that have a lower environmental impact, contributing significantly to the overall goal of achieving carbon neutrality. The utilization of renewable resources ensures a continuous and sustainable supply of feedstock for biogas production, mitigating the negative environmental and social impacts associated with the extraction and consumption of nonrenewable resources (Singh et al., 2023).

One of the noteworthy advantages of bioeconomy is its ability to facilitate carbon sequestration through the growth of biomass. As plants grow, they capture and store carbon from the atmosphere through photosynthesis (Shah et al., 2023). By incorporating biomass into the biogas production process, bioeconomy creates a closed-loop system that prevents the release of additional carbon. This dual benefit, capturing carbon during growth and utilizing it for energy production, aligns with the goal of achieving carbon neutrality by both reducing emissions and actively sequestering carbon, contributing to a more sustainable and balanced carbon cycle (Chen et al., 2022).

Bioeconomy aligns perfectly with circular economy principles, emphasizing resource efficiency and waste reduction. Biogas, derived from organic materials like agricultural residues and food waste, exemplifies circularity by converting organic waste into valuable energy (Lubańska & Kazak, 2023). This not only reduces the environmental burden of waste disposal but also minimizes the need for conventional energy sources, fostering a regenerative and closed-loop system. Circular economy practices in bioeconomy contribute directly to the overarching aim of achieving carbon neutrality by promoting sustainable resource utilization and minimizing the carbon footprint associated with waste (Stegmann et al., 2020). Fig. 5.1 outlines the key milestones and trends in carbon neutrality efforts and the impact of biogas in the context of achieving carbon neutrality.

5.2 Carbon neutrality in the biogas sector

The utilization of bioenergy through AD of biomass, coupled with carbon capture and storage (CCS) or carbon capture and utilization (CCU), has emerged as a prominent subject in discussions regarding negative emissions strategies (Al-Wahaibi et al., 2020). This approach involves the comprehensive collection of different organic wastes, subjecting them to the AD process to produce biogas. The potential impact of this process is substantial, with the possibility to mitigate GHG emissions by an estimated 3.29 to 4.36 billion tons of carbon dioxide equivalent ($CO_{2,\ eq}$) (Farghali et al., 2022).

This reduction might account for approximately 10%–13% of the overall global GHG emissions stemming from activities such as renewable bioenergy production, crop burning, evaded emissions management, landfill gas, deforestation, and fertilizer production emissions. By embracing bioenergy generation through AD in conjunction with CCS or CCU, there is a tangible opportunity to address a substantial portion of emissions associated with various sectors, thereby contributing to the broader efforts to combat climate change (Farghali et al., 2022).

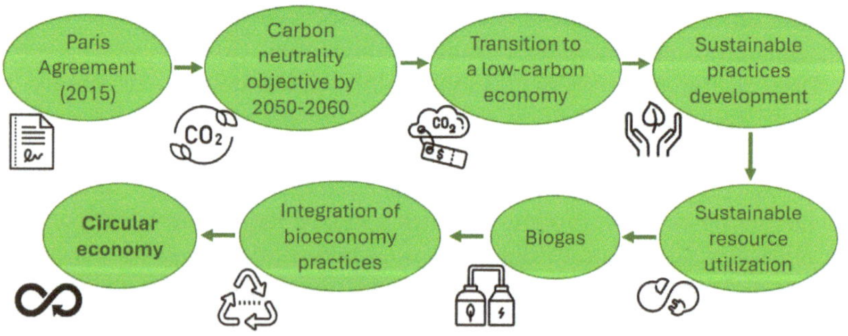

FIGURE 5.1 Outline of the key parameters in achieving carbon neutrality within the biogas sector.

Despite the current global operation of around 50 million microdigesters, 132,000 small-, medium-, and large-scale digesters, and 700 upgrading plants, the utilization of AD accounts for approximately 2% of its global potential (IRENA, 2019). This suggests an extraordinary growth potential for the biogas industry worldwide, with implications for every country (IRENA, 2019).

The potential for energy generation from currently available and sustainably grown or recovered major feedstocks, including livestock manure, food waste, sewage, crop residues, and energy crops, ranges from 10,100 to 14,000 TWh. This energy reservoir has the capacity to fulfill 6%–9% of the world's primary energy consumption or alternatively and 23%–32% of the global coal consumption (IRENA, 2019). When harnessed for electricity production, it holds the potential to satisfy 16%–22% of the world's total electricity consumption. Furthermore, if this energy is converted into biomethane, it could replace 993 to 1380 billion m^3 of natural gas, amounting to 26%–37% of the current global natural gas consumption. This highlights the significant role these feedstocks can play in addressing global energy needs and transitioning toward more sustainable alternatives (IRENA, 2019).

The removal and/or sequestration of CO_2 from raw biogas is crucial for enhancing the environmental impact of AD systems. This process not only reduces CO_2 emissions but also contributes to the overall GHG mitigation. Purifying or upgrading biogas becomes essential in this context, as it not only removes CO_2 but also improves the efficiency of biogas electricity transmission. Furthermore, it allows for standardization, enabling integrated incorporation into natural gas grids, and facilitates transportation and storage, thereby optimizing the role of renewable biogas as a substitute for fossil fuels (Khan et al., 2021).

In addition to biogas upgrading, methanation offers a sustainable approach to further reduce CO_2 emissions from biogas streams. Methanation involves reacting CO_2 with H_2 to produce methane and water through the Sabatier reaction ($CO_2 + 4H_2 \rightleftarrows CH_4 + 2H_2O$), thereby converting the captured CO_2 into additional renewable methane. This not only enhances the overall yield of biomethane from biogas but also effectively reduces the amount of CO_2 released into the atmosphere. The integration of methanation processes into biogas plants can significantly strengthen carbon neutrality efforts by providing a circular solution where CO_2 is recycled into the energy system rather than emitted as a waste product (González-Arias et al., 2024).

The conversion of biogas into biomethane has evolved into a well-established technology. Additionally, the process of capturing biomethane and CO_2 during biogas plant operations serves as an effective carbon sequestration tool. Notably, some plants harness the captured CO_2 for utilization in greenhouses, food processing plants, and beverage co-facilities.

Globally, there are 700 biogas upgrading plants, with a significant concentration in Europe (c. 80%). Beyond Europe, there are around 50 upgrading plants in the United States, 25 in China, and 20 in Canada.

Additionally, a few upgrading plants are present in Japan, India, Brazil, and South Korea. This distribution showcases the widespread adoption of biomethane upgrading and biogas separation technologies on a global scale, contributing to sustainable energy practices and carbon management (IEA Bioenergy, 2020; IRENA, 2019).

5.3 Challenges in achieving carbon neutrality

5.3.1 Technical challenges in biogas production

As aforementioned, the pursuit of carbon neutrality makes the production of biogas emerge as a promising solution, offering a renewable energy source derived from organic waste. Nonetheless, the journey toward achieving carbon neutrality possesses different challenges, particularly in the realm of biogas production. Several technical obstacles must be addressed to maximize the potential of this sustainable energy source.

AD viability relies on the availability of adequate feedstock. In small areas, this supply can be a significant challenge, impacting the project profitability and sustainability. In regions with low population density or without significant agro-industrial activity, the amount of organic waste may be insufficient, and transporting feedstock from afar increases logistical costs and GHG emissions, countering sustainability goals (Reynolds et al., 2022).

In this sense, one significant challenge is the variability and quality of feedstock used in biogas production. The composition of organic materials such as agricultural residues, municipal waste, and industrial by-products can fluctuate, impacting the efficiency and reliability of the biogas production process (Kulichkova et al., 2020)(Wang, 2014). Achieving a consistent and high-quality feedstock supply is crucial for ensuring optimal biogas yields.

To overcome this issue, there is a critical need for research and development in feedstock management and preprocessing technologies, as well as for political measurements (this last one will be explained in the next section). Innovative solutions, such as sorting and pretreatment techniques, can help standardize feedstock composition, ensuring a more consistent and reliable input for the AD process (Atelge et al., 2020). This might maximize biogas yield and minimize operational uncertainties associated with feedstock variability.

Moreover, addressing technical challenges extends to optimizing the AD process itself. Advanced reactor designs, improved microbial cultures, and enhanced process monitoring systems are essential components of this effort (Das et al., 2023; Rocha-Meneses et al., 2022). Developing more robust and efficient AD systems may allow for better control over the biogas production process, increasing its overall reliability and performance.

Another technical barrier is the need for continuous improvements in AD technology. Technological efficiency and regular upgrades are paramount to keep pace with advancements in the field (Brémond et al., 2021). Researchers and engineers must collaborate to develop cutting-edge technologies that enhance the overall efficiency of biogas production while minimizing energy losses and environmental impacts. Ongoing research and investment in upgrading existing infrastructure are essential for staying ahead of the curve in the rapidly evolving field of renewable energy (Obaideen et al., 2022). Furthermore, taking advantage of artificial intelligence, enhancing the operating conditions of the biogas reactor in the biogas production process has the potential to boost biogas yields and improve the overall economic viability of the operation (Asadi & McPhedran, 2021).

Digestate management poses another obstacle on the path to carbon neutrality (Lijó et al., 2017). The by-product of biogas production, digestate, contains valuable nutrients but requires proper handling to prevent environmental risks (Monfet et al., 2018). Developing efficient and environmentally responsible methods for digestate management is essential to mitigate potential negative impacts and maximize the overall sustainability of the biogas production cycle. Approaches such as covering and heating liquid digestate storage tanks, along with additional energy recovery from solid digestate through various posttreatment methods, are considered effective strategies for better digestate management (Bacenetti et al., 2016; Monlau et al., 2015).

Briefly, while the journey to carbon neutrality through biogas production is promising, different challenges must be tackled. Addressing challenges related to feedstock variability, digestate management, and technological efficiency is imperative to unlock the full potential of biogas as a sustainable energy source. The collective efforts of researchers, policymakers, and industry stakeholders are essential to surmount these obstacles and pave the way for a greener and more sustainable future.

5.3.2 Economic and policy challenges

To fully harness the complete potential of biogas in achieving sustainable energy goals, policy and regulatory support are essential. Decarbonizing energy production requires a level playing field with existing operators, often established in fossil fuel industries. The involved benefits of biogas, including waste treatment, energy generation, and fertilizer production, are not always considered in their true value, leading to a perception that biogas is more expensive than fossil fuels. Policymakers face challenges in understanding the intrinsic value, particularly when fossil fuels receive substantial incentives and tax breaks (IRENA, 2019).

Political support is critical at both global and national/regional levels. All policies should acknowledge the circular nature of the AD industry and recognize its full potential in energy generation, nutrient recovery, and recycling, as well as its role in waste treatment and transportation (IRENA, 2019).

The primary obstacle hindering the growth of the AD industry is the lack of financial recognition for its numerous services. While AD plays a pivotal role in addressing air pollution, which the Organization for Economic Co-operation and Development (OECD) estimates costs $2.6 trillion and results in 6–9 million premature deaths annually worldwide, its broader benefits are often overlooked. These include improvements in water quality and associated clean-up costs, the production of biofertilizers that enhance soil carbon capture and biodiversity, and the creation of skilled jobs that boost rural economies (World Biogas Association, 2021).

Despite these far-reaching health, environmental, and social impacts, AD is predominantly perceived as a renewable energy generator, neglecting its broader positive contributions. To rectify this, an international carbon tax and trading scheme would be the most efficient mechanism to financially value AD's services and recognize its central role in the circular economy. This approach, however, necessitates significant political will and global cooperation. The proposal envisions achieving this goal toward the end of the decade, aiming to position the AD industry for sustained success and independent viability, less susceptible to the changing priorities of national governments (World Biogas Association, 2021).

Investment and funding issues, policy frameworks and regulatory issues, and market dynamics and price fluctuations, among others, play pivotal roles in shaping the feasibility and success of biogas initiatives. Investment and funding issues are substantial barriers to the expansion of biogas production (Yousuf et al., 2016). The capital-intensive nature of AD facilities demands substantial financial backing for infrastructure development and technology upgrades (Zglobisz et al., 2010). Addressing this challenge requires innovative financing models, public–private partnerships, and a concerted effort to attract investments by showcasing the long-term economic and environmental benefits of biogas.

Policy frameworks and regulatory barriers notably influence the viability of biogas projects (Mateescu & Dima, 2022). Supportive policies are crucial to creating an enabling environment for sustainable energy initiatives (Capodaglio et al., 2016). Governments need to establish frameworks that simplify licensing procedures, provide incentives for biogas adoption, and ensure regulatory consistency (Capodaglio et al., 2016). Collaboration between policymakers, industry stakeholders, and environmental agents is necessary to promote a regulatory landscape favorable to carbon-neutral biogas production (Mateescu & Dima, 2022).

Market dynamics and price fluctuations further complicate the carbon-neutral biogas production landscape (Ottosson et al., 2020). The unpredictability of energy markets and fluctuations in the prices of renewable technologies can impact project economics (Capodaglio et al., 2016). Strategies such as long-term contracts, market stabilization mechanisms, and financial instruments are essential to mitigate risks associated with price fluctuations and create a stable market environment for biogas producers.

To sum up, reaching carbon neutrality through biogas production via AD requires a thorough approach that addresses investment challenges, establishes supportive policy frameworks, and considers the fluctuations of energy markets. By overcoming these obstacles collaboratively, stakeholders can fully utilize the potential of biogas as a sustainable and carbon-neutral energy source.

5.4 Bioeconomy case studies

5.4.1 Overview of the successful implementation of biogas in a circular economy

The scientific community is motivated to intensify the study of sustainable development policies, moving toward a more environmental-friendly energy consumption. In this sense, biogas production from biomass and organic waste arises as a suitable alternative. In this line, new energy policies aim at implementing biogas in a circular economy framework. To this end, the study of different alternatives including the improvement of environmental and economic benefits of the AD process, new ways of obtaining the biomass or waste to be digested, or the improvement in the process parameters is essential. Similarly, using biogas as feedstock for other processes aimed at producing various value-added chemicals and fuels should be also included in this line. In this section, some of these alternative works are presented and discussed. To summarize the diverse case studies discussed, Table 5.1 presents an overview including the country where the study was performed, the plant sizes considered, the feedstock used in each case, and the main findings of every study.

A study presented by Lubańska and Kazak investigates the alignment of biogas plants in Poland with the principles of the circular economy. They state that by emphasizing waste minimization through the reuse of apparently useless raw materials, the circular economy model promotes local systems, including supply chains, to maximize environmental benefits. In this context, this research focuses on biogas plants in Poland, to assess their adherence to the circular principles with a particular emphasis on locality (Lubańska & Kazak, 2023). The analysis employs questionnaire surveys categorizing biogas plants into three types: (1) those situated at sewage treatment plants, (2) those based on municipal waste, and (3) agricultural biogas plants. Statistical and spatial analyses of data on installed capacity, substrate distance, and biogas utilization were conducted to evaluate the extent of local operations. Results affirm that most biogas plants in Poland exhibit a predominantly local character, aligning with the goals of a circular economy. Particularly noteworthy are biogas plants located adjacent to sewage treatment plants,

TABLE 5.1 Overview of different case studies considered where biogas can be implemented in a circular economy scheme.

Country	Plant size	Feedstock used	Main findings	References
Poland	Various sizes	Sewage, municipal waste, agricultural waste	Biogas plants in Poland exhibit a predominantly local character, aligning with circular economy goals. Plants near sewage treatment plants have minimal transportation needs, showcasing potential for local sustainability and waste reduction.	Lubańska and Kazak (2023)
Switzerland	Different sizes of 61 biogas plants	Agricultural waste	Potential to increase biogas production tenfold by 2050, reducing mineral fertilizer use and GHG emissions. Economic value of digestates depends on type and mineral fertilizer prices.	Burg et al. (2023)
Lithuania	C. 1000 kW (8 different plants)	Agricultural waste and animal manure	Continuous application of digestate on low-fertility soil increases seminatural grass yield and methane production. Digestate as a valuable biofertilizer enhances biomass yields and biogas potential. Further research is needed on sustainability aspects of closed-loop digestate utilization.	Jurgutis et al. (2021)
Thailand	Three different scenarios	Cassava waste	Adoption of circular economy principles in cassava starch industry enhances demand for biogas systems, improves energy security, and addresses environmental concerns. Biogas systems utilizing wastewater and cassava pulp contribute to economic and environmental sustainability.	Lerdlattaporn et al. (2021)
Italy	—	Biomass	Biogas sector enhances resilience of energy systems, supports energy transition, and creates local business opportunities. Widespread adoption can reinforce energy security and decarbonization.	Sica et al. (2023)
Netherlands	5000, 20,000, and 100,000 t/year	Animal manure	Profitable cooperation for large-scale farms under specific conditions. Small-scale plants only profitable with low manure dry content and short transportation distances. Insights applicable for investors, suppliers, and transportation companies.	Yazan et al. (2018)

Greece	500 kW	Cow manure and additional feedstocks	Year-long monitoring shows stable digester performance with minimal pH control or trace metal supplementation. Slight variations in biogas and methane production. Successful example of growing biogas sector benefiting local farmers and livestock owners.	Spyridonidis et al. (2020)
India	—	Agricultural waste	Circular economy approach to agricultural waste management can produce biomethane, bio-CO_2, and digestate. Challenges include supply chain issues and lack of technical knowledge. Need for policy and regulatory support to enhance stakeholder knowledge and promote transition.	Kapoor et al. (2020)
Malaysia	Various	Organic fraction of municipal solid waste	Recovery of energy and biofertilizer through AD of OFMSW enhances environmental sustainability and economic development. Waste management shift needed toward sustainable practices.	Yong et al. (2021)
Finland	—	Biowaste and sewage sludge	Regional strategy for circular economy includes producing biogas and fertilizer from biowaste and sewage sludge. Challenges in financing and regulatory frameworks. Need for stakeholder collaboration and adaptable regulations.	Vanhamäki et al. (2020)

characterized by minimal transportation needs. These findings underscore the potential of biogas plants to exemplify circular economy principles, contributing to local sustainability and waste reduction objectives (Lubańska & Kazak, 2023).

Another study published by Burg et al. scrutinizes the current landscape of biomass utilization in agricultural AD plants in Switzerland, focusing on mass, nutrients, and energy flows. Emphasizing the pivotal role of biogas production and digestate as an organic fertilizer, the research evaluates their contribution to the circular economy and climate change mitigation. The key objective was to assess how these processes can substitute mineral fertilizers and fossil fuels, promoting a more circular approach within the agro-food system. The study highlights the potential for agricultural AD to amplify biogas production by 10-fold by 2050, concurrently delivering substantial savings in mineral fertilizer usage and GHG emissions. To enhance the commercialization of digestates, strategies such as improving marketing, pricing, and product quality, as well as expanding services, are essential. Mineral fertilizer prices play a significant role, making local organic fertilizers more appealing when mineral fertilizers become expensive or scarce. Economic values of digestates vary based on factors like type and mineral fertilizer prices. Fig. 5.2 shows the net theoretical economic value (TEV) calculated for the three typical agricultural digestates investigated in the work of Burg et al. based on the nutrient levels.

In essence, the study emphasizes the imperative for every country to fully harness its domestic resources, reinforcing the notion that underutilizing such resources is a luxury no nation can afford (Burg et al., 2023).

The research presented by Jurgutis et al. explores the economic value of digestates from agricultural biogas plants as biofertilizers and their impact on

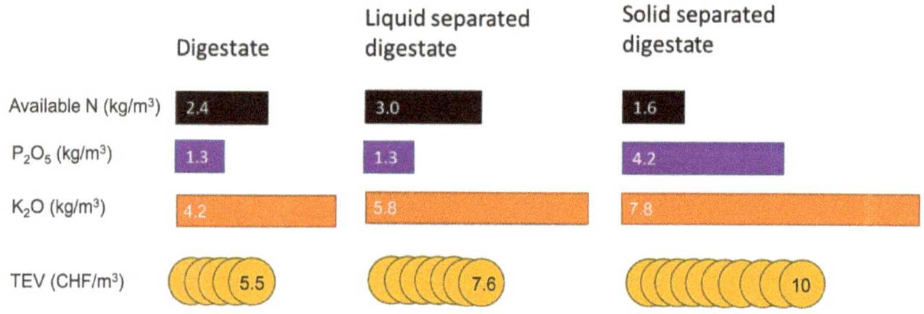

FIGURE 5.2 Net theoretical economic value calculated for three typical agricultural digestates based on nutrient levels.
Reproduced with minor modifications from Burg, V., Rolli, C., Schnorf, V., Scharfy, D., Anspach, V. & Bowman, G. (2023). Agricultural biogas plants as a hub to foster circular economy and bioenergy: An assessment using substance and energy flow analysis. Resources, Conservation and Recycling, 190. https://doi.org/10.1016/j.resconrec.2022.106770.

biomass yields and biogas potential when used on low-fertility soils. The study underscores the AD process as an exemplary model of a circular economy, promoting sustainable solutions for the utilization of organic biowaste through biogas production and the use of digestate as a valuable organic fertilizer. Moreover, digestate emerges as a potential income source for biogas power plants, leading to a more circular system. The research extends to the continuous application of digestate on low-fertility soil, demonstrating a notable increase in semi-natural grass yield. The subsequent biogas potential tests indicate that biomass cultivated on digestate-fertilized soil has the potential to generate significant methane yields. This study suggests that incorporating digestate into semi-natural grass biomass production areas near biogas plants could serve as an alternative strategy for diversifying the biogas plant feedstock portfolio. However, the authors emphasize the need for further research to comprehensively understand the sustainability aspects associated with the closed-loop digestate utilization concept (Jurgutis et al., 2021).

Another research presented by Lerdlattaporn et al. delves into the adoption of circular economy principles for utilizing wastes and by-products from the cassava starch industry to produce biogas, presenting a viable solution. This study aims to analyze the key drivers and challenges in implementing the circular economy concept in the cassava starch industry in Thailand, aiming to enhance demand for biogas systems, improve energy security, increase resource efficiency, and address environmental concerns associated with cassava wastes. They examined three different scenarios and the analysis covered economic feasibility, resource efficiency, water recovery, land use, and global warming potential. The results here obtained underscore the potential benefits of embracing circular economy principles in the cassava starch industry, specifically through the integration of biogas systems that utilize both wastewater and cassava pulp, contributing to economic, environmental, and resource sustainability (Lerdlattaporn et al., 2021).

In the work of Sica et al., a life cycle approach was followed to comprehensively analyze the socio-economic aspects and environmental implications of biogas production in Italy. Evaluating critical elements, opportunities, and limits within the biogas supply chain sustainability, the research explores the potential for a closed-loop approach, drawing insights from existing studies. The analysis identifies the strategic role of the biogas sector in enhancing the resilience of energy systems, particularly in the face of sudden changes. The study centers on the biogas sector's capacity to support Italy's energy transition, addressing structural vulnerabilities linked to fossil resource scarcity. This research offers policy, managerial, and social implications, emphasizing that the widespread adoption of biogas (a renewable, scalable, and flexible resource) can reinforce the security and decarbonization of the energy system. Additionally, it has the potential to create new local business opportunities, fostering a circular and bio-based economy. Overall, this study contributes to advancing the sustainable development of the biogas sector, illuminating pathways for its integration into resilient, low-carbon energy systems (Sica et al., 2023).

The study presented by Yazan et al. deals with the implementation of a circular economy using animal manure for biogas and fertilizers within a regional network. It assesses economic sustainability by examining key variables, that is, manure quantity, transportation distance, manure dry content, manure price, and discharge price. This research models both physical and monetary flows within the supply chain. Findings reveal profitable cooperation for large-scale farms under specific conditions. For instance, when the biogas producer pays the farmer to receive the manure up to 5 €/t or the farmer provides manure for free with disposal costs exceeding 10 €/t. Profitability for the biogas producer is consistent when the farmer pays for the supply of manure. Positive benefits are observed for medium-large-scale plants if the biogas producer receives manure for free. Small-scale plants are only profitable if manure dry content is lower than 12% and the transportation distance is lower than 10 km. This research offers a concise analysis of crucial variables, providing insights applicable to various contexts for investors, suppliers, and transportation companies (Yazan et al., 2018).

The work presented by Spyridonidis and colleagues describes a case study on the expansion of biogas plants in Greece, focusing on a year-long monitoring of a 500 kW mesophilic biogas plant using cow manure as the primary feedstock, assessing its performance in biogas and electrical energy production. The digester's stability allowed for minimal pH control or trace metal supplementation. Despite slight variations in organic loading rates (OLR) and biogas production rates (BPR), the specific BPR and biogas yield remained relatively consistent. However, the actual methane production was slightly lower than targeted due to the differences between laboratory tests and continuous operation. Electrical energy production correlated with the chemical oxygen demand (COD) introduced, with fluctuations attributed to maintenance and substrate composition changes affecting energy yields. This can be observed in Fig. 5.3.

The feeding mixture, primarily cow manure, included seven additional feedstocks based on availability, cost, and organic content. This case study exemplifies the success of the growing biogas sector in Greece, benefiting local farmers and livestock owners seeking environmentally friendly waste disposal methods. The sector shows promising economic potential but still has room for further growth (Spyridonidis et al., 2020).

In the work of Kapoor et al., the circular economy approach to agro-waste holds the potential for a lasting solution to agricultural waste management in India. Through AD, not only can biomethane be produced as a source of energy, but the by-products like bio-CO_2 and digestate also present opportunities for transformation into high-value products, facilitating the transition to a bio-circular economy.

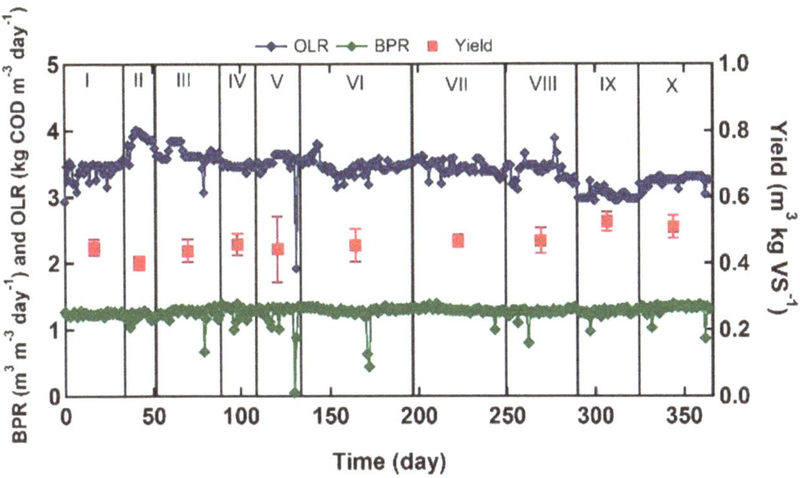

(A)

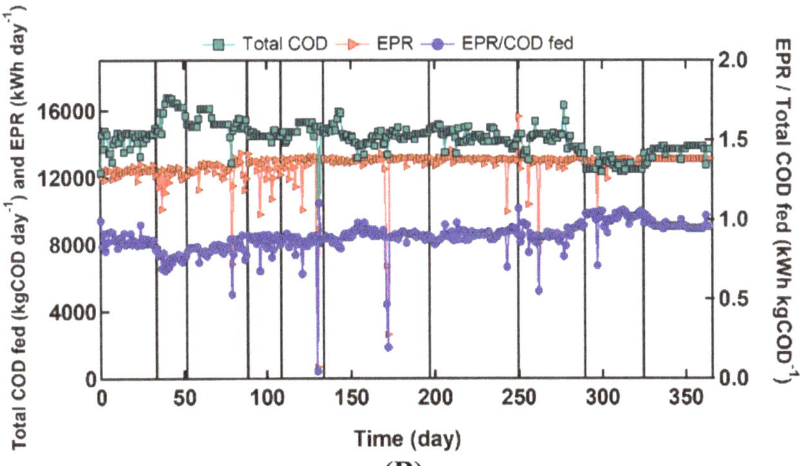

(B)

FIGURE 5.3 (A) The variation in organic loading rate, biogas production rate, and subsequent yield over time, and (B) the total chemical oxygen demand (COD) loading, electrical energy production rate, and electrical energy yield per COD fed throughout one year of operation. *From Spyridonidis, A., Vasiliadou, I.A., Akratos, C.S. & Stamatelatou, K. (2020). Performance of a full-scale biogas plant operation in Greece and its impact on the circular economy. Water (Switzerland), 12(11), 1–19. https://doi.org/10.3390/w12113074.*

However, challenges related to agro-waste supply chains and a lack of technical knowledge are hindering the growth of a circular economy based on agro-waste to biogas. Therefore there is a pressing need for knowledge enhancement among stakeholders and the development and implementation of policy and regulatory frameworks to support this transition (Kapoor et al., 2020).

In the work of Yong et al., the recovery of resources from the organic fraction of municipal solid waste (OFMSW) in Malaysia through AD holds substantial potential for enhancing environmental sustainability, expanding renewable energy sources, and fostering positive economic development. Waste management in Malaysia poses a persistent economic and environmental challenge, with over 80% of solid waste being disposed of in landfills and dumpsites. To address this issue, there is a pressing need for a shift toward sustainable solid waste management and resource recovery OFMSW. This study explores the feasibility of energy and biofertilizer recovery through AD of OFMSW, conducting economic and environmental benefit analyses. The study demonstrates that economic development can align with environmentally sound practices in waste management (Yong et al., 2021).

In Finland, the transition to a circular economy is actively pursued through a regional strategy. The strategy, developed through a stakeholder-engaged road map process, focuses on five key goals, including closing technical and biological loops, promoting sustainable energy technologies, fostering new consumption models, and establishing demonstration sites. This approach supports the systemic change required for a circular economy, as illustrated by a case study of a local bio-based industrial symbiosis producing biogas and fertilizer from biowaste and sewage sludge. Challenges in financing and regulatory frameworks are acknowledged, emphasizing the need for stakeholder collaboration and adaptable regulations to facilitate the transition to a circular economy society. The regional strategy and industrial symbiosis case serve as transferable good practices at the European level (Vanhamäki et al., 2020).

5.4.2 From biogas to valuable products in circular economy frameworks: the power of biogas upgrading plus CO_2 utilization

Biogas can be used within a circular economy framework not only as biofuel but also as feedstock for other valuable products such as other chemicals and fuels valorizing the CO_2 present in the biogas. Below, some examples can be found in the literature.

A recent study has explored an innovative approach to synthesize bio-methanol through the integration of biogas upgrading with CO_2 conversion, deviating from the traditional biogas-to-biomethane route. This process utilizes a $Cu/ZnO/Al_2O_3$ commercial catalyst to facilitate the reaction, in which CO_2 and H_2 are converted into CH_3OH and H_2O ($CO_2 + 3H_2 \rightarrow CH_3OH + H_2O$) at a temperature of 210°C and a pressure of 78 bars. This route is schematized in the process scheme involving biogas upgrading and bio-methanol production from CO_2, as shown in Fig. 5.4.

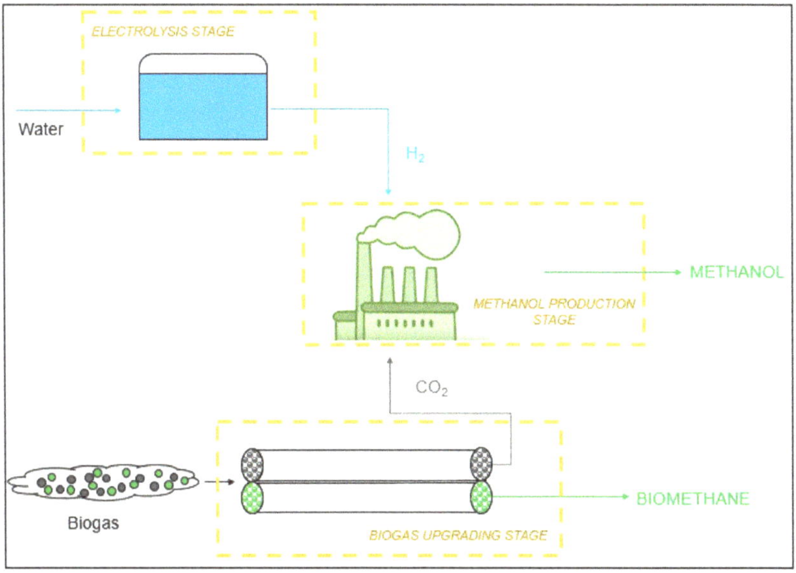

FIGURE 5.4 Process scheme of the biogas-to-methanol route proposed.
From Baena-Moreno, F.M., Pastor-Pérez, L., Wang, Q. & Reina, T.R. (2020). Bio-methane and bio-methanol co-production from biogas: A profitability analysis to explore new sustainable chemical processes. Journal of Cleaner Production, 265. https://doi.org/10.1016/j.jclepro.2020.121909.

Through a techno-economic analysis of 15 scenarios, including various biogas plant sizes, authors evaluate the profitability of this project for potential investors. Findings indicated that without subsidies, the process may lack economic viability. However, feed-in tariff subsidies render some biogas plants highly profitable. Sensitivity analysis highlights the critical role of electricity, natural gas, and bio-methanol prices in the overall profitability, emphasizing the dependence on subsidies for larger biogas production plants. They provide key insights into the complex dynamics shaping the feasibility and profitability of the proposed biogas upgrading and CO_2 conversion to bio-methanol process (Baena-Moreno et al., 2020).

Similarly, emphasizing the use of biomass within the circular economy principles, a recent work introduces an innovative approach of biogas utilization. These authors propose synergizing biogas upgrading to biomethane and formic acid production from CO_2. To evaluate its economic viability, a profitability analysis, focusing on net present value (NPV) and profitability index, was conducted. Despite the known environmental benefits, the process yields negative profitability. Reversing this requires formic acid prices outpacing market values to achieve a NPV of zero. Alternatively, biomethane subsidies are needed for reaching profitability, with consumables costs, notably catalytic conversion of the CO_2-rich feedstock, identified as a major factor. To improve the outcomes of this proposal, around 80% reduction in catalyst costs is required, suggesting the need for economically appealing catalysts. Fig. 5.5 shows the NPV results evolution with consumables catalysts costs reduction for the different plant sizes evaluated.

This study also explores biomethane subsidies as a percentage of investment, revealing encouraging outcomes for small-scale plants. This underscores the complex economic landscape and the necessity for further research to enhance the economic viability for this kind of innovative processes (Baena-Moreno et al., 2020).

The same research team investigated the technical and economic viability of another innovative approach for co-producing biomethane and urea from biogas, offering an alternative avenue to enhance the profitability of biogas upgrading plants. They assessed four different biogas plant sizes across four European countries (Spain, Italy, United Kingdom, and Germany), analyzing profitability within the current policy frameworks for biomethane production in each of those countries. The findings revealed that, under existing policy schemes, only medium- and large-scale plants in Italy demonstrate

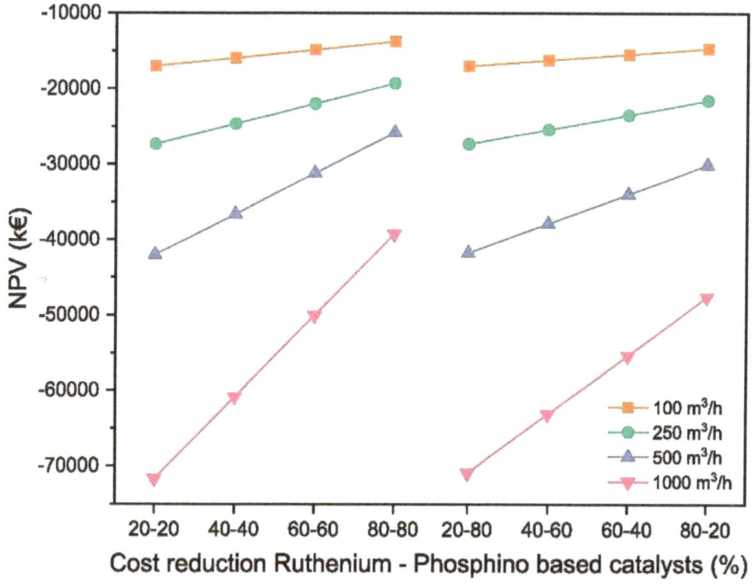

FIGURE 5.5 NPV variation with the catalyst cost reduction for the different plant sizes. *From Baena-Moreno, F.M., Pastor-Pérez, L., Zhang, Z. & Reina, T.R. (2020). Stepping towards a low-carbon economy. Formic acid from biogas as case of study. Applied Energy, 268. https://doi.org/10.1016/j.apenergy.2020.115033.*

profitability, attributed to the robust support for biomethane production through feed-in tariffs subsidies. This can be observed in Fig. 5.6 where the NPV values for the different countries studied are shown.

Exploring the potential impact of governmental incentives, particularly biomethane subsidies in the form of feed-in tariffs and investment percentage, the study unveiled feed-in tariffs as a viable solution for large plants, achieving profitability. Notably, southern EU countries exhibit a higher likelihood of achieving profitability with lower subsidies. The study also delved into the potential benefits of cost reduction, particularly ammonia price reduction, emphasizing its significance in minimizing the subsidies required. In summary, this study underscored the challenges that European policies face in advancing toward a bio-based economy, using biogas upgrading as a reference case, and emphasized the critical role of incentives and cost optimization in ensuring economic viability and sustainability in this transition (Baena-Moreno et al., 2020).

Other than valuable chemicals and fuels, biogas can also be used as feedstock for producing carbonates. These carbonates are valuable products

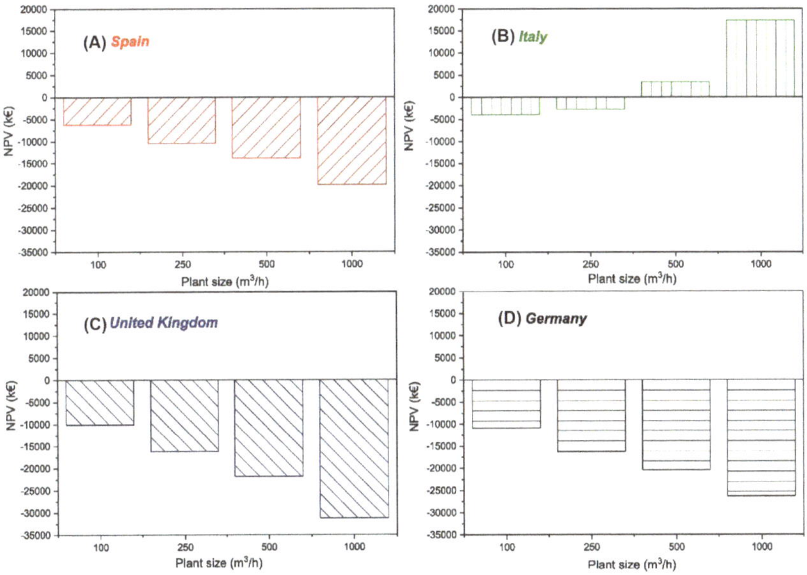

FIGURE 5.6 NPV results for the baseline scenario in different countries. (A) Spain, (B) Italy, (C) United Kingdom, and (D) Germany

From Baena-Moreno, F.M., Sebastia-Saez, D., Wang, Q. & Reina, T.R. (2020). Is the production of biofuels and bio-chemicals always profitable? Co-production of biomethane and urea from biogas as case study. Energy Conversion and Management, 220. https://doi.org/10.1016/j.enconman.2020.113058.

due to their wide range of industrial, agricultural, environmental, and pharmaceutical applications.

A novel study delved into a groundbreaking strategy for biomethane production and the valorization of magnesium chloride waste. Offering a potential alternative to existing biogas upgrading technologies, the innovative process involves the mineralization of the CO_2 removed from the upgraded biogas into valuable magnesium carbonate. Overall, the results affirm the technical feasibility of this innovative strategy, showcasing the synergy between CO_2 mineralization and renewable energy production (Baena-Moreno et al., 2019a).

Another similar work presents a novel strategy for biogas upgrading utilizing a continuous flow absorption unit with $CaCl_2$ as the capturing agent. The process scheme for this strategy is shown in Fig. 5.7.

Positioned as an alternative to conventional physical regeneration processes for capturing CO_2 in biogas effluents, this process aims to circumvent inherent high energy penalties. The study systematically investigates key parameters influencing calcium carbonate precipitation efficiency. The results obtained in this work underscore the technical feasibility of this alternative biogas upgrading process through $CaCO_3$ production, presenting a promising avenue for sustainable and energy-efficient CO_2 capture from biogas streams (Baena-Moreno et al., 2019b).

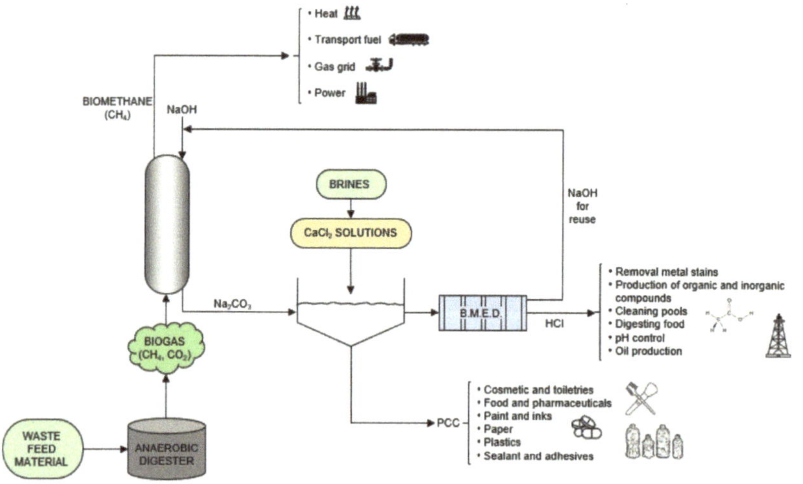

FIGURE 5.7 Suggested approach for upgrading biogas using a continuos flow absorption unit. *From Baena-Moreno, F.M., Rodríguez-Galán, M., Vega, F., Reina, T.R., Vilches, L.F. & Navarrete, B. (2019). Synergizing carbon capture storage and utilization in a biogas upgrading lab-scale plant based on calcium chloride: Influence of precipitation parameters. Science of the Total Environment, 670, 59–66. https://doi.org/10.1016/j.scitotenv.2019.03.204.*

5.5 Innovations and solutions

Industries aiming at reaching carbon neutrality by using sustainable energy sources such as biogas need to address different innovations and solutions. Three essential areas of advancement are technological innovations in biogas production (Nevzorova & Karakaya, 2020), policy innovations promoting carbon neutrality (Dahal et al., 2018), and the incorporation of the circular economy philosophy in biogas projects (Jain et al., 2022).

Current advances in technological innovations have considerably boosted the efficiency and viability of biogas production. Innovative reactor designs, novel monitoring systems, and improved microbial cultures are transforming the AD process, enhancing biogas yields (Tabatabaei et al., 2020a, 2020b). In addition, advanced preprocessing technologies are addressing feedstock variability issues by guaranteeing a more stable and reliable input, hence increasing the overall potential of biogas projects. The continuous innovation of these technological frontiers is crucial for achieving the complete potential of biogas as a sustainable energy source (Tabatabaei et al., 2020a, 2020b).

Governments and regulatory organizations also play a vital role in promoting carbon neutrality within the biogas sector through innovative policy frameworks, as aforesaid. Supportive regulations and/or incentives for renewable energy adoption are establishing an encouraging environment for biogas projects (Chen et al., 2022; Gustafsson & Anderberg, 2022). Furthermore, proactive policies that promote research and development, along with continuous commitments to reaching carbon reduction goals, play a key role in attracting investments and pushing the industry toward a more sustainable and carbon-neutral future (Chen et al., 2022; Gustafsson & Anderberg, 2022).

A paradigm shift toward circular economy principles is restructuring the model of biogas projects (Awasthi et al., 2019). The concept of closing the loop by using organic waste as feedstock not only addresses environmental concerns but also enhances the overall sustainability of biogas production (Lijó et al., 2017). Integrating circular economy principles involves maximizing resource efficiency, reducing waste, and creating a reformative system that aligns with the values of a circular economy (Awasthi et al., 2019). This general approach not only mitigates environmental impacts but also impacts to the economic and social magnitude of sustainability.

Briefly, the biogas industry is observing transformative advances through technological, policy, and circular economy fronts. These developments highlight a combined commitment to overcome challenges and pave the way for a more sustainable and carbon-neutral energy future. As these advancements progress, they have the capacity to reshape the role of biogas in our global energy portfolio.

5.6 Future outlook

The future of the biogas industry is very promising since it counts on emerging trends, innovative solutions, and a pivotal role in shaping carbon-neutral energy landscapes. Nonetheless, the current state of biogas production highlights several key points for consideration. Below, some of them are summarized.

The use of agricultural feedstocks is projected to continue. Sequential cropping is expected to be a potential solution to balance food and energy production (Garuti et al., 2022). In this sense, sustainable practices are crucial, requiring further development in determining suitable local practices, knowledge dissemination, and supervision to confirm a responsible use (Brémond et al., 2021). The rise of circular economy principles, where organic waste is utilized to its fullest potential, is not only addressing environmental concerns but also fostering sustainability and resource efficiency.

Continuous development and implementation of sustainability analyses, including life cycle assessment (LCA) and social LCA, are needed to support the ecological benefits of the biogas sector. Highlighting primary data from specific local contexts will boost the reliability of analyses and enable comparisons between different biogas systems (Ingrao et al., 2019).

As previously mentioned, the biogas and bioeconomy sectors are experiencing a dynamic shift linked to innovative trends in the field. Advanced technological solutions, such as improved AD and novel preprocessing techniques, are enhancing efficiency and scalability. In this line; investigating pretreatments and additives holds promise for enhancing biogas energy efficiency (Koch et al., 2020). Nonetheless, a thorough evaluation is required, considering limitations observed in batch tests and the potential ineffectiveness of some treatments over the long term (Janke et al., 2019). Semicontinuous pilot studies are advised to evaluate the most efficient solutions, considering factors such as feedstocks, pH, temperature, and gas flows (Fan et al., 2019). Additionally, the integration of digital technologies and data analytics is revolutionizing monitoring and control systems, further optimizing biogas production processes.

Existing and future biogas plants should study local opportunities to create additional revenue streams and territorial synergy. This involves considering CO_2 sinks and potential markets for nutrients/biomolecules valorization. Technology improvements and upscaling are vital for the success of these plans (Brémond et al., 2021). As countries and industries try to achieve carbon reduction targets, biogas stands out as a versatile and sustainable energy source. Its potential to provide renewable heat, power, and biofuels contributes significantly to decarbonizing various sectors.

European countries with a limited biogas sector require political incentives for development. This consists of understanding national and territorial specificities in terms of biomass and organic waste production, product demand, and existing infrastructure. The development of sustainable AD systems with local acceptance is crucial, and the "gas package" from the European Union may stimulate sector development in these countries (Bourdin & Nadou, 2020). Collaborative efforts between policymakers, industry stakeholders, and researchers are leading to the formulation of supportive policies that incentivize investment and streamline regulatory processes.

The future outlook for the biogas industry is marked by exciting trends, innovative solutions, and a pivotal role in advancing carbon-neutral energy landscapes. As technologies continue to evolve and collaborative efforts intensify, biogas can emerge as a keystone in the transition toward a more sustainable and resilient global energy system.

5.7 Conclusion

The global rise in GHG emissions due to industrialization and nonrenewable energy usage requires urgent action to address climate change and achieve carbon neutrality. In this chapter, we discuss the pivotal role of biogas production within the circular economy framework in mitigating GHG emissions and promoting sustainable development. Biogas, derived from organic waste through AD, offers a renewable energy source that aligns with principles of resource efficiency and waste reduction. Despite facing technical challenges such as feedstock variability and digestate management, biogas production presents significant potential for reducing GHG emissions globally. We explore various case studies highlighting innovative approaches to biogas production and upgrading, including the utilization of digestates as biofertilizers and the integration of biogas systems into industrial processes for additional valuable products. The importance of technological innovations, supportive policies, and the incorporation of circular economy principles in advancing biogas production toward carbon neutrality is also a topic covered. Collaboration between stakeholders is crucial to overcome economic and policy hurdles and fully harness the potential of biogas as a sustainable energy source. Future prospects for the biogas industry, driven by emerging trends, sustainable practices, and advancements in technology are also outlined. With continued efforts and supportive policies, biogas production has the potential to play a significant role in decarbonizing various sectors and shaping a resilient global energy system.

Acknowledgments

This work has been supported by the grants FJC2021–047672-I and RYC2023-043929-I co-financed by MCIN/AEI/10.13039/501100011033 and the European Union NextGenerationEU/PRTR funds, and the MICIU/AEI/10.13039/501100011033 and FSE+, respectively. This work has also been supported by the project PID2022–137299OB-I00 funded by MICIU/AEI/ 10.13039/501100011033/ FEDER, UE.

AI disclosure

During the preparation of this work, the author(s) used ChatGPT to proofread the document for grammar and spelling errors. After using this tool/service, the author(s) reviewed and edited the content as needed and take(s) full responsibility for the content of the publication.

References

Al-Wahaibi, A., Osman, A. I., Al-Muhtaseb, A. H., Alqaisi, O., Baawain, M., Fawzy, S., & Rooney, D. W. (2020). Techno-economic evaluation of biogas production from food waste via anaerobic digestion. *Scientific Reports, 10*(1). https://doi.org/10.1038/s41598-020-72897-5, https://www.nature.com/srep/index.html.

Ark Resources 2024 Climate Active Carbon Neutral - Ark Resources https://arkresources.com.au/climate-active-carbon-neutral/

Asadi, M., & McPhedran, K. (2021). Biogas maximization using data-driven modelling with uncertainty analysis and genetic algorithm for municipal wastewater anaerobic digestion. *Journal of Environmental Management, 293*, 112875. https://doi.org/10.1016/j.jenvman.2021.112875.

Atelge, M. R., Atabani, A. E., Banu, J. R., Krisa, D., Kaya, M., Eskicioglu, C., Kumar, G., Lee, C., Yildiz, Y., Unalan, S., Mohanasundaram, R., & Duman, F. (2020). A critical review of pretreatment technologies to enhance anaerobic digestion and energy recovery. *Fuel, 270*. https://doi.org/10.1016/j.fuel.2020.117494, http://www.journals.elsevier.com/fuel/.

Awasthi, M. K., Sarsaiya, S., Wainaina, S., Rajendran, K., Kumar, S., Quan, W., Duan, Y., Awasthi, S. K., Chen, H., Pandey, A., Zhang, Z., Jain, A., & Taherzadeh, M. J. (2019). A critical review of organic manure biorefinery models toward sustainable circular bioeconomy: Technological challenges, advancements, innovations, and future perspectives. *Renewable and Sustainable Energy Reviews, 111*, 115–131. https://doi.org/10.1016/j.rser.2019.05.017, https://www.journals.elsevier.com/renewable-and-sustainable-energy-reviews.

Bacenetti, J., Sala, C., Fusi, A., & Fiala, M. (2016). Agricultural anaerobic digestion plants: What LCA studies pointed out and what can be done to make them more environmentally sustainable. *Applied Energy, 179*, 669–686. https://doi.org/10.1016/j.apenergy.2016.07.029.

Baena-Moreno, F. M., Pastor-Pérez, L., Wang, Q., & Reina, T. R. (2020). Bio-methane and bio-methanol co-production from biogas: A profitability analysis to explore new sustainable chemical processes. *Journal of Cleaner Production, 265*. https://doi.org/10.1016/j.jclepro.2020.121909, https://www.journals.elsevier.com/journal-of-cleaner-production.

Baena-Moreno, F. M., Pastor-Pérez, L., Zhang, Z., & Reina, T. R. (2020). Stepping towards a low-carbon economy: Formic acid from biogas as case of study. *Applied Energy, 268*. https://doi.org/10.1016/j.apenergy.2020.115033, https://www.journals.elsevier.com/applied-energy.

Baena-Moreno, F. M., Rodríguez-Galán, M., Vega, F., Reina, T. R., Vilches, L. F., & Navarrete, B. (2019a). Converting CO_2 from biogas and $MgCl_2$ residues into valuable magnesium carbonate: A novel strategy for renewable energy production. *Energy, 180*, 457–464. https://doi.org/10.1016/j.energy.2019.05.106, https://www.elsevier.com/inca/publications/store/4/8/3/.

Baena-Moreno, F. M., Rodríguez-Galán, M., Vega, F., Reina, T. R., Vilches, L. F., & Navarrete, B. (2019b). Synergizing carbon capture storage and utilization in a biogas upgrading lab-scale plant based on calcium chloride: Influence of precipitation parameters. *Science of the Total Environment, 670*, 59–66. https://doi.org/10.1016/j.scitotenv.2019.03.204, www.elsevier.com/locate/scitotenv.

Baena-Moreno, F. M., Sebastia-Saez, D., Wang, Q., & Reina, T. R. (2020). Is the production of biofuels and bio-chemicals always profitable? Co-production of biomethane and urea from biogas as case study. *Energy Conversion and Management, 220*. https://doi.org/10.1016/j.enconman.2020.113058, https://www.journals.elsevier.com/energy-conversion-and-management.

Berndes, G., Abts, B., Asikainen, A., Cowie, A. L., Dale, V., Egnell, G., Lindner, M., Marelli, L., Paré, D., Pingoud, K., & Yeh, S. (2016). Forest biomass, carbon neutrality and climate change mitigation. *In From Science to Policy, 3*, 2016.

Bourdin, S., & Nadou, F. (2020). The role of a local authority as a stakeholder encouraging the development of biogas: A study on territorial intermediation. *Journal of Environmental Management, 258*, 110009. https://doi.org/10.1016/j.jenvman.2019.110009.

Brémond, U., Bertrandias, A., Steyer, J. P., Bernet, N., & Carrere, H. (2021). A vision of European biogas sector development towards 2030: Trends and challenges. *Journal of Cleaner Production, 287*. https://doi.org/10.1016/j.jclepro.2020.125065, https://www.journals.elsevier.com/journal-of-cleaner-production.

Burg, V., Rolli, C., Schnorf, V., Scharfy, D., Anspach, V., & Bowman, G. (2023). Agricultural biogas plants as a hub to foster circular economy and bioenergy: An assessment using substance and energy flow analysis. *Resources, Conservation and Recycling, 190*, 106770. https://doi.org/10.1016/j.resconrec.2022.106770.

Capodaglio, A. G., Callegari, A., & Lopez, M. V. (2016). European framework for the diffusion of biogas uses: Emerging technologies, acceptance, incentive strategies, and institutional-regulatory support. *Sustainability (Switzerland), 8*(4). https://doi.org/10.3390/su8040298, http://www.mdpi.com/2071-1050/8/4/298/pdf.

Chen, J. M. (2021). Carbon neutrality: Toward a sustainable future. *Innovation, 2*(3). https://doi.org/10.1016/j.xinn.2021.100127, https://www.cell.com/the-innovation.

Chen, L., Msigwa, G., Yang, M., Osman, A. I., Fawzy, S., Rooney, D. W., & Yap, P. S. (2022). Strategies to achieve a carbon neutral society: A review. *Environmental Chemistry Letters, 20*(4), 2277–2310. https://doi.org/10.1007/s10311-022-01435-8, http://springerlink.metapress.com/app/home/journal.asp?wasp=d86tgdwvtg0yvw9gvkwp&referrer=parent&backto=browsepublicationsresults,140,541.

Dahal, K., Juhola, S., & Niemelä, J. (2018). The role of renewable energy policies for carbon neutrality in Helsinki Metropolitan area. *Sustainable Cities and Society, 40*, 222–232. https://doi.org/10.1016/j.scs.2018.04.015.

Das, A., Das, S., Das, N., Pandey, P., Ingti, B., Panchenko, V., Bolshev, V., Kovalev, A., & Pandey, P. (2023). Advancements and innovations in harnessing microbial processes for enhanced biogas production from waste materials. *Agriculture (Switzerland), 13*(9). https://doi.org/10.3390/agriculture13091689, https://www.mdpi.com/journal/agriculture.

Fan, Y. V., Klemeš, J. J., Perry, S., & Lee, C. T. (2019). Anaerobic digestion of lignocellulosic waste: Environmental impact and economic assessment. *Journal of Environmental Management, 231*, 352–363. https://doi.org/10.1016/j.jenvman.2018.10.020, https://www.sciencedirect.com/journal/journal-of-environmental-management.

Farghali, M., Andriamanohiarisoamanana, F. J., Ahmed, M. M., Kotb, S., Yamashiro, T., Iwasaki, M., & Umetsu, K. (2019). Impacts of iron oxide and titanium dioxide nanoparticles on biogas production: Hydrogen sulfide mitigation, process stability, and prospective challenges. *Journal of Environmental Management, 240*, 160–167. https://doi.org/10.1016/j.jenvman.2019.03.089, https://www.sciencedirect.com/science/journal/03014797.

Farghali, M., Osman, A. I., Umetsu, K., & Rooney, D. W. (2022). Integration of biogas systems into a carbon zero and hydrogen economy: A review. *Environmental Chemistry Letters, 20*(5), 2853–2927. https://doi.org/10.1007/s10311-022-01468-z, https://www.springer.com/journal/10311.

Faster Capital 2024 Carbon Neutrality: Paving the Way for Green Shoots to Flourish - Faster Capital .https://fastercapital.com/content/Carbon-Neutrality–Paving-the-Way-for-Green-Shoots-to-Flourish.html

Feng, L., Aryal, N., Li, Y., Horn, S. J., & Ward, A. J. (2023). Developing a biogas centralised circular bioeconomy using agricultural residues – Challenges and opportunities. *Science of the Total Environment, 868*. https://doi.org/10.1016/j.scitotenv.2023.161656, http://www.elsevier.com/locate/scitotenv.

Garuti, M., Soldano, M., Mazzola, L., Fermoso, F. G., Rodriguez, A. J., Immovilli, A., & Dal Prà, A. (2022). Evaluation of triticale anaerobic digestion in a double cropping system: Relation between biomass yield, chemical composition, and biomethane production. *Biofuels, Bioproducts and Biorefining, 16*(6), 1599–1612. https://doi.org/10.1002/bbb.2393, http://onlinelibrary.wiley.com/journal/10.1002/(ISSN)1932-1031.

González-Arias, J., Gil, M. V., Fernández, R.Á., Martínez, E. J., Fernández, C., Papaharalabos, G., & Gómez, X. (2020). Integrating anaerobic digestion and pyrolysis for treating digestates derived from sewage sludge and fat wastes. *Environmental Science and Pollution Research, 27*(26), 32603–32614. https://doi.org/10.1007/s11356-020-09461-1, https://link.springer.com/journal/11356.

González-Arias, J., Torres-Sempere, G., Arroyo-Torralvo, F., Reina, T. R., & Odriozola, J. A. (2024). Optimizing biogas methanation over nickel supported on ceria-alumina catalyst: Towards CO_2-rich biomass utilization for a negative emissions society. *Environmental Research, 242*. https://doi.org/10.1016/j.envres.2023.117735, https://doi.org/10.1016/j.envres.2023.117735.

Gustafsson, M., & Anderberg, S. (2022). Biogas policies and production development in Europe: A comparative analysis of eight countries. *Biofuels, 13*(8), 931–944. https://doi.org/10.1080/17597269.2022.2034380.

IEA Bioenergy 2020 New – IEA Bioenergy Task 37 updated list of biogas upgrading plants | Bioenergy https://www.ieabioenergy.com/blog/publications/new-iea-bioenergy-task-37-updated-list-of-biogas-upgrading-plants/

Ingrao, C., Bacenetti, J., Adamczyk, J., Ferrante, V., Messineo, A., & Huisingh, D. (2019). Investigating energy and environmental issues of agro-biogas derived energy systems: A comprehensive review of life cycle assessments. *Renewable Energy, 136*, 296–307. https://doi.org/10.1016/j.renene.2019.01.023.

IRENA, Global Potential of Biogas Sponsors Global-Fossil-Fuel-Subsidies-Remain-Large-An-Update-Based-on-Country-Level-Estimates-46509. (2019), 2019.

Jain, A., Sarsaiya, S., Kumar Awasthi, M., Singh, R., Rajput, R., Mishra, U. C., Chen, J., & Shi, J. (2022). Bioenergy and bio-products from bio-waste and its associated modern circular economy: Current research trends, challenges, and future outlooks. *Fuel, 307*. https://doi.org/10.1016/j.fuel.2021.121859, http://www.journals.elsevier.com/fuel/.

Janke, L., Weinrich, S., Leite, A. F., Sträuber, H., Nikolausz, M., Nelles, M., & Stinner, W. (2019). Pre-treatment of filter cake for anaerobic digestion in sugarcane biorefineries: Assessment of batch versus semi-continuous experiments. *Renewable Energy, 143*,

1416–1426. https://doi.org/10.1016/j.renene.2019.05.029, http://www.journals.elsevier.com/renewable-and-sustainable-energy-reviews/.

Jurgutis, L., Šlepetienė, A., Šlepetys, J., & Cesevičienė, J. (2021). Towards a full circular economy in biogas plants: sustainable management of digestate for growing biomass feedstocks and use as biofertilizer. *Energies, 14*(14), 4272. https://doi.org/10.3390/en14144272.

Kapoor, R., Ghosh, P., Kumar, M., Sengupta, S., Gupta, A., Kumar, S. S., Vijay, V., Kumar, V., Kumar Vijay, V., & Pant, D. (2020). Valorization of agricultural waste for biogas based circular economy in India: A research outlook. *Bioresource Technology, 304*. https://doi.org/10.1016/j.biortech.2020.123036, https://www.elsevier.com/locate/biortech.

Khan, M. U., Lee, J. T. E., Bashir, M. A., Dissanayake, P. D., Ok, Y. S., Tong, Y. W., Shariati, M. A., Wu, S., & Ahring, B. K. (2021). Current status of biogas upgrading for direct biomethane use: A review. *Renewable and Sustainable Energy Reviews, 149*. https://doi.org/10.1016/j.rser.2021.111343, https://www.journals.elsevier.com/renewable-and-sustainable-energy-reviews.

Koch, K., Hafner, S. D., Weinrich, S., Astals, S., & Holliger, C. (2020). Power and limitations of biochemical methane potential (BMP) Tests. *Frontiers in Energy Research, 8*. https://doi.org/10.3389/fenrg.2020.00063, https://www.frontiersin.org/Energy_Research/.

Kulichkova, G. I., Ivanova, T. S., Köttner, M., Volodko, O. I., Spivak, S. I., Tsygankov, S. P., & Blume, Y. B. (2020). Plant feedstocks and their biogas production potentials. *Open Agriculture Journal, 14*(1), 219–234. https://doi.org/10.2174/1874331502014010219, https://benthamopen.com/contents/pdf/TOASJ/TOASJ-14-219.pdf.

Lerdlattaporn, R., Phalakornkule, C., Trakulvichean, S., & Songkasiri, W. (2021). Implementing circular economy concept by converting cassava pulp and wastewater to biogas for sustainable production in starch industry. *Sustainable Environment Research, 31*(1). https://doi.org/10.1186/s42834-021-00093-9.

Li, L., & Chen, Y. (2023). Intellectual property's role in achieving carbon neutrality through resource efficiency. *Resources Policy, 86*. https://doi.org/10.1016/j.resourpol.2023.104215 104215.

Lijó, L., Lorenzo-Toja, Y., González-García, S., Bacenetti, J., Negri, M., & Moreira, M. T. (2017). Eco-efficiency assessment of farm-scaled biogas plants. *Bioresource Technology, 237*, 146–155. https://doi.org/10.1016/j.biortech.2017.01.055, https://www.elsevier.com/locate/biortech.

Lijó, L., González-García, S., Bacenetti, J., & Moreira, M. T. (2017). The environmental effect of substituting energy crops for food waste as feedstock for biogas production. *Energy, 137*, 1130–1143. https://doi.org/10.1016/j.energy.2017.04.137, https://www.elsevier.com/inca/publications/store/4/8/3/.

Lubańska, A., & Kazak, J. K. (2023). The role of biogas production in circular economy approach from the perspective of locality. *Energies, 16*(9). https://doi.org/10.3390/en16093801, https://www.mdpi.com/journal/energies/.

Mateescu, C., & Dima, A. D. (2022). Critical analysis of key barriers and challenges to the growth of the biogas sector: A case study for Romania. *Biomass Conversion and Biorefinery, 12*(12), 5989–6002. https://doi.org/10.1007/s13399-020-01054-9, https://www.springer.com/journal/13399.

Monfet, E., Aubry, G., & Ramirez, A. A. (2018). Nutrient removal and recovery from digestate: a review of the technology. *Biofuels, 9*(2), 247–262. https://doi.org/10.1080/17597269.2017.1336348, http://www.tandfonline.com/loi/tbfu20?open=5&repitition=0#vol_5.

Monlau, F., Sambusiti, C., Ficara, E., Aboulkas, A., Barakat, A., & Carrère, H. (2015). New opportunities for agricultural digestate valorization: Current situation and perspectives. *Energy & Environmental Science, 8*(9), 2600–2621. https://doi.org/10.1039/c5ee01633a.

NASA 2024 Carbon Dioxide I Vital Signs – Climate Change: Vital Signs of the Planet .https://climate.nasa.gov/vital-signs/carbon-dioxide/.

Nevzorova, T., & Karakaya, E. (2020). Explaining the drivers of technological innovation systems: The case of biogas technologies in mature markets. *Journal of Cleaner Production, 259.* https://doi.org/10.1016/j.jclepro.2020.120819 120819.

Obaideen, K., Abdelkareem, M. A., Wilberforce, T., Elsaid, K., Sayed, E. T., Maghrabie, H. M., & Olabi, A. G. (2022). Biogas role in achievement of the sustainable development goals: Evaluation, challenges, and guidelines. *Journal of the Taiwan Institute of Chemical Engineers, 131.* https://doi.org/10.1016/j.jtice.2022.104207, https://www.elsevier.com/wps/find/journaldescription.cws_home/715607/description#description.

Ottosson, M., Magnusson, T., & Andersson, H. (2020). Shaping sustainable markets—A conceptual framework illustrated by the case of biogas in Sweden. *Environmental Innovation and Societal Transitions, 36*, 303–320. https://doi.org/10.1016/j.eist.2019.10.008.

Rabaey, K., & Ragauskas, A. J. (2014). Editorial overview: Energy biotechnology. *Current Opinion in Biotechnology, 27*, 5–6. https://doi.org/10.1016/j.copbio.2014.04.001, https://www.elsevier.com/locate/copbio.

Reynolds, J., Kennedy, R., Ichapka, M., Agarwal, A., Oke, A., Cox, E., Edwards, C., & Njuguna, J. (2022). An evaluation of feedstocks for sustainable energy and circular economy practices in a small island community. *Renewable and Sustainable Energy Reviews, 161.* https://doi.org/10.1016/j.rser.2022.112360 112360.

Rocha-Meneses, L., Zannerni, R., Inayat, A., Abdallah, M., Shanableh, A., Ghenai, C., Kamil, M., & Kikas, T. (2022). Current progress in anaerobic digestion reactors and parameters optimization. *Biomass Conversion and Biorefinery.* https://doi.org/10.1007/s13399-021-02224-z.

Salvia, M., Reckien, D., Pietrapertosa, F., Eckersley, P., Spyridaki, N. A., Krook-Riekkola, A., Olazabal, M., De Gregorio Hurtado, S., Simoes, S. G., Geneletti, D., Viguié, V., Fokaides, P. A., Ioannou, B. I., Flamos, A., Csete, M. S., Buzasi, A., Orru, H., de Boer, C., Foley, A., ... Heidrich, O. (2021). Will climate mitigation ambitions lead to carbon neutrality? An analysis of the local-level plans of 327 cities in the EU. *Renewable and Sustainable Energy Reviews, 135.* https://doi.org/10.1016/j.rser.2020.110253, https://www.journals.elsevier.com/renewable-and-sustainable-energy-reviews.

Shah, S. V., Lamba, B. Y., Sharma, R., Tiwari, A. K., Parveen, A., & Kumar, V. (2023). Effect of different light on microalgae growth and its integration with carbon dioxide sequestration technologies with emphasis on biofixation by microalgae. *Materials Today: Proceedings, 73*, 92–99. https://doi.org/10.1016/j.matpr.2022.09.370, https://www.sciencedirect.com/journal/materials-today-proceedings.

Sica, D., Esposito, B., Supino, S., Malandrino, O., & Sessa, M. R. (2023). Biogas-based systems: An opportunity towards a post-fossil and circular economy perspective in Italy. *Energy Policy, 182.* https://doi.org/10.1016/j.enpol.2023.113719, https://www.journals.elsevier.com/energy-policy/.

Singh, A. K., Pal, P., Rathore, S. S., Sahoo, U. K., Sarangi, P. K., Prus, P., & Dziekański, P. (2023). Sustainable utilization of biowaste resources for biogas production to meet rural bioenergy requirements. *Energies, 16*(14). https://doi.org/10.3390/en16145409, http://www.mdpi.com/journal/energies/.

Spyridonidis, A., Vasiliadou, I. A., Akratos, C. S., & Stamatelatou, K. (2020). Performance of a full-scale biogas plant operation in Greece and its impact on the circular economy. *Water (Switzerland), 12*(11), 1–19. https://doi.org/10.3390/w12113074, http://www.mdpi.com/journal/water.

Stegmann, P., Londo, M., & Junginger, M. (2020). The circular bioeconomy: Its elements and role in European bioeconomy clusters. *Resources, Conservation & Recycling: X, 6*, 100029. https://doi.org/10.1016/j.rcrx.2019.100029.

Tabatabaei, M., Aghbashlo, M., Valijanian, E., Kazemi Shariat Panahi, H., Nizami, A. S., Ghanavati, H., Sulaiman, A., Mirmohamadsadeghi, S., & Karimi, K. (2020a). A comprehensive review on recent biological innovations to improve biogas production, Part 1: Upstream strategies. *Renewable Energy, 146*, 1204–1220. https://doi.org/10.1016/j.renene.2019.07.037, http://www.journals.elsevier.com/renewable-and-sustainable-energy-reviews.

Tabatabaei, M., Aghbashlo, M., Valijanian, E., Panahi, H. K. S., Nizami, A. S., Ghanavati, H., Sulaiman, A., Mirmohamadsadeghi, S., & Karimi, K. (2020b). A comprehensive review on recent biological innovations to improve biogas production, Part 2: Mainstream and downstream strategies. *Renewable Energy, 146*, 1392–1407. https://doi.org/10.1016/j.renene.2019.07.047, http://www.journals.elsevier.com/renewable-and-sustainable-energy-reviews/.

United Nations 2015 United Nations ADOPTION OF THE PARIS AGREEMENT - Paris Agreement text English.

Vanhamäki, S., Virtanen, M., Luste, S., & Manskinen, K. (2020). Transition towards a circular economy at a regional level: A case study on closing biological loops. *Resources, Conservation and Recycling, 156*, 104716. https://doi.org/10.1016/j.resconrec.2020.104716.

Wang, F., Harindintwali, J. D., Yuan, Z., Wang, M., Wang, F., Li, S., Yin, Z., Huang, L., Fu, Y., Li, L., Chang, S. X., Zhang, L., Rinklebe, J., Yuan, Z., Zhu, Q., Xiang, L., Tsang, D. C. W., Xu, L., Jiang, X., ... Chen, J. M. (2021). Technologies and perspectives for achieving carbon neutrality. *Innovation. 2*(4). https://doi.org/10.1016/j.xinn.2021.100180, https://www.cell.com/the-innovation.

Wang, J. (2014). Decentralized biogas technology of anaerobic digestion and farm ecosystem: Opportunities and challenges. *Frontiers in Energy Research, 2*. https://doi.org/10.3389/fenrg.2014.00010.

Wei, Y. M., Chen, K., Kang, J. N., Chen, W., Wang, X. Y., & Zhang, X. (2022). Policy and management of carbon peaking and carbon neutrality: A literature review. *Engineering, 14*, 52–63. https://doi.org/10.1016/j.eng.2021.12.018, https://www.journals.elsevier.com/engineering/.

World Biogas Association, Biogas: Pathways to 2030 – Executive Summary. World Biogas Association. (2021), 2021.

Yang, M., Chen, L., Msigwa, G., Tang, K. H. D., & Yap, P. S. (2022). Implications of COVID-19 on global environmental pollution and carbon emissions with strategies for sustainability in the COVID-19 era. *Science of the Total Environment, 809*. https://doi.org/10.1016/j.scitotenv.2021.151657, www.elsevier.com/locate/scitotenv.

Yazan, D. M., Cafagna, D., Fraccascia, L., Mes, M., Pontrandolfo, P., & Zijm, H. (2018). Economic sustainability of biogas production from animal manure: A regional circular economy model. *Management Research Review, 41*(5), 605–624. https://doi.org/10.1108/MRR-02-2018-0053, http://www.emeraldinsight.com/Insight/viewContainer.do;jsessionid=746E76A1B81993BFFED6D07E37E888A6?containerType=Journal&containerId=15002370.

Yong, Z. J., Bashir, M. J. K., & Hassan, M. S. (2021). Biogas and biofertilizer production from organic fraction municipal solid waste for sustainable circular economy and environmental protection in Malaysia. *Science of the Total Environment, 776*. https://doi.org/10.1016/j.scitotenv.2021.145961, https://www.elsevier.com/locate/scitotenv.

Yousuf, A., Khan, M. R., Pirozzi, D., & Ab Wahid, Z. (2016). Financial sustainability of biogas technology: Barriers, opportunities, and solutions. *Energy Sources, Part B: Economics, Planning and Policy, 11*(9), 841–848. https://doi.org/10.1080/15567249.2016.1148084, http://www.tandf.co.uk/journals/titles/15567249.asp.

Yuhendra, A. P., Farghali, M., Mohamed, I. M. A., Iwasaki, M., Tangtaweewipat, S., Ihara, I., Sakai, R., & Umetsu, K. (2021). Potential of biogas production from the anaerobic digestion of Sargassum fulvellum macroalgae: Influences of mechanical, chemical, and biological pretreatments. *Biochemical Engineering Journal, 175*. https://doi.org/10.1016/j.bej.2021.108140, https://www.elsevier.com/locate/bej.

Zglobisz, N., Castillo-Castillo, A., Grimes, S., & Jones, P. (2010). Influence of UK energy policy on the deployment of anaerobic digestion. *Energy Policy, 38*(10), 5988–5999. https://doi.org/10.1016/j.enpol.2010.05.054.

Chapter 6

Valorization of digestate as a biofertilizer and its energy recovery using thermochemical conversion

Rafael Borja[1], D. de la Lama-Calvente[1], J.M. Mancilla-Leyton[2], and M.J. Fernández-Rodríguez[2]

[1]*Instituto de la Grasa (CSIC), Department of Food Biotechnology, Campus Universidad Pablo de Olavide, Sevilla, Spain,* [2]*Department of Vegetal Biology and Ecology, Faculty of Biology, University of Seville, Seville, Spain*

Abbreviation List

AD	Anaerobic digestion
AHL	Aqueous hydrothermal carbonization liquid
APC	Autogenous pressure carbonization
CMC	Component material category
COD	Chemical oxygen demand
Co-HTC	Co-pyrolisis/co-hydrothermal carbonization
DCN	Dolomite-supported g-C3N4
EAS	Electrochemical ammonia extraction
ED	Electrodialysis
EDt	Energy densification
EPA	Environmental Protection Agency
EU	European Union
EY	Energy yield
HC	Hydrocarbon
HHV	Higher heating value
HTC	Hydrothermal carbonization
MLHC	Microwave-assisted low-temperature hydrothermal carbonization
MLHT	Microwave-assisted low-temperature hydrothermal treatment
MSW	Municipal solid waste
PAH	Polyaromatic hydrocarbons
SBR	Sulfur-reducing bacteria
SCWG	Supercritical water gasification

Biogas in the Circular Economy. DOI: https://doi.org/10.1016/B978-0-443-29230-9.00008-5

SHS	Superheated steam
SS	Sewage sludge
TCs	Tetracyclines
USA	United States of America
VS	Volatile solids

6.1 Introduction

The increasing food demand of the global population is a constant challenge for the livestock and the agronomy sector. For instance, Zhou et al. (2024) predicted that the meat demand in China will significantly increase by the year 2030 which will translate into an increase in agricultural production. Furthermore, the diminishing supply of mineral fertilizers will be depleted faster than expected. This situation should therefore encourage an increase in arable land, improved agricultural efficiency, and the availability of fertilizers (Rizzioli et al., 2023), while taking into consideration the environmental impact of these measures and implementing the use of sustainable alternatives. In fact, Zhou et al. (2024) reported that by the 2030 year, the carbon emissions from meat, egg, and milk production and consumption in China will increase by a 40.1% when compared with the 2020 period.

Anaerobic digestion (AD) technology is a promising method, within a circular economy model, as it reduces the disposal of agro-industrial wastes by producing two valuable by-products: a methane-rich biogas and a stabilized digestate enriched with nutrient compounds (Chojnacka & Chojnacki, 2024). The potential of biogas production is widely assessed in the scientific literature and in other chapters of this book. However, studies on the applicability of anaerobic digestate are scarce, although they have increased significantly in recent years. Anaerobic digestates can be considered as a source of minerals, micro and macronutrients needed by crops (Carraturo et al., 2024). This application would reduce the use of fossil-based products by providing the main components of the mineral fertilizers, that is, nitrogen, phosphorus, and potassium (Bergstrand, 2022). In addition, the significant content of organic and mineral fractions in the anaerobic digestates suggests that this biomass could be a source of valuable chemicals and energy (Unyay et al., 2023).

This chapter reviews the current literature on the valorization of the anaerobic digestate as a fertilizer or as a secondary renewable energy source. The different technologies applied to the digestates to facilitate their use as biofertilizers due to legal or technical requirements are presented and discussed. Thermochemical conversion processes are also evaluated. Finally, future trends and research are recommended based on the current knowledge gaps identified during the review of the state of the art of anaerobic digestate valorization.

6.1.1 Main characteristics of anaerobic digestates

As discussed in detail in other chapters of this book, during the AD process, the easily degradable organic matter in the fed biomass is converted into methane-rich biogas. The remaining fraction, consisting mainly of recalcitrant materials (e.g., lignin and cellulose), together with other micro and macronutrients and the anaerobic microbial community, forms the semi-solid effluent known as the anaerobic digestate.

The characteristics and composition of anaerobic digestates vary widely due to several factors, the most influential of which are feedstock type and AD operating conditions (Sheets et al., 2015). Typical ranges of macronutrients, micronutrients, and heavy metals content found in anaerobic digestates by feedstock sources are shown in Table 6.1. The optimal pH for an efficient AD process has been established to be between 7 and 8.5 (Holliger et al., 2016). Thus, generally, digestates tend to be slightly alkaline, although there are some exceptions. For instance, this can vary significantly depending on the formation of ammonium (NH_4^+) or the accumulation of short-chain fatty acids during the AD performance (Rizzioli et al., 2023). In terms of organic matter content, digestates usually have a high C concentration (13%–50% of dry matter), but the usual macronutrients used to define the quality of a fertilizer (i.e., N, P, and K) can vary widely. For example, the digestate from the AD of food waste can have an N:P:K ratio ranging from 100:6:3 to 100:15:191 (Ries et al., 2023).

In any case, in addition to the availability of nutrients, the most important factor to consider is their actual form. For example, while digestate usually has a relatively high N content, the anoxic conditions of the AD process can reduce this to NH_4^+. In fact, this conversion typically represents 60%–80% of the total N, although it can reach 99% of the total N in NH_4^+ form (Ries et al., 2023), while it is generally accepted that a level below 40% could avoid ammonium toxicity to crops (Nelson, 2011). Phosphates are also an essential nutrient for plant growth, but these compounds can precipitate in the presence of organic matter in the form of insoluble salts, which can be exacerbated by alkaline pH and the presence of cations, such as Ca^{2+} or Mg^{2+} (Ries et al., 2023). Actually, P in the solid fraction can account for up to 90% of the total content (Christensen et al., 2009). S is another essential nutrient that enriches the fertilizers, but during the AD process, the sulfur-reducing bacteria produce the volatile hydrogen sulfide and provoke the precipitation of insoluble iron sulfides (Van den Brand et al., 2015). These typical drawbacks associated with anaerobic digestates and the various current techniques that could be applied to improve their valuable use are discussed in the following subsections.

TABLE 6.1 Overview of macro- and micronutrients and heavy metals in anaerobic digestates from different sources.

Parameter[a]	Animal manure	Food wastes	Agricultural wastes	Algae
pH	4.23–9.20	7–9.05	7.5–8.4	8.0–8.3
C (g kg^{-1})	159.4–506	128–488.6	300–400	344.7–390.9
N (g kg^{-1})	0–143.2	1.7–157	11.4–120	66–79.9
P (g kg^{-1})	0–13.1	2.2–49.1	2–66	0–7.49
K (g kg^{-1})	0–115.3	0.3–118.6	9–100	1.74–91.72
S (g kg^{-1})	4.1–7.2	0.05–27	2.9–14.7	1.01–1.09
Ca (g kg^{-1})	0–74.3	0.9–261	9–65.8	9–70.43
Mg (g kg^{-1})	0–11.8	0.04–56.2	4.1–24.6	0.3–8.44
Na (g kg^{-1})	0.69	0.43	0.68–24.6	1.7–11.43
Fe (g kg^{-1})	0–0.048	0.02–38	0.46–7.9	0.2
Cu (mg kg^{-1})	0–99	9.4–138	1–1373	1
Mn (mg kg^{-1})	0–124.6	33–481	240–1100	6
Zn (mg kg^{-1})	0–205.43	24.4–2465	43–3884	30
Cr (mg kg^{-1})	0–24.18	13–29	1.89–55	6
Ni (mg kg^{-1})	0–5.85	25–39	2–36	6
As (mg kg^{-1})	0–13	–	–	0.6
Cd (mg kg^{-1})	0–7.51	0.3–1	0.08–29	0.3
Hg (mg kg^{-1})	0–0.167	<0.1	<0.1	19
Pb (mg kg^{-1})	0–36.07	0.1–10	0–46	12
Co (mg kg^{-1})	2.23–3.57	0.06–2.8	0.4	1
Mo (mg kg^{-1})	2.5	0.11–15.2	9.1	0.7
Se (mg kg^{-1})	0–0.232	<0.07	<0.08	–
Source	Fu et al. (2024); Rao et al. (2024); Rittl et al. (2023); Shrestha et al. (2023); Wang, Sanusi, et al. (2023)	Golovko, et al. (2022); O'Connor et al. (2022); Sheets et al. (2015); Shrestha et al. (2023); Wang, Yu, et al. (2023)	Monlau et al. (2015); Fernández-Rodríguez et al. (2021); Sheets et al. (2015); Shrestha et al. (2023)	De la Lama-Calvente et al. (2023); Elalami et al. (2020)

6.1.2 The EU/USA legal framework for anaerobic digestates

Due to the widespread use of anaerobic digestate as a fertiliszer, the European Union (EU) and the United States of America (USA) government have regulated its use and definition. In Europe, Regulation (EU) 2019/1009 lays down the rules for making fertilizer products available on the market. With regard to anaerobic digestates, this regulation distinguished between two types, depending on the feedstock used for the AD system. Component material category (CMC) 4 is coded for fresh crop digestates and CMC 5 for digestates other than fresh crop digestates. There are no significant differences in terms of requirements; in fact, the technical criteria are almost the same in both cases, but the quality control system for CMC 5 digestates is stricter.

In both cases, there are certain requirements that apply directly to the AD process to minimize the potential presence of pathogens: AD should be carried out at thermophilic temperature (55°C) for 24 h and with a hydraulic retention time of 20 days, or it can be pasteurized or composted. However, if the AD is performed at mesophilic temperatures (37°C–40°C), the digestate must be either pasteurized or composted. The digestate itself (both the solid and the liquid fraction) shall not have an oxygen uptake rate higher than 25 mmol O_2 kg^{-1}VS h^{-1} or a residual biogas potential higher than 0.25 L biogas g^{-1} VS (where VS stands for volatile solids). CMC 5 digestates shall also not contain more than 6 mg kg^{-1}d.w. of the listed polyaromatic hydrocarbons (PAH) and not more than 5 g kg^{-1}d.w. of physical impurities of which 3 g kg^{-1}d.w. shall not exceed 2 mm.

In addition,to be considered an organic fertilizer, the anaerobic digestate may contain organic carbon and nutrients of exclusively biological origin and may have to comply with further limit values as shown in Table 6.2. The requirements for the minimum content in terms of N, P, or K would depend slightly on the nature of the digestate (solid or liquid), but in the case of a digestate that does not comply with any of these requirements, the regulation allows the digestate to be spiked with some inorganic fertilizer. In this case, however, the anaerobic digestate should be labeled as organo-mineral fertilizer.

In the United States, the regulatory framework also allows the direct use of anaerobic digestate. However, the restrictions are more relaxed than in the European regulation. For instance, the AD process can be run at 35°C–55°C if the average cell retention time is 15 d. It also allows lower temperatures (20°C), if the time is increased to 60 d. According to the US Environmental Protection Agency (EPA) regulation, these conditions should be sufficient to significantly reduce the potential presence of pathogens. Regarding the specific limit for certain elements, Table 6.2 shows those limits set for a single application, but the monthly, annual, and cumulative load is also regulated. This may limit the direct use of some digestates with high concentrations of

TABLE 6.2 EU and USA regulations on limits for toxic metals in fertilizers.

Element[a]	Europe[b]	USA[c]
Cr^{6+} (mg kg^{-1})	2	—
Ni (mg kg^{-1})	50	420
As (mg kg^{-1})	40 (as inorganic)	75
Cd (mg kg^{-1})	1.5	85
Hg (mg kg^{-1})	1	57
Pb (mg kg^{-1})	120	840
Co (mg kg^{-1})	—	4300
Mo (mg kg^{-1})	—	75
Se (mg kg^{-1})	—	100
Zn (mg kg^{-1})	—	7500

[a] *based on dry matter.*
[b] *Regulation (EU) 2019/1009.*
[c] *US EPA Regulation CFR40/503.13 Ceiling Concentrations (Version up to date as of 08/03/2024).*

these elements. However, as shown in Table 6.1, most of the digestates evaluated in the bibliography had concentrations significantly lower than those limited by the regulation. This supports the feasibility of AD as a biological process that facilitates not only the production of green energy but also the production of organic fertilizers.

6.2 Main technologies for the valorization of digestate as biofertilizer

6.2.1 Ammonia stripping

One of the methods of recovering nutrients present in the stabilized effluent from the AD process is known as stripping. As the digestate originates from an anaerobic atmosphere, NH_4^+ is the major nitrogen compound present. The concentration of ammonium/ammonia in the liquid digestate varies with the pH and temperature. The anaerobic process is easily inhibited by ammonium concentrations above 1700 mg L^{-1} (Yenigün & Demirel, 2013), which mainly affects the activity of methanogenic archaea. Therefore, NH_4^+ concentration in the liquid digestate typically ranges between 1 and 4 g N L^{-1} (Makian et al., 2024). The main approach to ammonium recovery is ammonia stripping, a process that separates nitrogen from liquid digestate using a gas stream within tower structures. This is achieved by controlling both pH and temperature, with pH being more crucial than temperature. A pH close to 10, high temperature, and subsequent extraction of ammonia using sulfuric acid result in the production of ammonium sulfate or other acids such as nitric acid to produce ammonium nitrate. The potential ammonia stripping efficiency can reach a theoretical maximum of 99%. However, for practical operational considerations, it is considered acceptable to achieve ammonia removal in the range of 80%–90% (Rizzioli et al., 2023). However, challenges arise due to its comparatively low nitrogen content compared with alternative fertilizers.

A recent research study on the energy utilization and efficiency of a full-scale ammonia stripping plant using a bubble reactor evaluated the stripping process in the liquid fraction of the digestate, analyzing the operating parameters of the plant both in its current state and in an optimized configuration (Abbà et al., 2023). Key factors influencing the process included temperature, pH, airflow, and feed characteristics. Notably, the pH increased to 10 during the experimental phase, eliminating the need for basifying agents. Elevated temperatures were found to enhance the stripping process, with the maximum temperature tested reaching 68°C. The airflow rate remained constant throughout the trials at 15 $Nm^3\ h^{-1}\ m^{-3}$ digestate in the predegassing reactors and 60 $Nm^3\ h^{-1}\ m^{-3}$ digestate in the degassing reactors throughout all trials. All energy requirements were met by the combined heat and power unit, which was fueled by biogas generated from the manure digestion. The results suggest that the combination of utilizing AD alongside stripping presents a feasible method for eliminating as much as 81% of NH_4^+, all achieved without necessitating external energy infusion or the inclusion of additional reagents (Abbà et al., 2023). Other systems have been also investigated; for instance, Aung et al. (2024) used a combination of electrodialysis (ED) and electrochemical ammonia extraction (EAS) to rapidly and efficiently recover ammonia from anaerobic digestate. To identify the most effective operational parameters for the novel method utilizing genuine anaerobic digestate, a sequence of experiments was undertaken, employing synthetic ammonia solutions. Key parameters were investigated for their influence on ammonia recovery. The results showed that cathodic feed had advantages over anodic feed by effectively converting ammonium to nitrate by raising the pH in the cathodic compartment. The increased current density provided sufficient OH^- ions to raise the pH in the cathodic compartment. The use of disodium phosphate (Na_2HPO_4) increased the pH due to its buffering capacity, resulting in an improved ammonia recovery. The experimental hybrid process for true anaerobic digestate concentrated ammonia to 3.775 g $NH_3\ L^{-1}$. With an ammonia recovery rate of up to 90.5%, the process consumed 11.6 kWh $(kg\text{-}NH_3)^{-1}$ (Aung et al., 2024).

6.2.2 Struvite precipitation

An alternative method of nutrient recovery involves adjusting the physicochemical conditions in the initial solution, such as pH and ionic strength, to promote the formation of salts with low solubility index. As a result, these salts precipitate, making it easier to recover them. A notable salt in this category is magnesium ammonium phosphate ($MgNH_4PO_4$–$6H_2O$) or struvite, which is widely used for the recovery of macro- and micronutrients, including P and N.

This process has a significant ability to remove P, achieving removal but failing short in the removal of NH_4^+-N, demonstrating only 10%–40% efficiency in this aspect (Lorick et al., 2020). However, it is possible to integrate ammonia stripping to remove NH_4^+ before the precipitation process. Struvite is a crystalline substance with a white appearance that is recognized as an environmentally friendly slow-release fertilizer. This characteristic aids in enhancing the plants' nutrient assimilation (Lorick et al., 2020). This recovery process has been tested on an industrial scale, particularly using municipal sludge digester effluents and industrial wastewater, as a P concentration of at least 100 mg L^{-1} is required (Rizzioli et al., 2023).

Another element commonly used for the caption of P is Ca. However, the process of calcium phosphate precipitation is complex and influenced by a variety of parameters. The outcome depends on factors such as the temperature, the calcium and phosphate ions content, the ionic strength, the types of ions present, the pH, and the duration of the reaction. Thus, it is necessary to introduce calcium hydroxide ($Ca(OH)_2$) into the liquid fraction, at a pH > 10 and at a temperature of 70°C, so phosphorus precipitates as either hydroxyapatite ($Ca_5(PO_4)_3OH$) or brushite ($CaHPO_4$–$2H_2O$) (Rizzioli et al., 2023).

It is also possible to extract K-struvite ($KMgPO_4$–$6H_2O$). This compound has a structure similar to that of struvite, the only difference being the substitution of NH_4^+ by K^+. Conclusions derived from their solubility constants suggest that if there is an excess of ammonia and potassium, struvite will precipitate before K-struvite. Thus, K-struvite will only form if the excess of K^+ is significantly higher than that of ammonium (Rizzioli et al., 2023).

Recently, Yazıcı et al. (2024) successfully isolated struvite, consisting of Mg^{2+}, NH_4+, and PO_4^{3-} from the liquid digestate produced during the processing of poultry manure in an anaerobic biogas plant. Yazıcı et al. (2024) concluded that maintaining a pH of 8.5 and using an overdose of either Mg^{2+} alone or a combination of Mg^{2+} and PO_4^{3-} with a molar ratio (NH_4^+/Mg^{2+}/PO_4^{3-}) of 1.0/1.3/1.0 gave the highest removal rates. The reductions in the chemical oxygen demand (COD), PO_4^{3-}, NH_4^+, and Mg^{2+} were 46%, 86%, 96%, and 67%, respectively. Despite these results, the authors recommend that future studies investigate other ions like SO_4^{2-}, K^+, and Na^+ and underscored the significance of enhancing the economic viability of the procedure by exploring the effects of affordable alternative external sources.

6.2.3 Ion exchange and adsorption

The ion exchange process involves the movement of cations away from a solid surface to be replaced by other ions present in the liquid within a densely packed column (SenGupta, 2017). This phenomenon, characterized by an equal exchange of electrical charge, is reversible between different ions (SenGupta, 2017). As ion exchange is a reversible reaction, the resins used in this process can be regenerated and reused. The size of the valence, ion diffusivity, and physical properties of the sorbents are factors that have collectively influenced the efficiency and effectiveness of ion exchange treatment for phosphorus and nitrogen recovery (SenGupta, 2017). The packed bed column is designed to increase the surface-to-volume ratio for a solid matrix of zeolites, resins, and clays (Dyer, 2000).

Sadrzadeh et al. (2023) used clinoptilolite, a widely used zeolite known for its high selectivity as a cation exchanger, specifically targeting ammonium and potassium ions. The primary focus of their investigation was nutrient recovery, with a particular emphasis on assessing the influence of initial loading on the ion exchange process for the removal of ammonium and potassium from the liquid phase of the digestate. The results of Sadrzadeh et al. (2023) indicate that increasing the initial loading ratio improves the removal of NH_4^+ and K^+ by clinoptilolite, but at the cost of a reduction in removal efficiency. They achieved removal efficiencies of up to 93% for NH_4^+ and 88% for K^+, with the optimum efficiency observed at an initial loading of 2.5 mg NH_4^+ g^{-1} zeolite. Sadrzadeh et al. (2023) experimented with higher concentrations and observed a significant decrease in removal efficiency, resulting in increased residual concentrations of NH_4^+ and K^+ in the liquid digestate.

6.2.4 Evaporation

In the evaporation process, the liquid portion of the digestate is evaporated at high-temperature evaporation. Before this process, the sample is acidified to facilitate the formation of a precipitate containing a significant amount of ammonium and volatile compounds. However, a notable challenge associated with evaporation is its significant energy requirements (Rizzioli et al., 2023).

A recent study by Wang & Yuan (2024) using a solar pilot system for the integrated treatment of the liquid part of the digestate demonstrated the potential for reducing the volume. This approach not only allowed for the recovery of nutrients but also facilitated the degradation of antibiotics. The use of dolomite-supported g-C3N4 (DCN) as a catalyst in the photocatalytic degradation of tetracyclines (TCs) in the liquid part of the digestate showed superior performance, particularly when the urea-to-dolomite doping ratio was 1:8. The optimum degradation efficiency for TCs in the liquid digestate reached 91% with a DCN/1:8 addition of 0.625 g mL^{-1}. Under dark adsorption at an initial pH of 4, the removal of TCs was 55% and the average degradation efficiency after 2 h of photocatalysis increased to 96%. Adjusting the initial pH to 4 proved to be beneficial for both liquid digestate concentration and antibiotic degradation. Even after five cycles, the efficiency of TC degradation remained stable. The introduction of DCN/1:8 into the solar evaporation system through a synergistic experiment increased the average TC degradation efficiency from 58% to 95% after 6 h of photocatalysis. The winter synergistic experiment achieved a 29% reduction in liquid digestate volume, with minimal impact from the addition of DCN/1:8. In terms of nutrient recovery, NH_4^+-N and soluble salts in the concentrated phase were recovered at rates of 93% and 83%, respectively. The pH levels in both the high concentrated and dilute phases were in line with standards for crop growth and agricultural irrigation water quality, respectively, falling within the pH range of 5.5–8.5. In addition, the use of solar panel not only streamlined the process but also contributed to cost reduction (Wang & Yuan, 2024).

6.2.5 Freeze concentration

Freeze concentration technology is engineered to selectively extract water as ice from a specified stream. Its objective is to diminish volume, generate a concentrated solution of superior quality, or reclaim purified water for reuse within an industrial operation. An important advantage of this approach is that freeze concentration requires less energy than evaporation.

The application of freeze concentration to digestate offers a means to treat and potentially recover valuable components in a more concentrated and efficient manner (Rizzioli et al., 2023). Research on the application of freeze concentration to digestate from AD is limited, with only one article found in scientific databases up to date. In their work, Uald-lamkaddam et al. (2021) used two different freeze concentration technologies—progressive and suspension freeze concentration. In both cases, the resulting concentrated fraction met standards for nutrient and metal content, rendering it appropriate for utilization as a nutrient-rich fertilizer. In their laboratory-scale investigation, Uald-lamkaddam et al. (2021) successfully recovered up to 71% of N, 90% of P, and 33% of K.

6.2.6 Membrane separation technologies

Membrane separation technology uses a membrane, which can be either organo-polymeric or ceramic, to facilitate the physical separation of liquid. The liquid is moved through the membrane by a flow, with the type of separation—microfiltration, ultrafiltration, or reverse osmosis—depending on the pore size of the membrane. While organo-polymer membranes are more cost-effective, they are difficult to clean and cannot operate efficiently at high pressures. Ceramic membranes, on the contrary, although typically more expensive, offer advantages such as cleanability, chemical resistance, and higher performance due to their ability to withstand higher pressures. It is important to note, however, that as separation efficiency increases, energy consumption also rises, which may be a major limitation to the widespread use of this technology (Camilleri-Rumbau et al., 2021).

Many experts claim that membrane filtration systems often face technical challenges, leading to uncertainties about their economic viability for digestate treatment (Rizzioli et al., 2023). Camilleri-Rumbau et al. (2021) reviewed the general physico-chemical characteristics of raw manure and digestate. They reviewed the technologies used for on-farm pretreatment for solid–liquid separation, outlined the essential membrane characteristics required for the treatment of liquid fractions of raw and digested manure, and examined membrane fouling mechanisms, together with purification strategies described in the existing literature. The review also provided valuable insights into nutrient recovery from manure, covering ammonia nitrogen and phosphorus, and the removal of heavy metals from these effluents by membrane processes. Camilleri-Rumbau et al. (2021) concluded that membrane fouling is a major challenge in the application of membrane technologies for the treatment of agricultural wastewater. However, they emphasized that a comprehensive understanding of the mechanisms involved, the composition of the fouling, and the implementation of effective membrane pretreatment and cleaning strategies can lead to excellent performance in terms of volume reduction and nutrient recovery.

6.2.7 Composting

Composting is a biological process involving the controlled decomposition and transformation of organic substrate into compost. Microorganisms organize this transformation by converting newly formed organic matter into a stable, nutrient-rich substance suitable for use as an organic soil improver or fertilizer. The composting process encourages microbial activity and environmental conditions are adjusted to facilitate the breakdown of organic matter. The elevated temperatures reached during composting help to reduce the presence of pathogens, ultimately improving the overall quality of the final product (Czekała et al., 2022).

European regulations (EU 2019/1009) restrict the direct application of digestate to soil and require it to be treated. One treatment required by the regulations is pasteurization of the solid fractions to eliminate pathogens. However, as the composting process reaches temperatures above 70°C and effectively eliminates pathogens, the prior pasteurization treatment is unnecessary.

Compost produced from the solid digestate has unique characteristics when compared to raw digestate. In addition, the solid fraction of the digestate has different characteristics from both the raw digestate and the compost derived from it. These differences are due to the inclusion of additional substrates commonly used in the composting process. Using substrates of known origin and free of contaminants in the AD process increases the chances of producing environmentally friendly compost with a stable composition. Throughout the composting stage, processes such as mineralization and humification take place, resulting in the conversion of certain organic components into mineral forms that can be absorbed by plants. In addition, odor emissions are significantly reduced and in some cases eliminated (Czekała et al., 2023).

While composting digestate has many advantages, it also has some disadvantages; in particular, the production of malodorous substances related to the N cycle. The thermophilic phase produces NH_3, which is volatile, odor-intensive, and toxic. The release of NH_3 into the atmosphere can lead to acid rain and the eutrophication of the water. In addition, microorganisms can develop during composting, converting ammonium to nitriteand finally nitrate, and reducing NO_x^- to N_2. These processes within the N cycle can lead to the emission of N_2O, a substance that contributes to global warming and ozone depletion. Other drawbacks include high moisture content, which reduces composting efficiency; low C/N ratio, which requires a longer composting period; and limited degradability of digestates, which results in insufficient temperature increase during composting and an inability to ensure the necessary disinfection of the compost (Czekała et al.,2022).

6.2.8 Microalgae

Legislation on nitrate-vulnerable zones and seasonal restrictions, as well as phosphorus soil overloading, can limit the application of digestate to agricultural land (Stiles et al., 2018). There are several approaches to microalgae cultivation, including photoautotrophic, mixotrophic, and heterotrophic methods. The first mentioned method uses sunlight as the primary energy source, which is converted into chemical energy through biochemical reactions. The second-mentioned method relies on organic compounds for both carbon and energy. In the last-mentioned approach, photosynthesis remains the primary energy source, with both organic compounds and CO_2 being essential. Typically, microalgae rely on a sufficient supply of carbon and sunlight to regulate photosynthesis for biomass growth. Microalgae have diverse metabolic pathways and the ability to change their metabolism in response to changes in their environment. Microalgae have the ability to extract macro- and micronutrients from the liquid fraction of the digestate, convert these elements into algal biomass, and produce a high-quality fertilizer. The liquid digestate is mainly composed of nitrogen, including NH_4^+, phosphorus, and organic matter, and their concentrations vary depending on the substrate used in the AD process (Alavianghavanini et al., 2024). Incorporating AD into the cultivation of a microalgal-bacteria consortium is a feasible option with additional benefits such as reducing the costs associated with microalgal biomass production and effectively managing digestate effluent (Kumar et al., 2023). However, the main limitations of microalgae cultivation are the high initial capital cost and adverse weather conditions, which can limit outdoor cultivation and reduce biomass productivity (Bele et al., 2023).

Once harvested, the microalgal biomass can act as a nitrogen fixer, facilitating the incorporation of atmospheric nitrogen into the soil and acting as a soil conditioner. In addition, microalgae can be further processed to produce more sophisticated biofertilizers and bio-stimulants (Kapoore et al., 2021). In addition, microalgal biomass can also be used in various sectors, including the production of biofuels, animal feed, pharmaceuticals, or as a human nutritional supplement (Stiles et al., 2018).

6.3 Thermochemical conversion processes

Following prior solid/liquid separation, thermal treatments are typically applied to the solid component of the digestate. The thermal processes involve subjecting the digestate to combustion by oxidation, which is achieved at temperatures in excess of 800°C and in the presence of oxygen above stoichiometric levels (Catenacci et al., 2022). As a result, only thermal energy will be recovered. In contrast, both energy and added-value materials are recovered in thermochemical conversion processes, which take place at much lower temperatures and predominantly in an inert environment. These thermochemical conversion technologies have a wide range of applications (Catenacci et al., 2021) and are based on the thermal decomposition of biomass (Hu et al., 2021). They are potentially able to process the entire organic matter of the material and have a higher efficiency, taking less time than required for biochemical conversion (Rizzioli et al., 2002). However, there are several challenges to scale-up, as thermochemical processes are multi-step and require high temperatures. Factors such as the quality of the substrate, the nature of the resulting products, compliance with local emission regulations, and economic considerations will influence the choice of a particular thermochemical process.

6.3.1 Pyrolysis

The pyrolysis involves a series of reactions that take place at rising temperatures in an inert environment. Depending on the temperature increase rate, pyrolysis is broadly classified as slow or fast pyrolysis ($< 100°C\ min^{-1}$ or $> 100°C\ min^{-1}$, respectively) (Xiang et al., 2020). While the primary emphasis in most studies has been on slow pyrolysis, only a limited number of articles have investigated the conditions that are typical for intermediate heating rates of pyrolysis (Careaga et al., 2020). Usually, pyrolysis is applied to substrates with high dry solids (> 90%), which requires an initial drying step to minimize energy consumption during carbonization (Tayibi et al., 2021). The untreated biomass, with a particle size between 5 and 50 mm, undergoes anaerobic decomposition at temperatures in the range 300°C–900°C, yielding higher quantities of biochar compared with alternative thermal processes (Jellali et al., 2021). Char yield, defined as char mass/initial feedstock mass ratio, is typically higher for slow pyrolysis compared with fast one (Chiappero et al., 2020; El Bari et al., 2024).

As highlighted by Sørmo et al. (2020), the pyrolysis process produces gaseous emissions containing various carbonaceous substances. In addition, there is a potential for the production of heavy metals and PAHs. However, a significant reduction in overall emissions has been achieved through the introduction of post-combustion of gaseous products as a result of advances in pyrolysis reactor technology. The feasibility of meeting EPA standards for CO, NO_x, and SO_2 without the need for advanced air pollution control systems has been demonstrated in studies carried out by Schwartz et al. (2020). Emission factors can be useful in estimating total emissions from pyrolysis processes, typically expressed as g kg^{-1} of biochar. Despite their usefulness, specific pyrolysis emission data remain scarce (Sørmo et al., 2020). Recent work has included the development of mathematical models to improve the assessment of hydrocarbon (HC), dioxin, and furan emissions during pyrolysis, considering input parameters such as waste composition and gas flow rates (Conesa et al., 2020).

Recently, Conesa et al. (2020) introduced and validated a model based on data from anaerobic digestate pyrolysis. This model includes three distinct stages of decomposition, each corresponding to different components that take place at three specific temperatures: easy-to-degrade organic matter (250°C), bacterial death organelles (350°C), and hard-to-degrade compounds (550°C). The first stage produces mainly CH_4, CO_2, acetic acid, and water. The second stage produces alcohols and hydrocarbons, while the last one produces H_2, CH_4, CO_2, alcohols, and hydrocarbons. Due to the predominance of the hard-to-degrade fraction as the primary char-forming component, anaerobic fermentation residues prove to be a suitable substrate for char production due to their relatively elevated content of difficult or nonbiodegradable organic substances (Conesa et al., 2020). At higher process temperatures, extensive molecular cracking converts biosolids into biochar, gases, and pyrolysis oil. This final step is known as catalytic cracking or thermal cracking, depending on whether a catalyst is used or not (Singh et al., 2020).

Currently, as highlighted by Basinas et al. (2023), pyrolysis stands out as the method of choice for digestate treatment to produce biochar suitable for soil applications. Temperature has a significant effect on the properties of biochar, and, thus, on its effectiveness in soil remediation, among the various parameters that influence pyrolysis. In general, high temperatures produce biochar with superior soil amendment potential. However, deviations from this norm have been observed for certain biomass materials. For example, the soil amendment capacity of the resulting biochar is reduced by a loss of microporosity when lignocellulose is pyrolyzed at temperatures above 700°C.

As described in Section 6.1.1, the physical and chemical composition of digestates could vary greatly. Consequently, the impact of temperature during the pyrolysis on the produced char may differ. In fact, biochar properties, such as pH, electrical conductivity, aromaticity, ash content, macro- and micronutrient composition, and polarity, among others, are primarily influenced by pyrolysis temperature (Stefaniuk & Oleszczuk, 2015). However, these effects are not well understood. Furthermore, there is a lack of understanding regarding the correlation between pyrolysis temperature and the effectiveness of digestate as a soil amender. (Basinas et al., 2023).

Catenacci et al. (2022) indicate that above 500°C the biochar yield increases moderately, suggesting that the charring process is close to being completed. Pyrolysis involves an oxygen-depleted environment to cautionary reduce the loss of organic matter at elevated temperatures (Singh et al., 2020). However, high-rise nitrogen flow rates can deplete the biochar production yield by reducing the residence time of the volatiles, thereby minimizing the secondary reactions and the formation of additional char (Catenacci et al., 2022).

The fundamental parameters to investigate the potential of char as an energy and fuel source are proximate and elemental analyses. An increase in carbon content, coupled with a decrease in hydrogen and oxygen content, indicates the formation of stronger C-C bonds, which provide more energy compared with C-O and C-H bonds. The energy yield (EY), the higher heating value (HHV), the ash content, and the energy densification (EDt) are the most commonly evaluated parameters related to the energy aspects. The ratio of the HHV of the biochar to the HHV of the feedstock, both on a dry basis, is used to calculate the EDt. To correlate the mass yield with the EDt, the parameter EY is defined as the product of the mass yield of the biochar and the EDt, which represents the energy content in the biochar (Akarsu et al., 2019). It is worth noting that high biochar HHVs can still be achieved by pyrolysis, although pyrolysis processes typically decrease biochar HHVs with increasing temperature. However, the use of dried digestate is preferable to its biochar as a solid fuel (Hung et al., 2017). Consequently, EY, HHV, and EDt are often insufficiently discussed in pyrolysis literature articles. The applicability of biochar as a soil amendment rather than their respective fuel sources is supported by their ash, nitrogen, and sulfur contents commonly found in digestate-derived biochar in relatively high amounts. The nitrogen, sulfur, and ash contents sum up with the fuel contribute to the emission of greenhouse gases, such as nitrogen oxides (NOx) and sulfur oxides (SOx), and particulates during combustion, thereby having a negative impact on quality of the fuel (Hung et al., 2017).

6.3.2 Hydrothermal carbonization

Biomass undergoes heating at temperatures lower than those utilized in pyrolysis, typically between 160°C and 250°C, in hydrothermal carbonization

(HTC). This process takes place in closed systems with water at saturation pressure (Catenacci et al., 2023). The pressure within the system is autogenous and increases proportionally with temperature, while the water containing the biomass serves as the reaction medium, and the temperature serves as the primary parameter to control the reaction mechanism. Substrates can be fed directly into the HTC process with high moisture content, typically in the range of 75% to 90%. This characteristic makes HTC particularly suitable for digestates which typically have a low total solids content. Reaction times in HTC vary widely, from a few minutes to several days, with common times in the range of 30 min to 72 h. Typically, the most common reaction times observed in practice are in the range of 1 to 8 h (Islam et al., 2021).

The HTC process produces sludge and synthesis gas (mainly CO_2 with small traces of CH_4, H_2, and CO). This slurry is then fractioned into a solid product (hydrochar) and a liquid phase (HTC aqueous liquid or process water). Typically, these fractions represent a 45%–70% and 5%–30% of the product mass, respectively, while the gases account for the smaller fraction (5%–25%) (Magdziarz et al., 2021). Atmospheric emissions associated with HTC are significantly lower than those associated with pyrolysis. They consist mainly of CO_2 and ammonia. In fact, HTC tends to produce chars with higher carbon concentration compared with those obtained by pyrolysis or gasification (Islam et al., 2021). One of the key advantages of HTC over pyrolysis or gasification is its ability to process wet biomass without the need for a predrying step. This feature enables direct processing of digestates, resulting in significant energy savings. In addition, HTC enhances mechanical sludge dewatering, improves feedstock stabilization and sanitization, and reduces digestate volume (Catenacci et al., 2022).

The HTC process involves several key reactions, including hydrolysis, dehydration, decarboxylation, condensation, polymerization, and aromatization, which occur sequentially and in parallel. The specific reaction pathways depend on the characteristics of the organic feedstock (Catenacci et al., 2022).

Similar to biochar yield, hydrochar yield tends to decrease with increasing process temperature due to increased solid destruction as process severity increases, leading to further decomposition and gasification of surface groups. Mild temperatures (180°C–210°C) and short residence times (30 min) are then recommended to improve the hydrochar formation. The biomass-to-water ratio also influences hydrochar yield and energy consumption, although the exact influence remains unclear and depends on the type of feedstock (Catenacci et al., 2022). Regarding the type of feedstock, agro-industrial digestate-derived hydrochars present higher HHVs compared with those derived from sewage sludge digestates and municipal solid waste (MSW) digestates. Furthermore, short residence times (30 min), mild temperatures (180°C–210°C), and moderate ash content in the hydrochar, together with a high COD concentration in the liquid fraction, are recommended for optimal energy recovery. This suggests the co-digestion of hydrochar and process water recirculation in biogas digesters as desirable applications (Catenacci et al., 2022). However, high-ash and low-lignin content in the feedstock could lead to lower HHV hydrochar, which would result in EDt ratios below 1. Sewage sludge digestates typically have, actually, a low degree of EDt compared with lignocellulosic feedstocks. Different feedstocks yield hydrochar with different EDt ratios, with higher values typically observed for hydrochar derived from MSW digestate. The effect of temperature and residence time on EY varies by feedstock, with increases in EY observed for certain digestates at higher temperatures (Akarsu et al., 2019).

Furthermore, hydrochar produced from food waste and its digestate exhibit comparable yields and similar characteristics for use as fuel. The main difference lies in the ash content, which tends to be higher in digestate-derived hydrochar, negatively affecting the HHV (Belete et al., 2021).

6.3.3 Gasification

Anaerobic digestate gasification for char production is an attractive technology for syngas production, but unlike pyrolysis and HTC, it is relatively under-explored. The quality of the syngas produced depends largely on its calorific value, which is primarily determined by the gasification temperature and the gasification agent used. Syngas, once purified from major pollutants such as tar, can be used either as a fuel or as a hydrogen source for biomethanization processes (Catenacci et al., 2022).

Gasification is emerging as a cost-effective and environmentally friendly method for hydrogen production from various biomass sources (Timofeeva et al., 2023). In this process, biomass is converted into syngas under oxygen-deficient conditions using various types of gasifiers. The syngas is used as combustion fuel in power plants and as feedstock in the chemical industry. The ratio of carbon monoxide (CO) to hydrogen (H_2) in the gas produced typically ranges from 1:1 to 1:3, depending on the gasification method and oxidants used. Oxygen, air, steam, and their mixtures are commonly used as oxidants. Steam gasification is characterized by its ability to maximize the yield of H_2 and CO, making it one of the most promising approaches for the thermochemical conversion of biomass into hydrogen-rich gas (Timofeeva et al., 2023).

As noted by Freda et al. (2019), the gasification of digestates produced after AD of biomass remains a relatively unexplored area. These authors investigated the use of digestates for the production of syngas by means of air gasification in a rotary drum. They also carried out steam gasification of food waste digestates at 850°C, resulting in a syngas with an average hydrogen content of 75%. The advantages of steam gasification over air gasification in terms of digestate utilization were demonstrated in comparative analyses. Steam gasification energy costs ranged from 93.4 to 113.8 €/MW-h, while air gasification costs ranged from 99.9 to 123.1 €/MW-h (Timofeeva et al., 2023).

Gasification is one of the most promising alternatives for biomass processing among direct thermochemical methods. It offers conversion of biomass to syngas with minimal harmful emissions compared with direct combustion and reduced resin formation compared with pyrolysis. Biomass gasification plays a crucial role in mitigating the greenhouse effect (Timofeeva et al., 2023), when combined with advanced CO_2 capture technology. The composition of the syngas components is controlled by selecting the type and flow rate of the gasifying agent. In particular, steam gasification facilitates the achievement of maximum hydrogen content in the syngas, making it one of the most attractive methods for thermochemical conversion of biomass to hydrogen-rich gas (Freda et al., 2019).

6.3.4 Torrefaction

In recent years, the torrefaction of biomass, also known as low-temperature pyrolysis, has received increasing attention. In this process, the properties and quantities of the individual product fractions are influenced by parameters, such as residence time, heating rate, temperature, and feedstock. Typical operating parameters include a temperature between 200°C and 350°C, a heating rate of 5 to 20 K/min, and a residence time of 5 to 20 min (Unyay et al., 2023). The primary products of torrefaction are charred solid biofuel, commonly referred to as biochar, and syngas, known as torgas. Valuable co-products such as 5-HMF, sodium acetate, furfural, and sodium formate in high purity can be obtained by condensing liquid products in the case of superheated steam torrefaction. Torrefaction facilitates the pretreatment of lignocellulosic biomass into lignite-like solid fuels, known as torrefied biomass or biochar. These fuels replace fossil fuels and contribute to the decarbonization of industrial processes when used in biomass boilers for industrial heat generation. In addition, process vapor condensate and torrefaction gas (torgas) serve as valuable co-products (Szufa et al., 2020). In rural areas, torrefaction offers significant potential for valorizing agro-industrial waste alongside forest biomass. Biomass torrefaction pellets, also known as black pellets, can be produced by further concentrating the biomass produced by torrefaction through pelletization processes. This provides a practical solution for biomass utilization (Unyay et al., 2023) by increasing transportability and improving combustion characteristics.

Superheated steam (SHS)-driven torrefaction is a thermochemical process that uses superheated steam as a heating medium instead of conventional gases such as torgas, N_2, or H_2. This process, which operates at atmospheric pressure, has significant economic efficiencies due to the elimination of additional gas costs. By using SHS torrefaction, precise control of the biomass upgrading process is achieved through the use of superheated steam in the reactor, improving the homogeneity of the resulting torrefied materials and optimizing reactor design procedures (Stelmach et al., 2021).

The benefits of SHS torrefaction can be outlined as follows: (1) increased HHV of the product; (2) recovery of valuable condensates such as acetic acid, propionic acid and ammonia as co-products for the chemical industry; (3) utilization of the torrefied biogas solid residue (solid phase of the digestate) as fertilizer; (4) reuse of heat from steam condensation, resulting in high overall process efficiency; and (5) prevention of reactor chamber fouling and resin oxidation during torrefaction, facilitated by the presence of SHS (Unyay et al., 2023).

6.3.5 Combined processes

A combination of processes holds promise for the full valorization of anaerobic digestates (Catenacci et al., 2022). HTC coupled with pyrolysis has been shown to effectively produce a char with enhanced soil applicability, that is, higher C content and lower PAH concentrations, compared with HTC or pyrolysis alone (Wang et al., 2022). In addition, the combination of HTC with supercritical water gasification (SCWG) on aqueous HTC liquid (AHL) or HTC with steam gasification on hydrochars has shown efficient conversions into valuable biofuels, char and H_2-enriched gas, while minimizing wastes and process costs (Taufer et al., 2021). In addition, promising methods for producing hydrochar with high HHV and favorable fuel applicability include microwave-assisted low-temperature hydrothermal treatment (MLHT, temperatures: 100°C–180°C) and microwave-assisted hydrothermal carbonization (MLHC—180°C–260°C) followed by pyrolysis. However, ash content is a limiting factor in these processes (Catenacci et al., 2022).

Autogenous pressure carbonization (APC), which favors the production of relatively high-purity biochemicals such as pyrenes and phenols, is another way to improve syngas yield and quality. In addition, APC can increase the nitrogen content of the char. While examples of catalytic processes in this context are limited, some studies have shown promising results. For example, Kim et al. (2020) investigated the effectiveness of using steel slag as a catalyst and CO_2 as a reactive gas medium during pyrolysis. Their study highlighted the potential for improved process outcomes with catalytic support, with a focus on evaluating carbon distribution. Similarly, Chakraborty et al. (2021) used a zeolite catalyst (ZSMe5) to accelerate process kinetics and increase the yield of aromatic hydrocarbons in bio-oil. Their results highlighted the efficiency of catalytic pyrolysis in reducing energy requirements compared to noncatalytic systems.

On the contrary, co-pyrolysis/co-hydrothermal carbonization (co-HTC) and physical/chemical activation or functionalization of chars have been widely studied to optimize the performance conditions and increase the final yields (Catenacci et al., 2022). The combination of different wastes in co-pyrolysis/co-HTC processes could address the drawbacks associated with limited feedstock availability, while product quality can be improved through synergistic effects. Studies have shown that co-pyrolysis of microalgae and swine manure digestate or digested MSW leads to improved kinetics and gas yields (Vuppaladadiyam et al., 2019; Vuppaladadiyam et al., 2022). Increasing the proportion of microalgae in these mixtures results in higher hydrogen production, likely because of the catalytic action of alkali metals present in the ash. Co-HTC has also been investigated to improve the combustion properties of hydrochar produced from digested sewage sludge (SS). The addition of digested cow dung was found to increase the carbon content and improve the HHV of the hydrochar. It also increased the activation energy of the hydrochar. During HTC, cellulose and hemicellulose from cow dung digestate interacted with proteins to form higher energy C-N bonds, resulting in an N-rich hydrochar (Xu et al., 2019).

6.3.6 Benefits of biochar and hydrochar in anaerobic digestion processes

There are significant benefits to the process from the use of biochar and hydrochar in anaerobic reactors (Habchi et al., 2024). Both types of char play a key role in enhancing hydrogen production in the initial stage of AD and facilitating methanogenesis in the subsequent stages. Theoretically, they act as process enhancers by reducing the lag phase of both steps, improving stability, promoting bacterial colonization, mitigating process inhibitors, increasing the buffering capacity of the system, and actively participating in methane production (Basinas et al., 2023). In addition, biomass-derived chars have remarkable abilities in the prevention of process acidification. These chars contain a significant amount of alkali functional groups and metal cations, which act as effective neutralizers of volatile fatty acids in an acidified reactor.

Among the morphological characteristics of biochar and hydrochar, surface area and microstructure are highlighted as critical factors for their efficacy in AD processes, providing favorable habitats for microorganisms and archaea to attach and proliferate (Basinas et al., 2023). In anaerobic systems, the colonization of microbial communities facilitates interspecies electron transfer and accelerates the acclimation rate of microbial cells, although it may increase the washout rate of active microorganisms (Catenacci et al., 2022).

Recent comparisons of the performance of bio and hydro chars in increasing the yield of biogas have shown that energy production is superior in digesters modified with hydrochar when using sewage sludge and pig manure as feedstock. Unlike biochar, the incomplete carbonization process of hydrochar results in the release of organic matter that can be degraded by the microbial communities involved in the AD process, increasing the overall methane yield. In addition, HTC promotes the formation of functional groups favoring electron transfer between microorganisms (Xu et al., 2020).

HTC effluents concentrate significant organic matter that can be co-digested with the main feedstock due to the solubilization of organic compounds; in fact, 15% to 20% of the initial carbon is present in the liquid fraction, resulting in a high COD (up to 90–110 g·L^{-1}), depending on HTC process conditions and feedstock type. Recycling HTC process waters to anaerobic digesters could potentially double the methane yield. Although ammonia inhibition from HTC liquor has not been observed in most studies, there is a risk under certain conditions and requires professional monitoring (Catenacci et al., 2022).

6.4 Future research

As highlighted above, the commercial use of HTC/pyrolysis technologies for digestate valorization is currently minimal. Despite an extensive scientific literature on the subject, critical challenges remain that hinder the full integration of AD with HTC/pyrolysis/gasification processes. To effectively translate scientific knowledge into practical applications, it is imperative to comprehensively address various aspects, ranging from substrate properties and techno-economic analysis to environmental and social impacts. In particular, careful monitoring of digestate characteristics is essential to ensure the production of biochar/hydrochar with standardized properties suitable for intended applications. This is particularly important given the higher variability of digestate characteristics in comparison with traditional lignocellulosic biomasses. In addition, to mitigate potential negative environmental and health impacts from the accumulation of contaminants, long-term monitoring of the application of biochar/hydrochar to agricultural soils is essential.

Digestate valorization, which allows the recovery and reuse of unused nutrients contained in the digestate, is a critical step in closing the loop of the AD process. By recovering resources and creating added-value products from digestate, the socio-economic benefits of bio-based industries can be secured and enhanced through sustainable development practices. The integrated biorefinery approach to AD and digestate management facilitates zero waste discharge from AD plants, thereby improving economic viability and sustainability.

References

Abbà, A., Domini, M., Baldi, M., Pedrazzani, R., & Bertanza, G. (2023). Ammonia recovery from livestock manure digestate through an air-bubble stripping reactor: Evaluation of performance and energy balance. *Energies, 16*(4), 1643. https://doi.org/10.3390/en16041643).

Akarsu, K., Duman, G., Yilmazer, A., Keskin, T., Azbar, N., & Yanik, J. (2019). Sustainable valorization of food wastes into solid fuel by hydrothermal carbonization. *Bioresource Technology, 292*, 121959. https://doi.org/10.1016/j.biortech.2019.121959.

Alavianghavanini, A., Shayesteh, H., Bahri, P. A., Vadiveloo, A., & Moheimani, N. R. (2024). Microalgae cultivation for treating agricultural effluent and producing value-added products. *Science of The Total Environment, 912*, 169369. https://doi.org/10.1016/j.scitotenv.2023.169369.

Aung, S. L., Choi, J., Cha, H., Woo, G., & Song, K. G. (2024). Ammonia-selective recovery from anaerobic digestate using electrochemical ammonia stripping combined with electrodialysis. *Chemical Engineering Journal, 479*, 147949. https://doi.org/10.1016/j.cej.2023.147949.

Basinas, P., Rusín, J. R., Chamrádová, K., & Kaldis, S. P. (2023). Pyrolysis of the anaerobic digestion solid by-product: Characterization of digestate decomposition and screening of the biochar use as soil amendment and as additive in anaerobic digestion. *Energy Conversion and Management, 277*, 116658. https://doi.org/10.1016/j.enconman.2023.116658.

Bele, V., Rajagopal, R., & Goyette, B. (2023). Closed loop bioeconomy opportunities through the integration of microalgae cultivation with anaerobic digestion: A critical review. *Bioresource Technology Reports, 21*, 101336. https://doi.org/10.1016/j.biteb.2023.101336.

Belete, Y. Z., Mau, V., Yahav-Spitzer, R., Posmanik, R., Jassby, D., & Iddya, A. (2021). Hydrothermal carbonization of anaerobic digestate and manure from a dairy farm on energy recovery and the fate of nutrients. *Bioresource Technology, 333*, 125164. https://doi.org/10.1016/j.biortech.2021.125164.

Bergstrand, K. J. (2022). Organic fertilisers in greenhouse production systems – A review. *Scientia Horticulturae, 295*, 110855. https://doi.org/10.1016/j.scienta.2021.110855.

Camilleri-Rumbau, M. S., Briceño, K., Søtoft, L. F., Christensen, K. V., Roda-Serrat, M. C., Errico, M., & Norddahl, B. (2021). Treatment of manure and digestate liquid fractions using membranes: Opportunities and challenges. *International Journal of Environmental Research and Public Health, 18*(6), 1–30. https://doi.org/10.3390/ijerph18063107.

Careaga, F. J. S., Porat, A., Briens, L., & Briens, C. (2020). Pyrolysis shaker reactor for the production of biochar. *The Canadian Journal of Chemical Engineering, 98*, 2417–2424. https://doi.org/10.1002/cjce.23771.

Catenacci, A., Boniardi, G., Mainardis, M., Gievers, F., Farru, G., Asunis, F., Malpei, F., Goi, D., Cappai, G., & Canziani, R. (2022). Processes, applications and legislative framework for carbonized anaerobic digestate: Opportunities and bottlenecks. A critical review. *Energy Conversion and Management, 263*, 115691. https://doi.org/10.1016/j.enconman.2022.115691.

Carraturo, F., Siciliano, A., Giordano, A., Di Capua, F., Barone, F., Casaletta, E., Cicotti, F., Guida, M., & Adani, F. (2024). Ecotoxicological assessment of waste-derived organic fertilisers and long-term monitoring of fertilized soils using a multi-matrix and multi-species approach. *Science of the Total Environment, 912*, 169341. https://doi.org/10.1016/j.scitotenv.2023.169341.

Chakraborty, S., Dunford, N. T., & Goad, C. (2021). A kinetic study of microalgae, municipal sludge and cedar wood co-pyrolysis. *Renewable Energy, 165*, 514–524. https://doi.org/10.1016/j.renene.2020.11.012.

Chiappero, M., Norouzi, O., Hu, M., Demichelis, F., Berruti, F., & Di Maria, F. (2020). Review of biochar role as additive in anaerobic digestion processes. *Renewable and Sustainable Energy Reviews, 131*, 110037. https://doi.org/10.1016/j.rser.2020.11003.

Chojnacka, K., & Chojnacki, M. (2024). Nutrient recovery from anaerobic digestate: Fertiliser informatics for circular economy. *Environmental Research, 245*, 117953. https://doi.org/10.1016/j.envres.2023.117953.

Christensen, M. L., Hjorth, M., & Keiding, K. (2009). Characterization of pig slurry with reference to flocculation and separation. *Water Research, 43*, 773–783. https://doi.org/10.1016/J.WATRES.2008.11.010.

Conesa, J. A., Ortuño, N., & Palmer, D. (2020). Estimation of industrial emissions during pyrolysis and combustion of different wastes using laboratory data. *Scientific Reports, 10*, 6750. https://doi.org/10.1038/s41598-020-63807-w.

Czekała, W., Jasiński, T., Grzelak, M., Witaszek, K., & Dach, J. (2022). Biogas plant operation: Digestate as the valuable product. *Energies, 15*, 8275. https://doi.org/10.3390/en15218275.

Czekała, W., Nowak, M., & Piechota, G. (2023). Sustainable management and recycling of anaerobic digestate solid fraction by composting: A review. *Bioresource Technology, 375*. https://doi.org/10.1016/j.biortech.2023.128813.

De la Lama-Calvente, D., Fernández-Rodríguez, M. J., García-Gómez, J. C., & Borja, R. (2023). Impact of natural degradation of the invasive alga *Rugulopteryx okamurae* on anaerobic digestion: Heavy metal pollution and kinetic performance. *Marine Pollution Bulletin, 192*, 115005. https://doi.org/10.1016/j.marpolbul.2023.115005.

Dyer, A. (2000). Ion exchange. *Encyclopedia of Separation Science*, 156–173.

El Bari, H., Fanezoune, C. K., Dorneanu, B., Arellano-Garcia, H., Majozi, T., Elhenawy, Y., Bayssi, O., Hirt, A., Peixinho, J., Dhahak, A., Gadalla, M. A., Khashaba, N. H., & Ashour, F. H. (2024). Catalytic fast pyrolysis of lignocellulosic biomass: Recent advances and comprehensive overview. *Journal of Analytical and Applied Pyrolysis, 178*, 106390. https://doi.org/10.1016/j.jaap.2024.106390.

Elalami, D., Monlau, F., Carrere, H., Abdelouahdi, K., Charbonnel, C., Oukarroum, A., Zeroual, Y., & Barakat, A. (2020). Evaluation of agronomic properties of digestate from macroalgal residue anaerobic digestion: Impact of pretreatment and co-digestion with waste activated sludge. *Waste Management, 188*, 127–136. https://doi.org/10.1016/j.wasman.2020.04.019.

Fernández-Rodríguez, M. J., Mancilla-Leytón, J. M., Jiménez-Rodríguez, A., Borja, R., & Rincón, B. (2021). Reuse of the digestate obtained from the biomethanization of olive mill solid waste (OMSW) as soil amendment or fertiliser for the cultivation of forage grass (Lolium rigidum var. Wimmera). *Science of Total Environment, 792*, 148465. https://doi.org/10.1016/j.scitotenv.2021.148465.

Freda, C., Nanna, F., & Villone, A. (2019). Air gasification of digestate and its co-gasification with residual biomass in a pilot scale rotary kiln. *International Journal of Energy and Environmental Engineering, 10*, 335e46. https://doi.org/10.1007/s40095-019-0310-3.

Fu, Z., Zhao, J., Guan, D., Wang, Y., Xie, J., Zhang, H., Sun, Y., Zhu, J., & Guo, L. (2024). A comprehensive review on the preparation of biochar from digestate sources and its application in environmental pollution remediation. *Science of the Total Environment, 912*, 168822. https://doi.org/10.1016/j.scitotenv.2023.168822.

Golovko, O., Ahrens, L., Schelin, J., Sörengard, M., Bergstrand, K. J., Asp, H., Hultberg, M., & Wiberg, K. (2022). Organic micropollutants, heavy metals and pathogens in anaerobic digestion based on food waste. *Journal of Environmental Management, 313*, 114997. https://doi.org/10.1016/j.envman.2022.114997.

Habchi, S., Lahboubi, N., Asbik, M., & El Bari, H. (2024). Enhancing biomethane production from food waste using olive pomace hydrochar: An optimization study. *Environmental Advances*. https://doi.org/10.1016/j.envadv.2023.100477.

Holliger, C., Alves, M., Andrade, D., Angelidaki, I., Astals, S., Baier, U., Bougrier, C., Buffiere, P., Carballa, M., de Wilde, V., Ebertseder, F., Fernandez, B., Ficara, E., Fotidis, I., Frigon, J. C., de Laclos, H. F., Ghasimi, D. S. M., Hack, G., Hartel, M., ... Wierinck, I. (2016). Towards a standardization of biomethane potential tests. *Water Science and Technology, 74*, 2515–2522. https://doi.org/10.2166/wst.2016.336.

Hu, M., Ye, Z., Zhang, H., Chen, B., Pan, Z., & Wang, J. (2021). Thermochemical conversion of sewage sludge for energy and resource recovery: Technical challenges and prospects. *Environmental Pollutants and Bioavailability, 33*, 145–163. https://doi.org/10.1080/26395940.2021.1947159.

Hung, C. Y., Tsai, W. T., Chen, J. W., Lin, Y. Q., & Chang, Y. M. (2017). Characterization of biochar prepared from biogas digestate. *Waste Management, 66*, 53–60. https://doi.org/10.1016/j.wasman.2017.04.034.

Islam, M. A., Limon, M. S. H., Romic, M., & Islam, Md. A. (2021). Hydrochar-based soil amendments for agriculture: A review of recent progress. *Arabian Journal of Geosciences, 14*(2). https://doi.org/10.1007/s12517-020-06358-8.

Jellali, S., Khiari, B., Usman, M., Hamdi, H., Charabi, Y., & Jeguirim, M. (2021). Sludge-derived biochars: A review on the influence of synthesis conditions on pollutants removal efficiency from wastewaters. *Renewable and Sustainable Energy Reviews, 144*, 111068. https://doi.org/10.1016/j.rser.2021.111068.

Kapoore, R. V., Wood, E. E., & Llewellyn, C. A. (2021). Algae biostimulants: A critical look at microalgal biostimulants for sustainable agricultural practices. *Biotechnology Advances, 49*, 107754. https://doi.org/10.1016/j.biotechadv.2021.107754.

Kim, J. H., Oh, J. I., Tsang, Y. F., Park, Y. K., Lee, J., & Kwon, E. E. (2020). CO_2-assisted catalytic pyrolysis of digestate with steel slag. *Energy, 191*, 116529. https://doi.org/10.1016/j.energy.2019.116529.

Kumar, V., Jaiswal, K. K., & Tomar, M. S. (2023). Production of high value-added biomolecules by microalgae cultivation in wastewater from anaerobic digestates of food waste: A review. *Biomass Conversion and Biorefinery, 13*, 9625–9642. https://doi.org/10.1007/s13399-021-01906-y.

Lorick, D., Macura, B., & Ahlström (2020). Effectiveness of struvite precipitation and ammonia stripping for recovery of phosphorus and nitrogen from anaerobic digestate: A systematic review. *Environmental Evidence, 9*, 27. https://doi.org/10.1186/s13750-020-00211-x.

Magdziarz, A., Mlonka-Mędrala, A., Sieradzka, M., Aragon-Briceño, C., Pozarlik, A., & Bramer, E. A. (2021). Multiphase analysis of hydrochars obtained by anaerobic digestion of municipal solid waste organic fraction. *Renewable Energy, 175*, 108–118. https://doi.org/10.1016/j.renene.2021.05.018.

Makian, M., Im, S., Mostafa, A., Prakash, O., Kim, J., Park, C., & Kim, D. H. (2024). Continuous production of high-concentrated ammonia broth through fermentation. *Bioresource Technology, 394*, 130217. https://doi.org/10.1016/j.biortech.2023.130217.

Monlau, F., Sambusiti, C., Ficara, E., Aboulkas, A., Barakat, A., & Carrère, H. (2015). New opportunities for agricultural digestate valorization: Current situation and perspectives. *Energy & Environmental Science, 8*(9), 2600–2621. https://doi.org/10.1039/c5ee01633a.

Nelson, P. V. (2011). *Greenhouse Operation and Management*. Boston, MA, USA: Prentice Hall.

O'Connor, J., Mickan, B. S., Rinklebe, J., Song, H., Siddique, K. H. M., Wang, H., Kirkham, M. B., & Bolan, N. S. (2022). Environmental implications, potential value, and future of food-waste anaerobic digestate management: A review. *Journal of Environmental Management, 318*, 115519. https://doi.org/10.1016/j.jenvman.2022.115519.

Rao, M., Zou, X., Huang, Z., Ye, J., Huang, C., Zhang, M., Chen, C., Kuang, C., Song, Y., Xin, K., Jia, D., Liu, Y., & Cheng, J. (2024). Purification of anaerobic fermentation digestate of dairy manure by Arthrospira biomass cultivated with biogas-CO_2 fixation. *Biochemical Engineering Journal, 203*, 109178. https://doi.org/10.1016/j.bej.2023.109178.

Ries, J., Chen, Z., & Park, Y. (2023). Potential applications of food-wate-based anaerobic digestate for sustainable crop production practice. *Sustainability, 15*(11), 8520. https://doi.org/10.3390/su15118520.

Rittl, T. F., Reidun, P., Anders, J., Havard, S., Hugh, R., & Anne-Kristin, L. (2023). Anaerobic digestion of dairy cattle slurry – long-term effects on crop yields and chemical soil characteristics. *Organic Agriculture, 13*(4), 547–563. https://doi.org/10.1007/s13165-023-00447-0.

Rizzioli, F., Bertasini, D., Bolzonella, D., Frison, N., & Battista, F. (2023). A critical review on the techno-economic feasibility of nutrients recovery from anaerobic digestate in the agricultural sector. *Separation and Purification Technology, 306*, 122690. https://doi.org/10.1016/j.seppur.2022.122690.

Sadrzadeh, F., Baykal, B. B., & Altinbas, M. (2023). An investigation on the importance of initial loading in the removal of nutrients from liquid digestate through sorption and nutrient recovery from the solid phase to be used as fertiliser. *Water, Air, and Soil Pollution, 234*(11). https://doi.org/10.1007/s11270-023-06717-7.

Schwartz, N. R., Paulsen, A. D., Blaise, M. J., Wagner, A. L., & Yelvington, P. E. (2020). Analysis of emissions from combusting pyrolysis products. *Fuel, 274*, 117863. https://doi.org/10.1016/j.fuel.2020.117863.

SenGupta, A. K. (2017). *Ion exchange in environmental processes: Fundamentals, applications and sustainable technology*. John Wiley & Sons.

Sheets, J. P., Yang, L., Ge, X., Wang, Z., & Li, Y. (2015). Beyond land application: Emerging technologies for the treatment and reuse of anaerobically digested agricultural and food waste. *Waste Management, 44*, 94–115. https://doi.org/10.1016/j.wasman.2015.07.037.

Shrestha, S., Pandey, R., Aryal, N., & Lohani, S. P. (2023). Recent advances in co-digestion conjugates for anaerobic digestion of food waste. *Journal of Environmental Management, 345*, 118785. https://doi.org/10.1016/j.envman.2023.118785.

Singh, S., Kumar, V., Dhanjal, D. S., Datta, S., Bhatia, D., & Dhiman, J. (2020). A sustainable paradigm of sewage sludge biochar: Valorization, opportunities, challenges and future prospects. *Journal of Cleaner Production, 269*, 122259. https://doi.org/10.1016/j.jclepro.2020.122259.

Sørmo, E., Silvani, L., Thune, G., Gerber, H., Schmidt, H. P., & Smebye, A. B. (2020). Waste timber pyrolysis in a medium-scale unit: Emission budgets and biochar quality. *Science of The Total Environment, 718*, 137335. https://doi.org/10.1016/j.scitotenv.2020.137335.

Stefaniuk, M., & Oleszczuk, P. (2015). Characterization of biochars produced from residues from biogas production. *Journal of Analytical and Applied Pyrolysis, 115*, 157–165. https://doi.org/10.1016/j.jaap.2015.07.011.

Stelmach, J., Kuncewicz, C., Adrian, Ł., Jirout, T., & Rieger, F. (2021). Change in mixing power of a two-PBT impeller when emptying a tank. *Processes, 9*(2), 341. https://doi.org/10.3390/pr9020341.

Stiles, W. A. V., Styles, D., Chapman, S. P., Esteves, S., Bywater, A., Melville, L., Silkina, A., Lupatsch, I., Fuentes-Grünewald, C., Lovitt, R., Chaloner, T., Bull, A., Morris, C., & Llewellyn, C. A. (2018). Using microalgae in the circular economy to valorise anaerobic digestate: challenges and opportunities. *Bioresource Technology, 267*, 732–742. https://doi.org/10.1016/j.biortech.2018.07.100.

Szufa, S., Wielgosiński, G., Piersa, P., Czerwińska, J., Dzikuć, M., Adrian, L., Lewandowska, W., & Marczak, M. (2020). Torrefaction of straw from oats and maize for use as a fuel and additive to organic fertilisers—TGA analysis, kinetics as products for agricultural purposes. *Energies, 13*(8), 2064. https://doi.org/10.3390/en13082064.

Taufer, N. L., Benedetti, V., Pecchi, M., Matsumura, Y., & Baratieri, M. (2021). Coupling hydrothermal carbonization of digestate and supercritical water gasification of liquid products. *Renewable Energy, 173*, 934–941. https://doi.org/10.1016/j.renene.2021.04.058.

Tayibi, S., Monlau, F., Marias, F., Cazaudehore, G., Fayoud, N. E., & Oukarroum, A. (2021). Coupling anaerobic digestion and pyrolysis processes for maximizing energy recovery and soil preservation according to the circular economy concept. *Journal of Environmental Management, 279*, 111632. https://doi.org/10.1016/j.jenvman.2020.111632.

Timofeeva, S. S., Karaeva, J. V., Kovalev, A. A., Kavalev, D. A., & Litti, Y. V. (2023). Steam gasification of digestate after anaerobic digestion and dark fermentation of lignocellulosic biomass to produce syngas with high hydrogen content. *International Journal of Hydrogen Energy, 48*, 7559–7568. https://doi.org/10.1016/j.ijhydene.2022.11.260.

Uald-lamkaddam, I., Dadrasnia, A., Llenas, L., Ponsá, S., Colón, J., Vega, E., & Mora, M. (2021). Application. *Sustainability, 13*, 13769. https://doi.org/10.3390/su132413769.

Unyay, H., Piersa, P., Perendeci, N. A., Wielgosinski, G., & Szufa, S. (2023). Valorization of anaerobic digestate: Innovative approaches for sustainable resource management and energy production— Case studies from Turkey and Poland. *International Journal of Green Energy*. https://doi.org/10.1080/15435075.2023.2276158.

Van den Brand, T. P. H., Roest, K., Chen, G. H., Brdjanovic, D., & van Loosdrecht, M. C. M. (2015). Potential for beneficial application of sulfate reducing bacteria in sulfate containing domestic wastewater treatment. *World Journal of Microbiology and Biotechnolgy, 31*, 1675–1681. https://doi.org/10.1007/s11274-015-1935-x.

Vuppaladadiyam, A. K., Liu, H., Zhao, M., Soomro, A. F., Memon, M. Z., & Dupont, V. (2019). Thermogravimetric and kinetic analysis to discern synergy during the co-pyrolysis of microalgae and swine manure digestate. *Biotechnology and Biofuels, 12*, 170. https://doi.org/10.1186/s13068-019-1488-6.

Vuppaladadiyam, A. K., Antunes, E., Vuppaladadiyam, V. S. S., Shehzad, F., Somasundaram, M., & Memon, M. Z. (2022). Discernment of synergy during the copyrolysis of lipid-extracted microalgae and digested municipal solid waste: a thermogravimetric–mass spectrometric study. *Journal of Chemical Technology and Biotechnology, 97*(2), 490–500. https://doi.org/10.1002/jctb.6702.

Wang, P., & Yuan, Q. (2024). Synergistic experimental study by a solar pilot evaporation system for the volume reduction and antibiotics photocatalytic degradation of liquid digestate. *Journal of Water Process Engineering, 59*. https://doi.org/10.1016/j.jwpe.2024.105023.

Wang, S., Wen, Y., Shi, Z., Niedzwiecki, L., Baranowski, M., & Czerep, M. (2022). Effect of hydrothermal carbonization pretreatment on the pyrolysis behavior of the digestate of agricultural waste: A view on kinetics and thermodynamics. *Chemical Engineering Journal, 431*, 133881. https://doi.org/10.1016/j.cej.2021.133881.

Wang, Y., Yu, G., Xie, S., Jiang, R., Li, C., & Xing, Z. (2023). Pyrolysis of food waste digestate residues for biochar: Pyrolytic properties, biochar characterization, and heavy metal behaviours. *Fuel, 353*, 129185. https://doi.org/10.1016/j.fuel.2023.129185.

Wang, Z., Sanusi, I. A., Wang, J., Ye, X., Kana, E. B. G., Olaniran, A. O., & Shao, H. (2023). Developments and prospects of farmland application of biogas slurry in China—A review. *Microorganisms, 11*(11), 2675. https://doi.org/10.3390/microorganisms11112675.

Xiang, W., Zhang, X., Chen, J., Zou, W., He, F., & Hu, X. (2020). Biochar technology in wastewater treatment: A critical review. *Chemosphere, 252*, 126539. https://doi.org/10.1016/j.chemosphere.2020.126539.

Xu, Z. X., Song, H., Zhang, S., Tong, S. Q., He, Z. X., & Wang, Q. (2019). Co-hydrothermal carbonization of digested sewage sludge and cow dung biogas residue: Investigation of the reaction characteristics. *Energy, 187*, 115972. https://doi.org/10.1016/j.energy.2019.115972.

Xu, S., Wang, C., Duan, Y., & Wong, J.-C. (2020). Impact of pyrochar and hydrochar derived from digestate on the co-digestion of sewage sludge and swine manure. *Bioresource Technology, 314*, 123730. https://doi.org/10.1016/j.biortech.2020.123730.

Yazıcı, H., Kumbul, B. S., & Ekinci, K. (2024). Assessment of the effectiveness of process-induced struvite crystallization process on the recovery of nutrients from high-strength liquid digestate of an anaerobic biogas plant. *Journal of Environmental Chemical Engineering, 12*(2). https://doi.org/10.1016/j.jece.2024.112129.

Yenigün, O., & Demirel, B. (2013). Ammonia inhibition in anaerobic digestion: A review. *Process Biochemistry, 48*, 901–911. https://doi.org/10.1016/j.procbio.2013.04.012.

Zhou, K., Chen, L., Wu, J., Li, H., Wang, H., & Li, J. (2024). The carbon emissions calculation, decomposition and peak scenario simulation of meat, egg, and milk in China: from a production and consumption perspective. *Clean Technologies and Environmental Policy*. https://doi.org/10.1007/s100988-024-02765-7.

Chapter 7

Biomethane production and utilization in the energy and transportation sector

Mirko Barz[1], Hartmut Wesenfeld[2], Asnakech Lass-Seyoum[3], Carla Wunsch[3], and Zinnabu Tassew Redda[4]
[1]*Faculty 1, Engineering, Hochschule für Technik und Wirtschaft Berlin (HTW Berlin), Berlin, Germany,* [2]*Faculty II, Mathematics - Physics - Chemistry, BHT Berliner Hochschule für Technik, Berlin, Germany,* [3]*Faculty of Engineering – Energy and Information, SC Renewable Energy, HTW Hochschule für Technik und Wirtschaft, Berlin, Germany,* [4]*Addis Ababa University, Institute of Technology, Addis Ababa, Ethiopia*

7.1 Biogas technologies and utilization pathways—A general overview

Anaerobic digestion (AD) is a biochemical process in which microorganisms break down organic material in anaerobic environments and produce biogas as a combustible gas mixture. The process is widely disseminated in nature and takes place in the rumen of ruminants, in wetlands or swamps, at the bottom of lakes or in slurry tanks.

The formation of methane in AD processes involves various groups of bacteria and archaea that work together in four complicated and interdependent biochemical reactions, namely hydrolysis, acidogenesis, acetogenesis, and methanogenesis, which lead to the degradation of organic material and the production of methane (CH_4), carbon dioxide (CO_2), and as well as trace amounts of hydrogen (H_2), hydrogen sulfide (H_2S), ammonia (NH_3), and nitrogen (N_2) and various volatile organic compounds (Harirchi et al., 2022). A simplified scheme of the decomposition of organic matter during AD is shown in Fig. 7.1. Each of the shown steps is carried out by different groups of microorganisms.

During the first hydrolysis step, complex organic compounds such as carbohydrates, proteins, and fats are broken down into simpler molecules by enzymes produced by hydrolytic bacteria. Large molecules like starch, cellulose, and proteins are converted into smaller molecules like simple

Biogas in the Circular Economy. DOI: https://doi.org/10.1016/B978-0-443-29230-9.00006-1

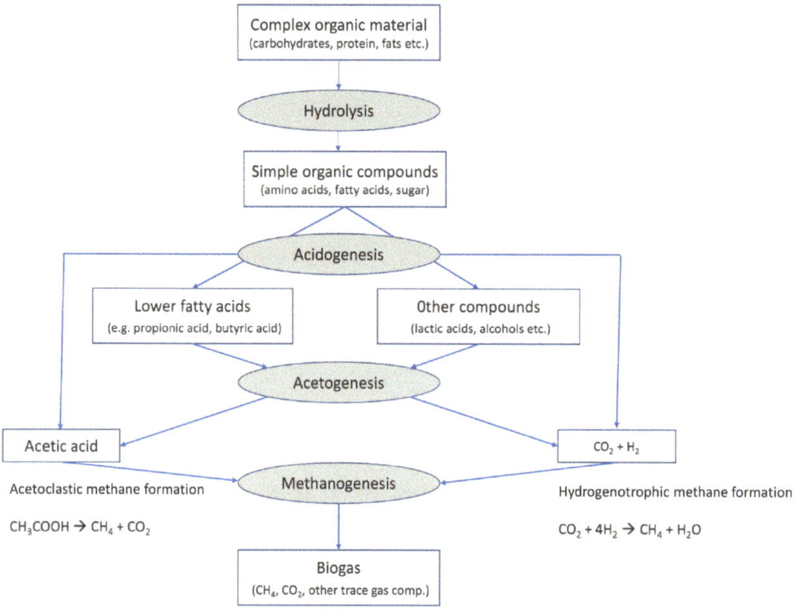

FIGURE 7.1 Scheme of the decomposing process of organic matter within AD. *Adapted from FNR: Agency for Renewable Resources, (2013).*

sugars, amino acids, and fatty acids and make the organic matter more accessible to other groups of microorganisms in the subsequent steps.

In the second acidogenesis step, acid-forming bacteria further break down the simpler molecules produced during hydrolysis into volatile fatty acids (VFAs) and alcohols, such as acetic acid, propionic acid, and butyric acid. Acidogenesis results in the production of organic acids and other intermediate compounds, along with hydrogen and carbon dioxide.

In the third acetogenesis step, acetogenic bacteria utilize the VFAs and other organic acids produced in the acidogenesis step to convert these intermediate products into acetate, hydrogen, and carbon dioxide through biochemical reactions. Acetogenesis is an important step as it generates acetate, which serves as a precursor for the methane production by acetoclastic methanogenesis in the next step.

During the fourth and last step, called methanogenesis, specialized microorganisms like methanogenic archaea utilize the acetate, hydrogen, and carbon dioxide which are produced in the previous steps for biochemical reactions to produce methane (CH_4) and carbon dioxide (CO_2) as metabolic byproducts. Methanogenesis is the final step in AD and is responsible for the production of biogas, which mainly consists of methane along with some carbon dioxide and traces of other gases.

These four steps work together in a complex biochemical process to decompose organic matter and produce biogas as a renewable energy source. Each step relies on specific groups of microorganisms that sequentially break down the organic matter into simpler molecules until methane is produced.

The biogas itself is a mixture of various gas components, consisting mainly of methane (45%–75%), carbon dioxide (25%–45%), water vapor, and other gases such as hydrogen sulfide, ammonia, nitrogen, and oxygen. The relative content of CH_4 and CO_2 in biogas mainly depends on the substrate being digested, used digestion technologies and process parameters of the digestion system such as hydraulic retention time (HRT), process temperature, organic loading rate (OLR), pH values in the digester, or mixing systems used in the fermenter (Barz et al., 2021). The produced biogas can be used directly as a fuel to generate heat and electricity in combined heat and power (CHP) plants or processed into biomethane, which can be fed into gas grids and used for various purposes, for example, as a fuel for the transportation sector or as a chemical raw material (Barz et al., 2022). The use of biogas and the production of biomethane plays an important role in the transformation of the energy sector toward climate protection and sustainability and also contributes to increasing the energy security of national energy economies, as biogas can replace conventional fossil fuels, which are often imported from politically unstable regions.

Various concepts are applied to the installations used for biogas production and the type of AD technology, and the design of the biogas reactor depends on the intended application (small or large scale, CHP plant, or biomethane production) and the properties of the available substrates (Lisowyj & Wright, 2020). The concepts differ according to feedstock type and process characteristics, such as the dry-matter content, process temperature in the digester, the way in which the substrates are fed into the digester, or the number of process phases implemented into the design (FNR: Agency for Renewable Resources, 2013). A classification of various applicable biogas design concepts is shown in Table 7.1.

In organic waste-based biogas plants, the biogas is produced from various organic waste materials such as agricultural residues (crop residues and animal manure), municipal solid waste, sewage sludge, food waste, and organic industrial waste. In energy crop-based biogas plants, the biogas is produced from dedicated energy crops grown specifically for biogas production, such as maize, sorghum, sugarcane, and energy grasses like switchgrass or miscanthus. The co-digestion process involves the simultaneous digestion of multiple feedstocks, often combining different organic wastes with energy crops or other substrates to optimize biogas yield and improve process stability. A further classification can be made by the dry-matter (dm) content of the used substrates, whereby wet fermentation systems use organic material with a

TABLE 7.1 Classification of biogas processes according to different criteria.

Criterion	Design concept
Feedstock type	• Organic waste-based biogas plants • Energy crop-based biogas plants • Co-digestion biogas plants
Dry-matter (dm) content of the substrate	• Wet fermentation (dm < 20%) • Dry fermentation (dm > 20% < 40%)
Process temperature	• Psychrophilic (T < 25°C) • Mesophilic (25°C–40°C) • Thermophilic (41°C–55°C)
Type of feeding	• Discontinuous (batch) • Continuous • Semicontinuous
Number of process phases	• Single-phase process • Two-phase process
End use of the biogas	• Heat generation • Electricity generation • Combined heat and power generation • Biomethane as vehicle fuel

consistency of 10%–20% dry matter or less and dry fermentation systems for organic matter with consistency of 20%–40% dry matter or more (Stolze et al., 2015). Dry fermentation systems often use batch or plug-flow digesters with airtight chambers to contain the solid substrate. These digesters may have different configurations to facilitate gas collection and substrate loading. Wet fermentation systems can utilize various digester designs, including continuous stirred tank reactors (CSTRs), plug-flow digesters, or anaerobic lagoons, depending on the specific characteristics of the substrate and operational requirements. A further important parameter for the implementation is the process temperature for the AD. Three different temperature ranges of AD are defined, depending on the suitable range for microorganisms to grow and metabolize the substrates. Psychrophilic conditions typically range from 10°C to 25°C, mesophilic conditions from 25°C to 40°C with an optimal performance at around 37°C, and thermophilic conditions in the range between 41°C and 55°C. The use of higher temperatures above 55°C, for example, at 70°C is a relatively new trend, but until now without practical relevance. Worldwide, especially in developing countries, most biogas plants are operated under psychrophilic conditions in a temperature range below 25°C. Most

European biogas plants are operated under mesophilic conditions and moderate temperatures typically ranging from 25°C to 40°C, because of low microbial activity and biogas production rates under psychrophilic conditions. Thermophilic conditions are mostly used for the use of critical organic waste components (e.g., the organic fraction of municipal solid waste or slaughterhouse waste due to it being a pathogen-free process in addition to the enhanced biogas yield and short HRT (Singh et al., 2023).

Biogas can be utilized in various end-use applications. Some common end-use options for biogas include the following:

Electricity generation: Biogas can be used to generate electricity through CHP systems or gas turbines. The electricity generated can be used onsite or fed into power grids.
Heat generation: Biogas can be used directly for heating purposes in industries, residential buildings, or other facilities. It can be utilized in boilers, furnaces, or combined with other fuels in heating systems.
Combined heat and power: CHP systems, also known as cogeneration systems, utilize biogas to generate both electricity and heat simultaneously. This integrated approach increases overall energy efficiency compared with-separate generation methods.
Vehicle fuel: Biogas can be processed and upgraded to biomethane, suitable for use as a vehicle fuel to replace natural gas used in compressed natural gas (CNG) vehicles or as a substitute for diesel in trucks and buses.
Injection into natural gas grid: Biomethane produced from biogas can be injected into the existing natural gas grid, blending with fossil natural gas. This enables wider distribution and utilization of renewable energy in various sectors.
Biochemicals and biofuels: Biogas can be converted into bio-based chemicals or biofuels through further processing steps such as gas cleaning, upgrading, and conversion. This can include production of ethanol, hydrogen, or other valuable chemicals.
Combustible gas for cooking and lighting: In regions where access to traditional energy sources, like electricity or clean cooking fuels, is limited or unreliable, biogas can be used for lighting and cooking purposes.
Remote and off-grid applications: Biogas systems can provide decentralized energy solutions for remote or off-grid areas where access to conventional energy sources is limited. They can power community facilities, farms, or small-scale industries.

The implementation of biogas technologies covers a very wide spectrum of components and combinations. However, it should be noted that an analysis of the plant and system suitability and a capacity adjustment is required in each individual case for the design of a biogas plant depending on the intended use (FNR: Agency for Renewable Resources, 2010).

Household biogas plants are typically small-scale systems designed to cater to the energy needs of a single household or a small community. Design considerations include simplicity, ease of maintenance, and the ability to process household organic waste effectively.

Community biogas plants are designed to serve larger communities or clusters of households. They require higher processing capacities and more sophisticated systems to handle an increased waste input. Considerations include scalability, community participation, and integration with existing waste management infrastructure.

Agricultural biogas plants are medium- and large-scale applications that use organic materials from agricultural (e.g., special grown energy plants like maize or grasses, often in combination with agricultural organic waste and residues from animal breeding) to produce biogas for various end-use purposes.

Industrial biogas plants are larger-scale facilities typically integrated with industrial processes or agricultural operations. Design criteria include high processing capacities, advanced waste pre-treatment systems, and compatibility with industrial waste streams. Gas purification may also be necessary for certain industrial applications.

Further design criteria are the availability of suitable feedstocks (substrates), the climate and environmental conditions at the intended plant site to ensure that the chosen technology is suitable for the local climate and complies with relevant environmental standards, as well as the technical feasibility and the economic viability of the biogas projects. There is a wide range of AD system designs primarily due to different feedstock types and applications in specific environments (IEA: International Energy Agency, 2020). Most common digester types are as follows:

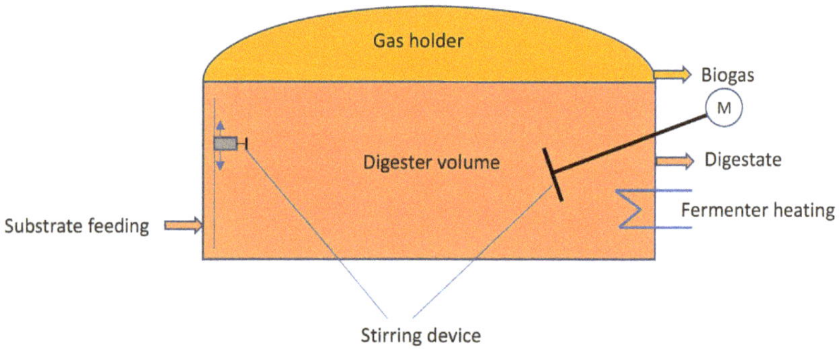

FIGURE 7.2 Scheme of a continuous-flow stirred tank reactor.

Continuous-flow stirred-tank reactor: In CSTR systems (Fig. 7.2), the biomass fed into the reactor is constantly mixed with stirring devices to ensure a perfect homogenization at controlled temperatures. Depending on the feedstock and its properties, different mixing devices can be installed. Mixing prevents floating and sinking layers in the digesters and ensures uniform conditions of availability of nutrients to microorganisms and uniform temperature conditions in the system. The dry-matter contents in CSTR systems are in the range between 5% and 12% dm. The retention time is variable and depends on the types of biomasses used to feed the plant. The CSTR shows high process stability and an effective retention time which is lower than theoretical average retention times (Barchmann et al., 2016; Lindmark et al., 2014).

Plug-flow digester: In the most cases, plug-flow digesters (Fig. 7.3) are lying tanks (round or rectangular) with several horizontal agitators that mix the substrate in different zones but also move it forward in the digester. The digester is a horizontal performing system and fed on one side of the digester. The feedstock moves through the digester as a "plug" and leaves the digester on the other side (Rosenwinkel et al., 2015). Dry-matter content of the substrate might be higher compared with conventional CSTR systems up to 20% and more since the systems are suitable for dry fermentation processes.

Upflow anaerobic sludge blanket digester: In UASB systems (Fig. 7.4), a blanket of granular sludge is formed and kept in suspension in the reactor. UASB digester systems are mostly used for substrates with low dm contents such as for wastewater treatment (e.g., wastewater from breweries or municipal wastewater). A blanket of granular flocs containing an assortment of anaerobic microorganisms is suspended in the digester. The treated wastewater should not contain particles and should have a maximum concentration of 3% total suspended solids. The reactor has a simple structure and does not require any mixing device and is equipped with a set of gas–liquid solid separator to separate solids (granules) from the effluent (Lim & Kim, 2014).

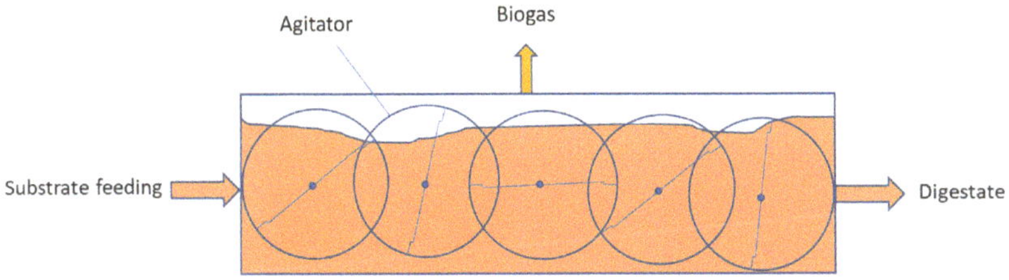

FIGURE 7.3 Schematic of a plug-flow reactor.

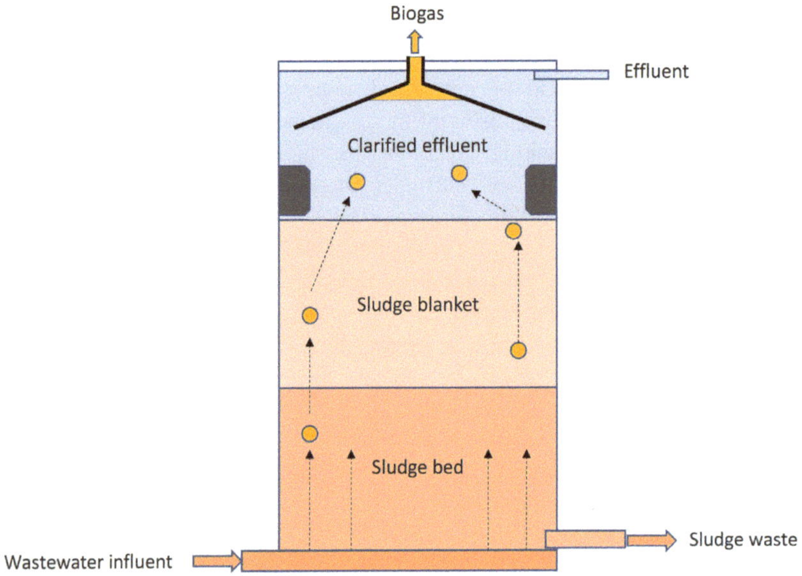

FIGURE 7.4 Schematic of an UASB digester.

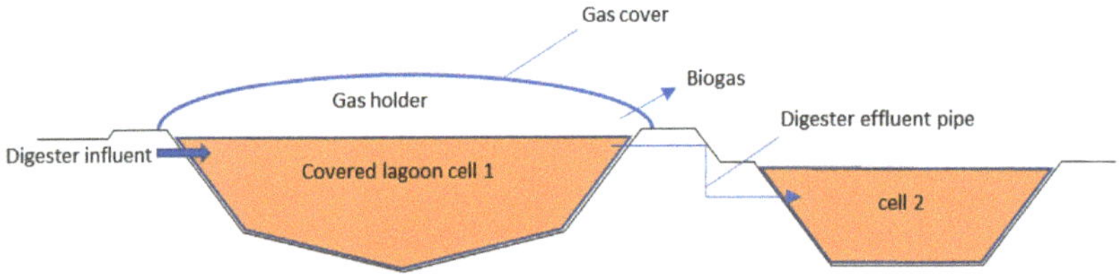

FIGURE 7.5 Scheme of a covered lagoon digester.

Covered lagoon digesters (Fig. 7.5) are the simplest and cheapest solutions for AD and widely distributed in developing countries. The digester itself is an impermeable pit or basin where organic waste with low solid matter content (usually less than 4%), such as liquid animal manure, or wastewater from industrial processing of agricultural goods (e.g., palm oil mill effluent), is digested. A durable cover made of flexible materials like polyethylene (HDPE) is installed over the lagoon to create an airtight environment and facilitate the capture of biogas, caused by the simplicity they are operated under psychrophilic conditions at the temperature of the ambient soil without any heating device and the retention times is longer compared with other solutions, ranging from 20 to 180 days depending on the location and the respective climate and temperature conditions (O'Connor et al., 2021).

7.2 Status quo of biogas and biomethane utilization in Europe and worldwide

Biogas production began to grow in the 1990s especially in Europe and has been rising since then. Over the last 25 years, the global biogas production has increased enormously, increasing almost sixfold and reached 445 TWh in 2022. Almost half of the production is based in Europe (Fig. 7.6), with Germany alone meeting almost 20% of global consumption. Other big biogas-consuming countries are China (21%), the United States (12%), and India (9%) (IEA: International Energy Agency, 2024).

The installed capacity of biogas energy worldwide reached about 21.5 GW in 2022, but there are major differences in the production and use of biogas between developed and developing countries. During in the so-called developed countries, biogas is mainly produced from large-scale agricultural or industrial biogas plants and used for CHP production or the production of biomethane, which can be injected in natural gas grids; the biogas production in developing countries occurs often in simple and small-scale household or community-based biogas plants to provide gas for cooking and lighting and/or to reduce firewood consumption (Brémond et al., 2021).

With about 18,774 biogas plants in 2020 (Motola et al., 2022), Europe is the world leader in biogas electricity production and biomethane production. The implementation of the highly developed biogas plants in Europe has been

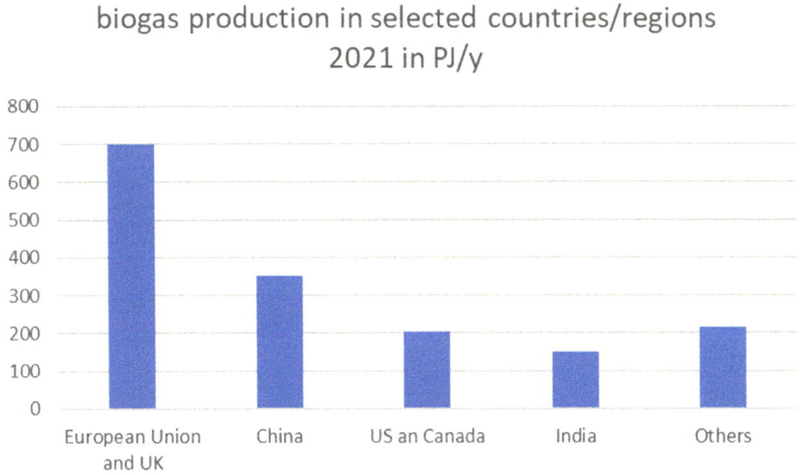

FIGURE 7.6 Main biogas-producing regions 2021.

driven mainly by Europe's renewable energy-related policies and regulations, comprehensive agricultural regulations and waste management-related policies (Vasco-Correa et al., 2018). The EU has established the ambitious goal to reach 80%–95% greenhouse gas (GHG) emissions reduction by 2050, and some EU countries have integrated favorable policies toward renewable sources into their general energy policy and planning framework. With around 200 incentives, the EU has created favorable conditions for the promotion of biogas electricity and 18 EU countries have introduced a feed-in tariff system for biogas electricity (Banja et al., 2019).

Besides using biogas for electricity production, the production of biomethane is a new and increasing trend in the European biogas sector. Especially since Europe has felt the effects of an extremely volatile market for fossil natural gas in recent years, the production of biomethane is becoming increasingly important. Biomethane is defined as all methane of organic/renewable origin, and it can be obtained by means of separation of carbon dioxide from raw biogas using conventional biogas upgrading plants or through methanation of hydrogen and carbon dioxide (Mignogna et al., 2023). According to the European Biogas Association, Europe reached a total of 1,323 biomethane-producing facilities by April 2022 with a continuously expected increase for the future. The development of biogas production volume is shown in the typical composition of biogas and biomethane in shown in Table 7.2.

7.3 Production of biomethane as flexibility option for the biogas industry

Upgrading biogas to biomethane generates new possibilities for its use since it can replace natural gas, which is used extensively in many countries worldwide. Europe is the world's leading producer of biomethane for the use as a

TABLE 7.2 Typical composition of biogas and biomethane.

Gas component	Biogas	Biomethane
Methane (CH_4)	45%–75%	96%–99.9%
Carbon dioxide (CO_2)	25%–45 %	0.1%–4%
Nitrogen (N_2)	< 2%	< 1%
Oxygen (O_2)	< 2%	< 1%
Hydrogen (H_2)	< 1%	Traces
Hydrogen sulfide (H_2S)	20–20,000 ppm	< 10 ppm

vehicle fuel or for injection into the natural gas grid and the number of biogas plants equipped with biomethane upgrade facilities is rapidly increasing in Europe, with 1,323 plants in 2022 (European Biogas Association, 2023).

The total combined biogas and biomethane production in Europe reached 21 bcm in 2022 with a share of 20% (4.2 bcm) of biomethane (Fig. 7.7).

According to the REPowerEU Plan of May 18, 2022, the EU's biomethane production, either as biogas or its upgraded version, needs to reach 35 billion cubic meters (bcm) per year by 2030 and the estimated investment need for the period amounts to €37 billion.

The future perspective of biomethane production appears promising for several reasons. As countries worldwide commit to reducing greenhouse gas emissions and transitioning to renewable energy sources, biomethane is gaining attention as a versatile and sustainable alternative to fossil fuels. Its ability to be produced from various organic feedstocks makes it a valuable component of the renewable energy mix and contributes to the renewable energy transition. Furthermore, biomethane can play a crucial role in decarbonizing various sectors, including transportation, heating, and industry. Its use as a low-carbon fuel can help reduce emissions of greenhouse gases and pollutants. Biomethane production also promotes the concept of a circular economy by utilizing organic waste streams such as agricultural residues, food waste, and wastewater to generate renewable energy. This not only helps in waste management but also creates economic value by turning waste into a valuable resource.

Many governments and international organizations are implementing policies and incentives to support biomethane production and its integration into energy systems. These policies include renewable energy targets, feed-in

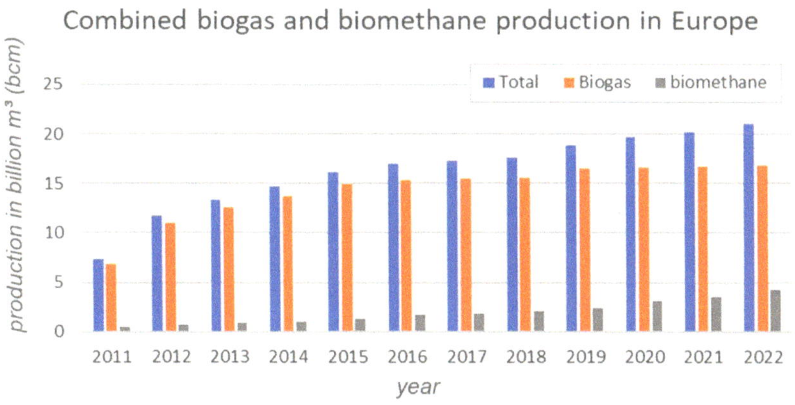

FIGURE 7.7 Combined biogas and biomethane production in Europe 2011–2022. *Adapted from European Biogas Association, (2023).*

tariffs, tax incentives, and sustainability standards, which create favorable conditions for investment and development in the biomethane sector. Last but not least, the increasing demand for renewable energy, coupled with growing awareness of the environmental and social benefits of biomethane, is creating new market opportunities for biomethane producers, suppliers, and technology providers. Biomethane can also enhance energy security by diversifying the energy sector and reducing dependence on imported fossil fuels (Sulewski et al., 2023).

Overall, the future of biomethane production looks promising as it aligns with the goals of sustainability, decarbonization, and resource efficiency. Continued technological innovation, supportive policies, and market developments are expected to drive further growth and adoption of biomethane as a renewable energy solution in the coming years.

7.4 Analysis of the role of biomethane production for a climate-friendly energy industry

Biogas and biomethane production play significant roles in fostering a climate-friendly energy industry since they are promising forms of renewable energy that can be obtained from a wide variety of biogenic energy sources. Residues from agriculture, animal husbandry, the biogenic portion of municipal waste, various industrial and food waste, and special grown energy plants such as maize are excellent substrates for the production of clean and sustainable energy as part of a circular economy. Their conversion into biogas and biomethane through the AD process is an efficient solution for the treatment of waste from different origins and can contribute to build a circular economy through the valorization of organic residues and wastes. Compared with the traditional combustion of biogas in cogeneration units (CHP), upgrading biogas to biomethane can bring both environmental benefits (reduction in greenhouse gas emissions) and economic benefits (higher efficiency of energy use contained in biomass) (Sulewski et al., 2023). The production and use of biomethane bring important environmental benefits, such as the reduction of GHG emissions compared with those generated by using traditional fossil fuels. The prospects for biomethane, as well as their methods of generation, have increased exponentially, always aiming at sustainable energy development in support of the green transition (Mignogna et al., 2023). The European Union is currently the largest producer of biomethane in the world and has significant potential for further development (Sulewski et al., 2023). In addition, biogas and biomethane production plays an important role in promoting several sustainable development goals (SDGs), defined by the United Nations by promoting renewable energy, sustainable resource management, climate protection, and innovation and collaboration for a more sustainable future. The following SDGs can be addressed by biogas and biomethane generation and utilization pathways:

SDG 7: Affordable and clean energy: Biogas and biomethane are renewable energy sources that can be used for electricity generation, heating, and transportation. By replacing fossil fuels with biomethane, countries can reduce the dependency from imports and reduce greenhouse gas emissions. The use will further improve air quality, contributing to the goal of ensuring access to affordable, reliable, sustainable, and modern energy for all.
SDG 9: Industry, innovation, and infrastructure: Biogas and biomethane production requires investment in infrastructure, technology, and innovation. Developing biogas and biomethane facilities promotes industrial growth and innovation, while also enhancing infrastructure for sustainable energy production and distribution.
SDG 11: Sustainable cities and communities: Biogas and biomethane production helps address waste management challenges in urban areas by diverting organic waste from landfills and reducing methane emissions. Biogas plants can be integrated into urban waste management systems (e.g., AD of the solid fraction of municipal solid waste) and contributing to cleaner and more sustainable cities.
SDG 12: Responsible consumption and production: Biogas and biomethane support responsible consumption and production by utilizing organic waste streams that would otherwise be disposed of in landfills or incinerated. By converting waste into energy, biomethane production promotes resource efficiency and reduces negative environmental impacts.
SDG 13: Climate action: Biogas and biomethane production mitigates greenhouse gas emissions by capturing methane from organic waste and converting it into a renewable energy source. By reducing methane emissions and displacing fossil fuels, biomethane contributes to efforts to combat climate change and its impacts.
SDG 15: Life on land: Biogas and biomethane production promotes sustainable land use by utilizing agricultural residues, energy crops, and organic waste as feedstocks. This reduces pressure on land resources and ecosystems, contributing to the conservation and sustainable management of terrestrial ecosystems.
SDG 17: Partnerships for the goals: Achieving the SDGs requires collaboration and partnerships among governments, businesses, civil society, and other stakeholders. Biogas and biomethane projects often involve partnerships between public and private entities, as well as collaboration across sectors such as agriculture, energy, and waste management.

One of the key actions proposed with the REPowerEU plan was the creation of a Biomethane Industrial Partnership (BIP) to promote the sustainable production and use of biomethane. The BIP is an industrial partnership in which policymakers, industry, and other stakeholders team up with the goal to support the achievement of the target of 35 bcm annual production and use of sustainable biomethane by 2030, and to create the preconditions for a further ramp-up of its potential toward 2050 (Biomethane Industrial Partnership, 2022).

7.5 Potential for substitution of natural gas by biomethane

Biomethane is becoming a promising energy source across the world and will be a key factor in achieving net-zero GHG emissions (Marconi & Rosa, 2023) and achieving sustainability in the energy sector. It can be fed into the existing gas infrastructure and thus be made available to a large number of conventional natural gas applications. In the context of rising gas prices caused by the Russian war in Ukraine, biomethane can potentially play several roles in mitigating the impacts such as price stability, price moderation, and energy security. In general, biomethane [also known as renewable natural gas (RNG)] can be produced through several pathways, such as AD followed by biogas upgrade, thermochemical conversion of organic matter, for example, thermochemical gasification of biomass followed by a synthesis with hydrogen or by the direct chemical synthesis of gaseous carbon sources, such as CO_2 with hydrogen. Today, about 90% of the biomethane production results from biogas upgrade plants, but new developments of power to gas (PtG) technologies are showing huge potentials for a future increase of these technologies. Besides technical development based on innovative research and development results, many regions worldwide are implementing policies and incentives to promote biomethane/RNG production and utilization pathways. These include feed-in tariffs, renewable energy credits, carbon pricing mechanisms, and renewable fuel standards, which facilitate market penetration and investment in biomethane infrastructure. Europe is the leading producer of biomethane, and it has been increasingly injecting it to the current natural gas network (de Carvalho et al., 2023).

According to the European Biogas Association, Europe is producing today 21 bcm of biogas and biomethane combined with a share of 20% (4.2 bcm) of biomethane. Calculations show that reaching the 35 bcm biomethane production capacity goal by 2030 (as outlined in the REPowerEU plan) an annual growth rate of about 30% in biomethane production is required. Several studies show that enough sustainable feedstocks are available in the EU to meet the 35 bcm target until 2030 with a further increase up to 153 bcm (equivalent to 1.5 thousand TWh) estimated for 2050 (Sulewski et al., 2023). Globally, biomethane can offset 67% (6415 TWh per year) of natural gas net imports worldwide. Furthermore, replacing fossil natural gas with biomethane will decrease emissions from natural gas systems by 11%, equivalent to 1.1 Gt CO_2-eq per year (Marconi & Rosa, 2023).

7.6 Required biomethane product quality and utilization requirements

Biomethane (also known as RNG) is a purified form of biogas that meets the quality standards for natural gas. Since the chemical composition and energy content of biomethane are close to natural gas, it can likewise be used in the same way:

- Gas grid injection and used as a natural gas substitute,
- Vehicle fuel.

To allow injection of biogas into the natural gas grid or the use as a vehicle fuel, it must be upgraded which means that carbon dioxide and other impurities are removed, whereas the share of methane is increased to usually above 96% (Savickis et al., 2020) so that it meets the quality standards for natural gas. Usually, the gas standards on grid injection and transport use are defined by natural gas properties. The CEN/TC 408 (European Committee for Standardization, Project Committee—Natural gas and biomethane for use in transport and biomethane for injection in the natural gas grid) developed specifications for biomethane (EN16723-1 for injection into natural gas grids and EN16723-2 for transport fuel), but the cited test methods are not harmonized and validated (CEN: European Committee for Standardization, 2023a, 2023b).

7.6.1 EN 16726

This standard defines the gas quality standards for natural gas. It establishes technical rules for gas pressure equipment and operations to improve the interoperability of gas systems in EU Member States.

7.6.2 EN 16223-1 and EN 16223-2

These standards focus specifically on the injection, integration, and safe use of biomethane in the natural gas network and for transport. They consider the specific requirements of biomethane and contribute to standardization.

Related to biomethane, the EU Reference Standards and other technical specifications vary from country to country. All countries injecting biomethane into the grid have developed their own national quality standards. For example, the Netherlands let biomethane with methane content of only 85% be injected, but Switzerland and Sweden require 96% and 97% of methane content (Savickis et al., 2020).

Germany has a standard for biogas injection (G262) (DVGW: German Technical & Scientific Association for Gas and Water, 2011) that has been elaborated in cooperation between the German Water and Gas Association and the German Biogas Association. The standard is based on the German standard for natural gas (G260) (DVGW: German Technical & Scientific Association for Gas and Water, 2021), see Table 7.3.

7.7 Concepts for biomethane production and utilization

All concepts for biomethane production and utilization need to contain the steps of biogas *upgrade*, *conversion*, *distribution* and *utilization*. The sequential arrangement of the different process steps varies between the different concepts. Also, the efficiency becomes less if more conversion steps are needed. The aim of every concept is reaching a high efficiency under technical feasible and economic profitable conditions.

Processes for biogas upgrade, conversion, and utilization will be discussed in the following Sections 7.8 to 7.10.

The distribution of the raw biogas or the products biomethane or electricity should also be regarded. Besides the distribution the storage also has an important role, biogas and biomethane can be stored in natural caverns, artificial tanks, or within the distribution network. Energy produced from biomethane can be electricity and/ or heat that can be stored as electrical or mechanical energy as well as thermal or chemical energy. Therefore, a wide range of potential storage devices is applicable (Sterner and Stadler, 2017). In principle, *centralized* and *decentralized* systems can be implemented.

Centralized systems have a high efficiency, and central plants can be technically more complex and advanced. The effort for service and

TABLE 7.3 Requirements for gas injection into the national gas grid according to German standard G260/G262.

Term	RNG quality requirement according G260/G262
Caloric value	8.4–13.1 kWh/m³
Wobbe Index	H-Gas: 13.6–15.7 kWh/m³ L-Gas: 11.0–13.0 kWh/m³
Relative density	0.55–0.75
Water content	200 mg/m³ (MOP ≤ 10 bar) 50 mg/m³ (MOP > 10 bar)
Dust	Technically free
CH_4	H-Gas: ≥95% L-Gas: ≥90%
CO_2	< 6% (regulated by min CH_4 content)
O_2	MOP < 16 bar, max 3% MOP ≥ 16 bar, max 0.01%
H_2	< 10%
H_2S	max. 5 mg/m³

maintenance is concentrated at one location saving operation costs. However, central plants require large distribution systems, either for the feed (biogas) or for the products (biomethane or electricity and heat). The investment for distribution systems can be higher than the investment for the central plant. This disadvantage can be compensated if existing infrastructure can be used for distribution.

An example for a centralized system is the application of the PtG concept. Fig. 7.8 illustrates the basic concept.

If the location of the biogas plant allows a connection to a central natural gas distribution network, biogas can be upgraded by separation and/or methanation of the CO_2 share. This allows feeding into the distribution network and utilization in central combined gas and steam power plants which have a high electrical efficiency of 60%. If the biogas upgrade is performed by methanation of the CO_2 share with green hydrogen, produced by electrolysis with power from renewable sources, the system becomes a circle in which power from renewable energy sources can be stored as chemical energy in the form of CH_4 that can be utilized as power again if there is a high power demand (Zapf, 2022). The overall efficiency of this power storage system is about 40%.

Decentralized systems are not so efficient because the technical installations are divided into small sections and widespread distributed. Therefore, synergetic effects cannot be generated and used as well as in central facilities. The effort for service and maintenance is distributed to many locations producing higher operation costs because for example the specialized technical

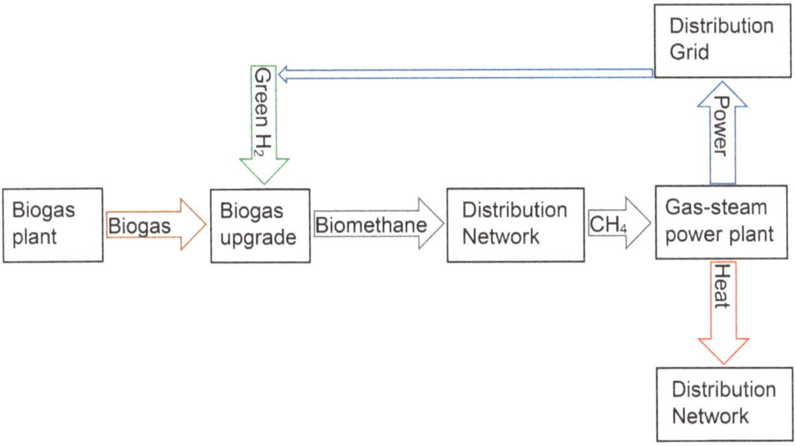

FIGURE 7.8 Power to gas concept—centralized biomethane utilization for power supply.

experts are not available at any facility and no central procurement for working materials and operation supplies is possible.

Depending on the structure of a country, biogas is produced in plants with a wide range of scale, with sizes containing a few households requiring *decentralized* systems up to large agricultural or wastewater treatment plants with 100,000 to 1,000,000 connected inhabitants requiring *centralized* systems. In any case, a large flexibility in the concepts for managing the output of biogas plants is necessary (Barz et al., 2021).

Decentralized systems will be described in the following passage by the example of direct biogas utilization in decentral CHP units, illustrated in Fig. 7.9.

The most common system for biogas plants with about 10,000 connected inhabitants or above is the direct power generation from the produced biogas in CHP units. The biogas plants are mostly located in rural areas. Therefore, distribution networks for power and heat are seldom reachable, and the energy will be used by the producing biogas plant. The electrical efficiency of biogas-driven motors used in CHP units is about 40% (Zahoransky, 2022). A demand for electricity exists around the whole year, whereas the demand for heat in summertime is less than the produced heat. This means that less biogas can be utilized during the warm season, or the heat must be destroyed by electrical coolers, both reducing the efficiency of the biogas utilization.

There are restraints and some disadvantages in the conventional centralized and decentralized systems as described above. The more flexible the system and utilization options are, the higher is the utilization efficiency and the less is the direct carbon dioxide impact of the produced biogas to the environment. Replacement of conventional fuels by biomethane or secondary products is one more option to expand the variety of biogas utilization. As later described in Section 7.10, biomethane can directly be used as fuel for cars. Biofuel filling stations could be established close to biogas plants for a decentralized utilization. Another possibility is feeding the produced biomethane into the natural gas distribution network and transporting it to refineries for conversion into other synthetic fuels, for example, methanol or bio-dimethyl ether (DME). This flexible utilization option is illustrated in the following Fig. 7.10.

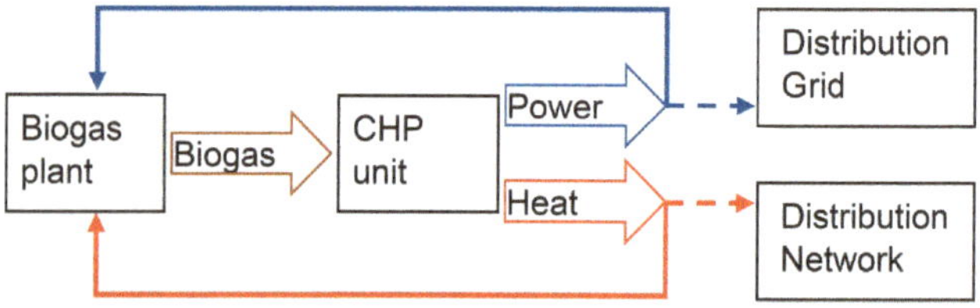

FIGURE 7.9 Conventional utilization of biogas in decentralized CHP units.

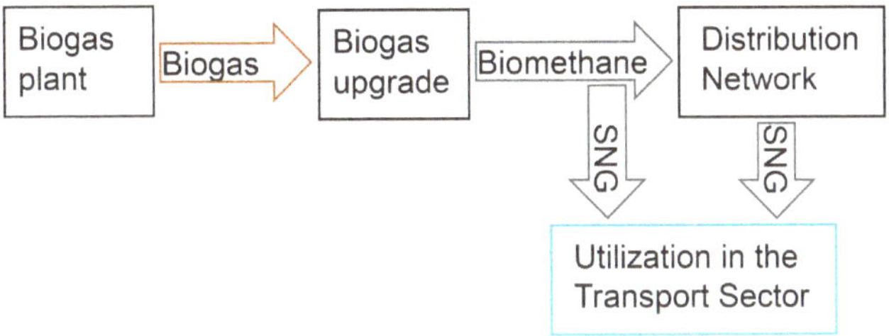

FIGURE 7.10 Flexible use of biomethane in the transportation sector.

Utilization concepts for biomethane in the transportation sector can cover the following options:

- Combustion in vehicle motors,
- Conversion to other fuels,
- Direct or indirect use in fuel cells for power production.

The utilization of direct combustion in conventional vehicle motors is possible and only needs additional CH_4 tanks. This option will be described later in Section 7.10.

The other two options will be described below.

Conversion to other fuels

CH_4 can be converted to H_2 via pyrolysis or steam reforming. Conventional steam reforming of CH_4 produces H_2 in a mixture with carbon monoxide, so-called synthesis gas (Syn'gas). Steam reforming with addition of CO_2 leads to the reactant mixture for methanol synthesis, see Table 7.4. Besides the use in fuel cells, methanol can be also used in the process and product chain within the methanol to gasoline process. Methane pyrolysis for hydrogen production is a process that is still investigated. Currently, plants for methane pyrolysis exist in pilot plant scale. The other processes are well known and have been applied in industrial scale (Arpe & Weissermel, 2003).

There are various alternative processes available that lead to the products in Table 7.4. For example, hydrogen can be obtained by electrolysis and methanol can be produced more directly by the gasification of biomass. The processes listed in the table assume methane being the originator having the advantage that it is partly produced from renewable biomass in biogas plants and transported via the existing natural gas infrastructure. Therefore, CH_4 offers a storage and transportation option for hydrogen that can be regained from CH_4, because the storage and transportation of CH_4 is state of the art and not so challenging as the comparable processes for hydrogen.

Fuel cells

TABLE 7.4 Possible reaction pathways for the conversion of CH_4 to other fuels. CH_4 can be converted to H_2 via pyrolysis or steam reforming. Conventional steam reforming of CH_4 produces H_2 in a mixture with carbon monoxide, so-called synthesis gas (Syn'gas). Steam reforming with addition of CO_2 leads to the reactant mixture for methanol synthesis.

Reaction	Reaction equation	Product
Methane pyrolysis	$CH_4 \rightarrow C + 2\ H_2$	H_2
Steam reforming	$CH_4 + H_2O \rightarrow CO + 3\ H_2$	Syn'Gas, H_2
Steam reforming + CO_2 a) Steam reforming b) Water gas shift c) CO_2 addition	 $3\ CH_4 + 3\ H_2O \rightarrow 3\ CO + 9\ H_2$ $3\ CO + 3\ H_2O \rightarrow 3\ CO_2 + 3\ H_2$ $CO_2 \rightarrow CO_2$	
Total reaction (a + b + c) *ratio*	$3\ CH_4 + 6\ H_2O + CO_2 \rightarrow 4\ CO_2 + 12\ H_2$ 1 : 3	Syn'Gas for methanol production
Methanol production	$CO_2 + 3\ H_2 \rightarrow CH_3OH + H_2O$	Methanol

An overview about the different kinds of fuel cells working with fuel and oxygen and the applicable fuels is given in Table 7.5. The molten carbonate fuel cell (MCFC) does not directly work with CH_4 (CH_4 will first be converted to synthesis gas) and needs a high temperature. The solid oxide fuel cell (SOFC) also needs a high operation temperature and is in an early stage of development. Therefore, additional equipment is necessary, and it is not applicable for the use in small vehicles like cars.

7.8 Technologies for conventional biogas upgrade to biomethane

Depending on the feedstock and process conditions (Section 7.1), biogas can contain various portions of carbon dioxide that has to be converted to methane or removed for the biomethane applications described in Sections 7.2 to 7.4.

The conversion of the carbon dioxide can be performed by biological and catalytic processes that will be described in Section 7.9 and that are still at the research stage or at the transfer stage to an industrial application.

Removing or separation of the carbon dioxide can be performed by industrial state-of-the-art processes. These processes can be mechanical processes, such as adsorption, physical absorption, or membrane separation, chemical processes, such as chemical absorption, or thermal processes, such as

TABLE 7.5 Overview fuel cells [adapted from Watter, (2022)].

Name	Electrolyte	Fuel	Output (kW)	Temperature (°C)	Efficiency (%)
Alkaline fuel cell (AFC)	KOH	H_2	10 … 100	< 80	60–70
Polymer electrolyte membrane fuel cell (PEMFC)	Polymer membrane	H_2	0.1 … 500	60–80	35
Direct methanol fuel cell (DMFC)	Polymer membrane	CH_3OH	< 0.001 … 100	90–120	40
Phosphoric acid fuel cell (PAFC)	H_3PO_4	H_2	$< 10,000$	200	38
Molten carbonate fuel cell (MCFC)	Alkali carbonate melt	H_2, CH_4	100,000	650	48
Solid oxide fuel cell (SOFC)	Oxidic ceramic electrolyte	H_2, CH_4	$< 100,000$	800–1000	47

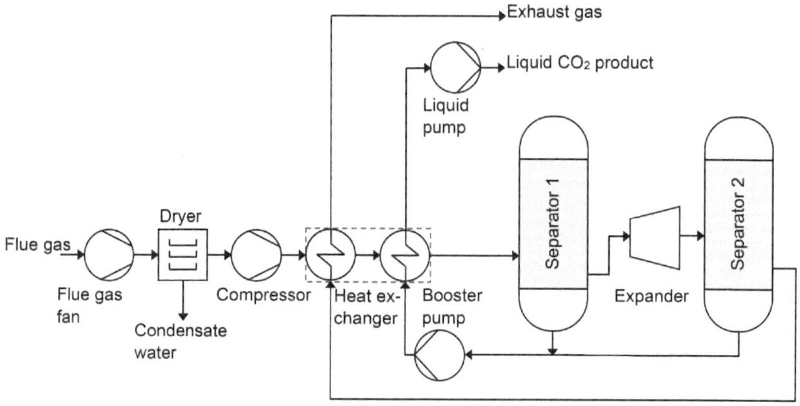

FIGURE 7.11 Schematic of CO_2 capture via low-temperature process.
Adapted from Inamuddin et al., 2019.

cryogenic separation. The mentioned separation processes will be explained in the following passages.

7.8.1 Pressure swing adsorption—adsorption of CO_2

Adsorption is the adherence of a fluid (e.g., carbon dioxide or methane gas) to a solid layer that can be released by a variation of the conditions. Different conditions like the size of the molecules or a different pressure have influence on the adsorption capacity of the layer for special molecules. These effects are used in the pressure swing adsorption.

7.8.2 Physical absorption

Physical absorption is the dissolution of CO_2 in an absorption liquid. The absorption capacity depends on the gas solubility of the applied scrubbing solution. Scrubbing solution water can be used, if the pressure is increased. Other scrubbing solutions can consist of organic compounds.

7.8.3 Chemical absorption

Chemical absorption is the dissolution of CO_2 in an absorption liquid that forms a chemical bonding to the carbon dioxide. Applicable scrubbing liquids are amines, such as monoethanolamine (MEA), diethanolamine (DEA), or methyl-diethanolamnine (MDEA).

7.8.4 Membrane separation

A membrane acts like a sieve or a filter. The membrane material is only permeable for specific gas compounds. Separation is performed by the pore size of the membrane material and/or by adsorption to the membrane material.

7.8.5 Cryogenic separation

The cryogenic separation puts the temperature down in a mixture of gases so that one gas condenses to a liquid or solid and can easily be separated from the gaseous compound. Under atmospheric pressure, carbon dioxide directly transitions from the gaseous to the solid aggregation state if the temperature is put down. Under higher pressure, carbon dioxide can transition from the gaseous to the liquid aggregation state. A schematic of CO_2 separation is shown in Fig. 7.11.

Adsorption, physical absorption, and membrane separation processes result in biogas quality up to 96%–99% CH_4. Higher purity of more than 99% can be reached by chemical absorption. The highest purity of more than 99.9% CH_4 can be achieved by a cryogenic separation processes. An overview about the different separation processes is given by Kaltschmitt (2019).

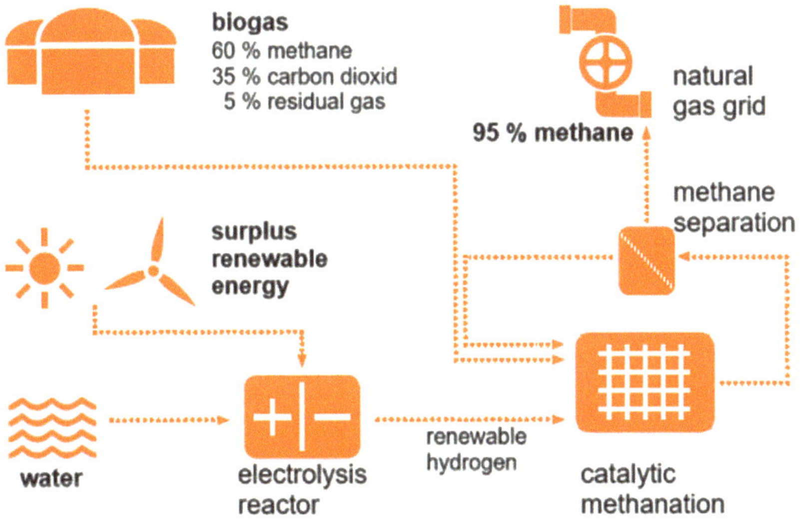

FIGURE 7.12 Process scheme of the catalytic methanation of biogas.

7.9 Direct methanation of raw biogas to biomethane

The methanation of carbon dioxide with hydrogen to methane and water is known as the Sabatier reaction:

$$CO_2 + 4H_2 \rightleftharpoons CH_4 + 2H_2O \ \Delta H = -165.0 kJ/mol \tag{7.1}$$

This equilibrium reaction is exothermic. Following the Le Chatelier's principle, an increase in the temperature or the presence of the product methane will shift the equilibrium to the left side, which is the side of the reactants.

To accelerate the reaction without shifting the equilibrium, a catalyst is used. To avoid the presence of products or the presence of catalyst poison, the state-of-the-art methanation is performed with pure carbon monoxide that can be gained by separation from the other compounds. Possible separation technologies are complex and described in Section 7.8. Therefore, several researchers have risen to the occasion of a direct methanation of raw biogas to biomethane which only requires comparatively plant-engineering efforts. There are two possible ways, the biological/biochemical methanation in a fermenter or the thermochemical catalytic methanation of the biogas in a downstream reactor using a catalyst. Both approaches have successfully led to operating facilities in pilot plant scale being functional.

Biological/biochemical methanation is based on the methanogenesis by specialized microorganisms like methanogenic archaea that utilize carbon dioxide and convert it into methane as described in Section 7.1 (see Fig. 7.1, fourth step). These archaea act as biocatalysts and the chemical reaction is analogous to the Sabatier reaction. Two different concepts exist for biological methanation, the integrative methanation, and the selective methanation. Integrative methanation is performed in the fermenter of the biogas plant, whereas selective methanation requires an additional fermenter. Both concepts work under mesophilic or thermophilic conditions. The biological methanation reaches a methane partition of above 98% in only one step without additional gas purification. Therefore, the produced biomethane could be directly fed into the natural gas distribution network (Zapf, 2022).

Various nickel- or ruthenium-based catalysts are available for thermochemical catalytic methanation. The catalyzed reaction works with raw dried and desulphurized biogas in a fixed bed reactor in a temperature range between 250°C and 450°C to achieve high reaction rates and reaching the equilibrium conversion in a comparatively small reactor. This equilibrium conversion is between 80% and 90% carbon dioxide conversion. To reach methane shares of above 96% for feeding into the natural gas distribution network, a circular flow (sweep) is used with a separation of the biomethane by membranes (Fig. 7.12).

7.10 Biomethane as a short- and long-term contribution to transport, distribution, and storage

7.10.1 Road vehicles powered by fuel cells

An overview about different fuel cells is given in Table 7.5. Because of the large size of SOFC fuel cells, cells directly powered by methane are not installed in comparatively small cars. However, CH_4 can be converted to H_2 or methanol as described in Section 7.7, and these compounds can be used in fuel cells installed in cars.

7.10.2 Road vehicles directly powered by biomethane

Road vehicles powered by methane (as CNG) are available as passenger vehicles for private transport, as buses for public transport or as other commercial vehicles. Road vehicles can be powered by H-Gas and L-Gas. H-Gas has a higher methane share (95.5%) and net calorific value (NCV, 11.1 kWh/m³) than L-Gas (79.8% CH_4 and 8.2 kWh/m³ NCV) with gas quality refering to G 260 (DVGW: German Technical & Scientific Association for Gas and Water, 2021). The difference when used in cars derives from the methane share that leads to a lower consumption and higher operating distance if H-Gas is used.

The benefits of using methane for vehicles include the following:

- lower and reduced overall emissions, especially if the CO_2 used for methanation comes from a biogas unit,

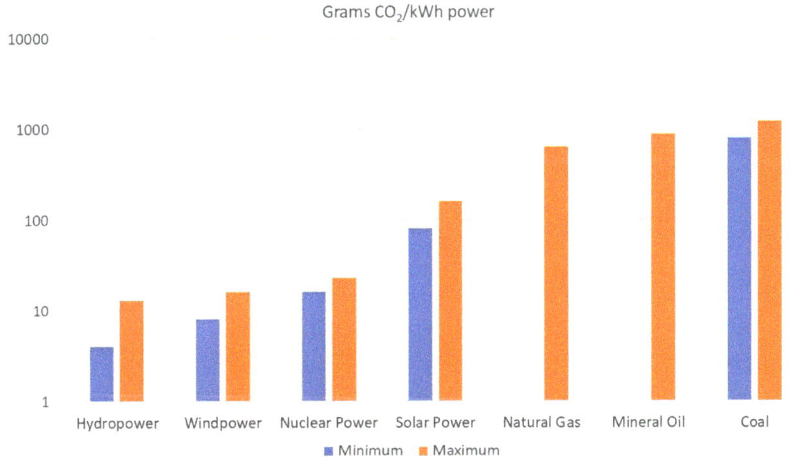

FIGURE 7.13 Overview about CO_2-greenhouse-gas emissions by different energy sources.

TABLE 7.6 Reactions within the carbon dioxide cycle via biomass/biomethane.

Reaction	Reaction equation	Product
d) Photosynthesis	$6\ CO_2 + 6\ H_2O \rightarrow C_6H_{12}O_6 + 6\ O_2$	Biomass, oxygen
e) Anaerobic digestion	$C_6H_{12}O_6 \rightarrow 3\ CH_4 + 3\ CO_2$	Biogas
f) CH_4 combustion	$3\ CH_4 + 6\ O_2 \rightarrow 6\ H_2O + 3\ CO_2$	Water, CO_2
Total reaction (d + e + f)	$6\ CO_2 + 6\ H_2O \rightarrow 6\ H_2O + 6\ CO_2$	Cycle

- absence of particulate emissions in comparison with a diesel motor, and
- less operation noise (3 dB less) (Boudellal, 2018).

Natural gas is an environmentally friendly alternative to conventional gasoline or diesel. Nevertheless, German car producers have stopped CNG-car production for the private sector because of the following reasons:

- lacking demand by the customers,
- less than 800 CNG filling stations in Germany with decreasing tendency, and
- clear commitment to electric cars by public policy and economy (ADAC General German Automobile Association, 2023).

In the commercial vehicles sector, the situation is slightly better. For example, 35,000 buses were used for short-distance traffic in Germany in 2018. A portion of 95% was powered by diesel, and a portion of 3% was powered by CNG. Taking into account in Germany, gas buses have a high availability on the market and a high technical maturity, the share of only 3% is comparatively small (DVGW: German Technical & Scientific Association for Gas and Water, 2019).

7.10.3 Decarbonization potential in the transport sector

Fig. 7.13 shows the carbon dioxide emissions of different energy sources. Renewable energy sources like hydropower, wind power, and solar power as well as nuclear power have the lowest CO_2 emissions per produced power unit. Methane has a higher carbon dioxide emission, but the emissions are lower than those of every other fossil energy source, especially if the methane is utilized to electrical power in a combined gas-steam powerplant with 60% electrical efficiency (Lübbert, 2007).

Biomethane is methane produced from biomass. Biomass has originally been produced by crops and trees, consuming exactly the same amount of carbon dioxide that is produced by biogas formation and biomethane combustion as shown in Table 7.6, which makes biomass a renewable energy source in contrast to fossil fuels not being regenerated in the period of a couple of generations.

In several cities in Germany, methane buses are in operation. A study of the German Technical and Scientific Association for Gas and Water (DVGW), a recognized standardization body for the gas and water industry in Germany, comes to the result, that with comparable technical specifications, a conventional diesel bus has well-to-wheel emissions of about 1.21 kg CO_2-eq/km, whereas a CH_4 bus only produces emissions of 0.23 kg CO_2-eq/km (DVGW: German Technical & Scientific Association for Gas and Water, 2019).

In Germany, 854 (Status 2018) SNG filling stations had an output of 1,900 Mio kWh/year of SNG. In 2018, a number of 212 biogas plants were connected to the central SNG-distribution network with a feeding capacity of total 132,000 m^3/h. With a calorific value of 12.5 kWh/kg for SNG and a density of 0.840 kg/m^3 for H-Gas, this leads to an energy output of the connected plants of about 1,390,000 kWh/h (estimated factor 7500 h/year) or 10,245 Mio kWh/year. The potential for decarbonization in the public transport sector by the use of biomethane is enormous.

7.11 Conclusion

Biomethane is a very flexible energy source that can be used for energy production and for powering vehicles in the transport sector. The flexibility is based on the various options for decentralized and centralized use as well as for direct utilization and indirect utilization. Indirect utilization is performed by conversion to other fuels or to power generation. In the long-term perspective, hydrogen will be used as carbon-free energy source and fuel. However, the technology for hydrogen utilization is still in development. The processes and technologies for biomethane utilization mostly exist in several different options and can be combined to tailor-made concepts. The infrastructure for transporting biomethane also exists and can be used; existing facilities will become more flexible. The decarbonization potential for biomethane being the shortest hydrocarbon chain is high in comparison with conventional energy sources and fuels based on coal or other hydrocarbons. Due to the production from biomass, biomethane is a renewable energy source. Theoretically, the same amount of CO_2 will be consumed for formation as it will be emitted again when being utilized. Therefore, the utilization of biomethane has the potential for being an immediately available bridging technology in the current ongoing energy transition.

References

ADAC General German Automobile Association 2023 "Erdgas: CNG-Antrieb ohne Zukunft." https://www.adac.de/verkehr/tanken-kraftstoff-antrieb/alternative-antriebe/erdgas/.

Arpe, H.-J., & Weissermel, K. (2003). *Industrial Organic Chemistry*. Wiley-VCH Verlag GmbH Weinheim, 3–527.

Banja, M., Jégard, M., Motola, V., & Sikkema, R. (2019). Support for biogas in the EU electricity sector— A comparative analysis. *Biomass and Bioenergy, 128*, 105313. https://doi.org/10.1016/j.biombioe.2019.105313.

Barchmann, T., Mauky, E., Dotzauer, M., Stur, M., Weinrich, S., Jacobi, H. F., Liebetrau, J., & Nelles, M. (2016). Erweiterung der Flexibilität von Biogas-anlagen—Substratmanagement, Fahrplan-synthese and ökonomische Bewertun. *Landtechnik. 71*(6), 233–251. Available from: https://doi.org/10.15150/LT.2016.3146, https://www.landtechnik-online.eu/landtechnik/issue/archive.

Barz, M., Laß-Seyoum, A., Kadow, S., Wesenfeld, H., Jiawei, H., Ovsitser, O., Knist, S., Geppert, P., Boitin, A., D. (2022). Hoffmann, Proceedings of the 8th International Conference on Sustainable Energy and Environment (SEE 2022): The Road to Net-Zero: Energy Transition Challenges and Solutions Catalytic conversion of biogas to biomethane proceedings of the 8th International Conference on Sustainable Energy and Environment. JGSEE of KMUTT. Bangkok, Thailand, JGSEE of KMUTT. Bangkok, Thailand, 2022.

Barz, M., Wesenfeld, H., Laß-Seyoum, A., Meibohm, A., & Knist, S. (2021). In E. Koukios, & A. Sacio-Szymanska (Eds.). *Bio#Futures Increasing Flexibility of Biogas Plants Through the Application of Innovative Concepts* Switzerland: Springer Nature 10.1007/978-3-030-64969-22021.

Biomethane Industrial Partnership 2022 2024 6 27 "Teaming up to achieve 35 bcm of sustainable biomethane by 2030." https://bip-europe.eu/.

Boudellal, M. (2018). *Power-to-Gas: Renewable Hydrogen Economy for the Energy Transition*. Berlin: De Gruyter 978–3.

Brémond, U., Bertrandias, A., Steyer, J.-P., Bernet, N., & Carrere, H. (2021). A vision of European biogas sector development towards 2030: Trends and challenges. *Journal of Cleaner Production, 287*, 125065. https://doi.org/10.1016/j.jclepro.2020.125065.

CEN: European Committee for Standardization 2023a EN 16723-1: Natural gas and biomethane for use in transport and biomethane for injection in the natural gas network— - Part 1: Specifications for biomethane for injection in the natural gas network. https://standards.cencenelec.eu/dyn/www/f?p=205:22:0::::FSP_ORG_ID,FSP_LANG_ID:853454,25&cs=186593F9198C956B5C9607B613F5F36B9.

CEN: European Committee for Standardization 2023b EN 16723-2: Natural gas and biomethane for use in transport and biomethane for injection in the natural gas network — Part 2: Automotive fuels specification. https://standards.cencenelec.eu/dyn/www/f?p=205:7:0::::FSP_ORG_ID:853454&cs=1D4652CA6A534598910AD4A58144F72C2.

de Carvalho, F. S, dos Santos Reis, L. C. B, Lacava, P. T., de Araújo, F. H. M, & de Carvalho, J. A (2023). Substitution of natural gas by biomethane: Operational aspects in industrial equipment. *Energies, 16*(2). 10.3390/en16020839, http://www.mdpi.com/journal/energies/.

DVGW: German Technical and Scientific Association for Gas and Water 2011 Technische Regel G 262 0176-3490 DVGW Bonn 09/2011 Nutzung von Gasen aus regenerativen Energiequellen in der öffentlichen Gasversorgung. https://shop.wvgw.de/leseprobe/108420_lp_G_262_2011_09.pdf.

DVGW: German Technical and Scientific Association for Gas and Water 2019 DVGW Bonn "Bewertung von Gasbussen für den öffentlichen Personennahverkehr und Vergleich mit Alternativkonzepten (Busstudie)." https://www.dvgw.de/medien/dvgw/forschung/berichte/g201810-busstudie-abschlussbericht.pdf.

DVGW: German Technical and Scientific Association for Gas and Water 2021 Technische Regel G 260 DVGW Bonn 09/2021 Gas Quality. https://shop.wvgw.de/G-260-Arbeitsblatt-09-2021/310700.

European Biogas Association 2023 "EBA Statistical Report 2023: Tracking biogas and biomethane deployment across Europe." https://www.europeanbiogas.eu/eba-statistical-report-2023/.

FNR: Agency for Renewable Resources, "Guide to Biogas - from production to use". 5, Fachagentur Nachwachsende Rohstoffe e.V. Gülzow, Fachagentur Nachwachsende Rohstoffe e.V. Gülzow, (2010), 2010.

FNR: Agency for Renewable Resources 2013 Fachagentur Nachwachsende Rohstoffe e.V. Gülzow Biogas an introduction. https://www.fnr.de/fileadmin/Projekte/2021/Mediathek/brosch.biogas-2013-en-web-pdf.pdf.

Harirchi, S., Wainaina, S., Sar, T., Nojoumi, S. A., Parchami, M., Parchami, M., Varjani, S., Khanal, S. K., Wong, J., Awasthi, M. K., & Taherzadeh, M. J. (2022). Microbiological insights into anaerobic digestion for biogas, hydrogen or volatile fatty acids (VFAs): A review. *Bioengineered, 13*(3), 6521–6557. https://doi.org/10.1080/21655979.2022.2035986.

IEA: International Energy Agency. (2020). International perspectives report of the symposium "Drivers for Successful and Sustainable Biogas Projects. *IEA Bioenergy, 37* 978–1.

IEA: International Energy Agency 2024 "Renewables 2023 Analysis and forecast to 2028", Revised version. https://iea.blob.core.windows.net/assets/96d66a8b-d502-476b-ba94-54ffda84cf72/Renewables_2023.pdf.

Inamuddin, Asiri, A.M., & Lichtfouse, E. (2019). Carbon Sequestration Vol. 2 Materials and Chemical Methods. In Inamuddin, Asiri A.M., & Lichtfouse E. (Eds.), Sustainable Agriculture Reviews. 10.1007/978-3-030-29337-6.

Kaltschmitt, M. (2019). *Energy from Organic Materials (Biomass)*. New York: Springer Nature. 978–1 10.1007/978-1-4939-7813-72019.

Lim, S. J., & Kim, T. H. (2014). Applicability and trends of anaerobic granular sludge treatment processes. *Biomass and Bioenergy, 60*, 189–202. https://doi.org/10.1016/j.biombioe.2013.11.011.

Lindmark, J., Thorin, E., Fdhila, R. B., & Dahlquist, E. (2014). Effects of mixing on the result of anaerobic digestion: Review. *Renewable and Sustainable Energy Reviews, 40*, 1030–1047. https://doi.org/10.1016/j.rser.2014.07.182.

Lisowyj, M., & Wright, M. M. (2020). A review of biogas and an assessment of its economic impact and future role as a renewable energy source. *Reviews in Chemical Engineering, 36*(3), 401–421. https://doi.org/10.1515/revce-2017-0103.

Lübbert, D. (2007). CO_2-Bilanzen verschiedener Energieträger im Vergleich—INFO-BRIEF WD 8 - 056. *Wissenschaftliche Dienste des Deutschen Bundestages*. https://www.bundestag.de/resource/blob/504060/d408ca51555a813c5b3a750c4c0c1fa1/co2-bilanzen-verschiedener-energietraeger-im-vergleich-data.pdf.

Marconi, P., & Rosa, L. (2023). Role of biomethane to offset natural gas. *Renewable and Sustainable Energy Reviews, 187*, 113697. https://doi.org/10.1016/j.rser.2023.113697.

Mignogna, D., Ceci, P., Cafaro, C., Corazzi, G., & Avino, P. (2023). Production of biogas and biomethane as renewable energy sources: A review. *Applied Sciences, 13*(18), 10219. https://doi.org/10.3390/app131810219.

Motola, V., Scarlat, N., Hurtig, O., Buffi, M., Georgakaki, A., Letout, S., Mountraki, A., & Joanny, G. (2022). *Clean Energy Technology Observatory: Bioenergy in the European Union – 2022 Status Report on Technology Development, Trends*. Value Chains and Markets, Publications Office of the European Union 10.2760/5771042022.

O'Connor, S., Ehimen, E., Pillai, S. C., Black, A., Tormey, D., & Bartlett, J. (2021). Biogas production from small-scale anaerobic digestion plants on European farms. *Renewable and Sustainable Energy Reviews, 139*, 110580. https://doi.org/10.1016/j.rser.2020.110580.

Rosenwinkel, K.-H., Kroiss, H., Dichtl, N., & Seyfried, C.-F. (2015). In K.-H. Rosenwinkel, H. Kroiss, N. Dichtl, C.-F. Seyfried, & P. Weiland (Eds.). *Anaerobtechnik - Abwasser-, Schlamm- und Reststoffbehandlung, Biogasgewinnung*Springer-Verlag 10.1007/978-3-642-24895-52015.

Savickis, J., Zemite, L., Zeltins, N., Bode, I., Jansons, L., Dzelzitis, E., Koposovs, A., Selickis, A., & Ansone, A. (2020). The Biomethane injection into the natural gas networks: The EU's gas synergy path. *Latvian Journal of Physics and Technical Sciences, 57*(4), 34–50. https://doi.org/10.2478/lpts-2020-0020.

Singh, R., Hans, M., Kumar, S., & Yadav, Y. K (2023). Thermophilic anaerobic digestion: An advancement towards enhanced biogas production from lignocellulosic biomass. *Sustainability, 15*(3), 1859. https://doi.org/10.3390/su15031859.

Sterner, M., & Stadler, I. (2017). *Energiespeicher - Bedarf, Technologien, Integration* (2). Springer-Verlag GmbH Deutschland 10.1007/978-3-662-48893-5.

Stolze, Y., Zakrzewski, M., Maus, I., Eikmeyer, F., Jaenicke, S., Rottmann, N., Siebner, C., Pühler, A., & Schlüter, A. (2015). Comparative metagenomics of biogas-producing microbial communities from production-scale biogas plants operating under wet or dry fermentation conditions. *Biotechnology for Biofuels, 8*(1). 10.1186/s13068-014-0193-8.

Sulewski, P., Ignaciuk, W., Szymańska, M., & Wąs, A. (2023). Development of the biomethane market in Europe. *Energies, 16*(4), 2001. https://doi.org/10.3390/en16042001.

Vasco-Correa, J., Khanal, S., Manandhar, A., & Shah, A. (2018). Anaerobic digestion for bioenergy production: Global status, environmental and techno-economic implications, and government policies. *Bioresource Technology, 247*, 1015–1026. https://doi.org/10.1016/j.biortech.2017.09.004.

Watter, H. (2022). *Regenerative Energiesysteme* (6). Springer Fachmedien Wiesbaden. 978–3 10.1007/978-3-658-35868-6.

Zahoransky, R. (2022). *Energietechnik* (9). Springer Fachmedien Wiesbaden. 978-3 10.1007/978-3-658-34831-12022.

Zapf, M. (2022). *Stromspeicher und Power-to-Gas im deutschen Energiesystem: Rahmenbedingungen, Bedarf und Einsatzmöglichkeiten* (2). Springer Fachmedien Wiesbaden 10.1007/978-3-658-37129-6.

Chapter 8

Biomethane in the water–energy–food nexus context

Ouahid El Asri[1], Fatima Safa[1], and Hassan El Bari[2]
[1]*Laboratory of Microbial Biotechnology and Vegetal Protection, Faculty of Sciences, Ibn Zohr University, Agadir, Morocco,* [2]*Laboratory of Electronic Systems, Information Processing, Mechanics and Energetics, Faculty of Sciences, Ibn Tofail University, Kenitra, Morocco*

8.1 Introduction

Our daily lives are currently faced with increased water, energy, and food needs to satisfy our new way of life. This improvement in living standards is reflected in the Earth's population. Our planet is subject to an increase in the number of inhabitants, reaching 8.5 billion in 2030 (Al-Obadi et al., 2022). Ravindran et al. (2022) estimated that the number will increase to 9.6 billion by 2050. At the same time, the demand for water supply is increasing. Niccolucci et al. (2011) reported that the volume of bottled water increased from 2.35 to 15.20 GL between 1980 and 2011. Another study by Hussien et al. (2016) showed that water consumption per capita increases according to the standard of living of each citizen, that is, the level of low- to high-income households ranging from 241 to 290 L/inhabitant/day, respectively. However, several researchers have confirmed that only 0.014% of this freshwater reserve is directly accessible to meet our needs (Shatat et al., 2013). Additionally, worldwide wastewater has been estimated at around 380 billion m^3, that is, 380 trillion liters per year (Kundu et al., 2022). In front of these modernization types, we have witnessed a significant increase in drinking water consumption and wastewater generation.

In addition to the demand for water, we assisted in an exponential increase in global energy consumption, from 8588 Mtoe in 1995 to 13,147 Mtoe in 2015 (Dong et al., 2020). The energy currently consumed annually is phenomenal, estimated to be more than 176,000 TWh per year worldwide. Kohli et al. (2023) and Rempel & Gupta (2022) have also described fossil fuels comprising more than 79% of the global primary energy demand. Their intensive exploitation leads to the progressive exhaustion of these resources, and certain regions are already

Biogas in the Circular Economy. DOI: https://doi.org/10.1016/B978-0-443-29230-9.00003-6

experiencing signs of a reduction in their exploitable reserves. Several researchers state that we will see the physical depletion of conventional oil by 2030 (Sorrell et al., 2010). So, fossil energy types progress toward depletion, producing an energy deficit in several countries. Thus, we are faced with the need to ensure a growing energy supply despite the depletion of fossil resources and to minimize their use because of their danger to natural ecosystems.

The food sector did not escape this influence and was also affected by this improvement in living standards. For example, between 1960 and 2011, wheat production tripled from 1 to 3.5 tons/ha (Erenstein et al., 2022). Meat consumption will increase from 75% to 145% in 2050. Blair et al. (2023) and Otles et al. (2015) have described that the food industry presents a gigantic increase in food supply chain demands that will significantly increase food production in the upcoming 50 years. Therefore, we are witnessing an explosion in the production of feedstock to sustain our existence.

We face a pressing triad of crises today: water, energy, and food. Additionally, these sectors are intricately linked, with the impact of one sector reverberating directly and indirectly on the others, a phenomenon known as the nexus. Many factors, including improved purchasing power, evolving work dynamics, social issues, food controversies, industrialization, and commercialization, further influence this nexus. To address and alleviate these daily crises, anaerobic digestion product (ADP), biomethane, and digestate are a beacon of hope, a sustainable energy solution poised to tackle the contemporary challenges of water, energy, and food. This chapter integrates biomethane into the energy–food–water nexus, offering a holistic approach to our era's most pressing contemporary challenges.

This work is structured into three main axes: In the first axis, we discussed all the characteristics of the water–energy–food nexus, such as the definitions, concepts, and tools of the nexus; then, we presented the state of world art currently in water, energy, and food. In the second axis, we dissected the process of anaerobic digestion (AD) and innovative biotechnology-producing biomethane. Then, we presented all the characteristics of this gas and an emerging opportunity to integrate and improve the water–energy–food nexus in the whole world. In the last part, we discussed the different solutions provided by ADP for the three sectors of this nexus, highlighting recent advances, technological innovations, and practical applications of ADP in this nexus approach in a sustainable development context. At the end of this axis, we presented some strategic development plans for specific countries by adapting ADP in their water, energy, and food sectors.

8.2 Presentation of the water–energy–food nexus

8.2.1 Nexus: Concept and meaning

Nexus often refers to a connection or link between resources, things, concepts, or ideas. In another way, it is the central point where these various elements converge, intersect, or come together significantly. The nexus concept makes it possible to reveal the existence of interdependencies and substantial relationships between the sectors that constitute it. Brouwer et al. (2018) described the nexus as a coherent system of interconnected natural resource management. Therefore, the nexus concept represents the exercise of pressure on the part of the nexus, producing pressure on the other components of the nexus. Many connections can be evaluated in interaction with the WEF nexus, for instance, climate change, food security, water management, energy security, and others. The energy/climate nexus is a bidirectional system between the two sectors, that is to say, energy consumption directly affects greenhouse gas production and directly impacts the planet's climate. So, the nexus concept has become a powerful approach to highlighting the social and environmental challenges produced by economic and technological development. Huckleberry & Potts (2019) described it as an analytical approach that makes it possible to quantify the links between the components of nexus systems to assess the impact of the governance adopted by politicians. So, this nexus approach helps push political decision-makers, entrepreneurs, and researchers to secure and sustainably manage the components of this nexus. Thus, this concept strongly contributes to achieving sustainable development goals (Tashtoush et al., 2019). To adopt this concept and achieve sustainable practices and a fulfilling economy, we must know the nexus tools.

Several works have studied the tools adopted by the nexus. González-Rosell et al. (2023) described four types of tools of the nexus methodology: The first type is tools for analyzing intersectoral policies between the components of nexus, the second type is of quantitative and qualitative characters, the third type is based on spatial evaluation, and the last type is based on temporal evaluation using tools for simulating future scenarios. On the contrary, Taguta et al., 2022added six other tools to the previous ones: tools based on analyzing returns, entry conditions, exit conditions, analytical capabilities, and optimization and visualization. So, we have less than 10 connection tools of the nexus concept with multicentric, multi-dimensional, and spatiotemporally characteristics. After wastewater treatment, the result is sewage sludge, a combination of organic and inorganic substances, and is very rich in water, the main component. This sewage sludge as a biomass can be primary, which occurs in primary clarifiers, or secondary or activated sludge obtained after purification in secondary treatment bioreactors (Demirbas et al., 2017). In European countries, yearly sewage sludge generation is more than 50 million tonnes (Chen, Chaudhary, et al., 2020; Demirbas et al., 2017). Currently, the global quantity of sludge biomass is around 462 billion tons (Zhu et al., 2023). Thus, we face a large amount of biomass that needs to be treated and valorized before it reaches nature.

8.2.2 Global context of water, energy, and food challenges

The energy supply from different resources like oil, gas, and coal continues to increase. A guaranteed energy supply is considered as much as a guarantee for the establishment of sustainable economic growth. For example, from 1990 to 2020, global energy use increased enormously by more than 69% of the energy consumed (Chanthakett et al., 2024). Over the past 20 years (1995 to 2015), we have witnessed an exponential increase in global energy consumption from 8588.9 to 13,147.3 Mtoe. Dong et al. (2020) and Moodley & Trois (2021) predicted that global energy demand will increase by 48% in the next 20 years. Nevertheless, Ahmad & Zhang (2020) described that global energy demand will reach 17,487 Mtoe in 2040. So, we continuously increase yearly, which is linked mainly to population growth and economic development.

With the improvement in living standards and the development of our civilization, we have witnessed an increase in global energy consumption, reaching more than 176,000 TWh of energy consumed annually today (Kohli et al., 2023). The industrial and transport sectors are the first energy consumers, accounting for 28% and 38% of global energy consumption, respectively (Semrau & Engell, 2023). Additionally, the majority of this energy demand (80%) comes from fossil fuels (Jamshidi & Askarzadeh, 2019). On the contrary, the major nations, except for Europe, demonstrated a 50% increase in natural gas production (Ahmad & Zhang, 2020). This excessive consumption leads directly to an accelerated depletion of fossil energy resources and their direct adverse effects on the environment.

Water is essential for the survival of living beings on our planet and for human development. More than 70% of the Earth's surface is covered by water, but most are unsuitable for human drinking (Musie & Gonfa, 2023). Freshwater represents only 3% of all global water reserves. It is generally in the form of glaciers and ice caps, but the rest (97%) is in salt water (Skuse et al., 2021). These data show that we have a low fresh water supply. Several researchers have confirmed that only 0.014% of this freshwater reserve is directly accessible to meet our needs (Shatat et al., 2013). Therefore, a small quantity is unsalted and unavailable for our consumption. This tiny quantity can be reduced due to different forms of pollution. So, we must protect and reason for using this tiny fraction of water.

We are witnessing a significant increase in food consumption throughout our history. The demands for alimentation requirements and agroindustry production are increasing (P. et al., 2020). The milk production and cattle slaughtered worldwide reflect a linear trend between 1961 and 2019, illustrated by a factor of more than 2% (Faverdin et al., 2022). Blair et al. (2023) have described increasing meat consumption from 75% to 145% in 2050. Morita (2021) described the increase in global production of agricultural food products from 1980 to 2012 as 2.25% per year. Otles et al. (2015) have described that the food industry presents a gigantic increase in food supply chain demands that will significantly increase food production in the upcoming 50 years. This growing production translates into an increase in financial income of around 1180 billion US dollars to 2411 billion US dollars. Therefore, we are witnessing a continuous flow of food production, which continues to increase.

This food production flow generates a large quantity of food waste, which is found in our trash bags every day. Approximately 600 million tons of food waste in China are generated yearly (Meng et al., 2015). In some countries like the United States, Bangladesh, Canada, and Afghanistan, the waste food sector produces 59, 65, 79, and 82 kg/capita/year (Sahoo et al., 2023; Sinha & Tripathi, 2021). Currently, food waste is produced from a third of the food generated globally and intended for human consumption, estimated at around 1.3 billion tonnes annually (Morone et al., 2019). Generally, all humans on our planet can produce 65 kg of food waste daily: 25% from vegetables, 24% from cereals, and 12% from fruits (Chen, Zhao, et al., 2020). Therefore, we are witnessing a continuous flow of food waste, which continues to increase.

8.3 Anaerobic digestion and biomethane

AD is a biological process that allows the microbiological conversion of organic matter into a mixture of gases without oxygen (El Asri et al., 2022b; El Gnaoui et al., 2020). This mixture of gases produced by AD is called biogas; it comprises methane, carbon dioxide, hydrogen, ammonia, water vapor, siloxanes, and sulfur compounds (Calbry-Muzyka et al., 2022; Ourradi et al., 2022). The composition of biogas depends on several parameters, such as the microbiological and physicochemical characteristics of the substrate intended for AD (El Asri et al., 2020; Ouahid et al., 2019), the temperature of incubation within the mesophilic (30°C–40°C) or thermophilic (50°C–60°C) (Serrano et al., 2014), pH, which helps promote maximum bacteria activity mostly methanogens (Lay et al., 1997), type of pretreatment adopted (Hamraoui et al., 2020), organic load rate, and hydraulic retention time in this process (Elasri et al., 2018); inoculum quality used influences the proportion of methane in the biogas (El Asri et al., 2022a; Gnaoui et al., 2020). Thus, an adequate understanding and control of these parameters makes it possible to optimize the production of methane-rich biogas, thus improving its efficiency from an energy point of view.

Generally, AD occurs in four main stages: hydrolysis, acidogenesis, acetogenesis, and methanogenesis. These stages are characterized by organic compounds, microorganisms, enzymes, biochemical pathways, and metabolic conditions (Bakraoui et al., 2019). The progress of these stages in a standard, continuous manner without any inhibition is essential for the final conversion of organic matter into biomethane. In addition to biomethane production, AD produces a solid residue called digestate. This AD residue is a soil fertilizer (Cristina et al., 2020). Among these gases produced, we distinguish biomethane from methane because it is produced by the biological process of AD compared with other techniques for producing methane. We distinguish the production of synthetic methane by purely chemical processes such as alkaline electrolyzers, proton exchange membrane electrolyzers, and solid oxide electrolyzers (Biswas et al., 2020). Thus, biomethane is a purely microbiological generation by AD.

Biomethane is considered a flammable gas that generates electricity and heat. Among the main properties of this gas is that it is odorless and easily flammable. The energy value of biomethane is estimated at approximately 9.9 kWh per 1 m^3 of biomethane produced (Eswari et al., 2023). It is purified from other contaminating gases to be compressed and packaged to meet the quality standards required for injection into natural gas networks or use as fuel for vehicles (El Asri et al., 2015). For these reasons, we have witnessed a wave of estimation of biomethane potential by several researchers and the integration of this potential into the development strategies of each country. For example, in Italy, Noussan et al. (2024) estimated an increase in biomethane adoption from 2.3 to 7.6 billion m^3 in 2030. Silva et al. (2024) described an increase in global demand for biomethane of around 40% in the transport sector from 2018 to 2019 to 1.5 billion m^3, with an estimated 3% increase in demand for world gas by 2030. Therefore, biomethane production is a sustainable energy source that ensures decentralized energy with low-carbon emissions to partially reduce the importation of fossil energy and the environmental danger caused by these types of energy.

The digesters are the birthplace of biomethane production. Hence, their construction and adoption must be well-designed. Biomethane digesters are classified into three main generations: (1) domestic biomethane reactors, which are simple small-scale structures generally adopted in rural regions; currently, portable type digesters are the most demanded, and they are in different designs (Ajay et al., 2021); (2) optimized bioreactors, which are medium and large-scale structures characterized by combined with technological advances such as heating systems, insulation, mixing, sensors, and gas composition analyzer; these installations are adapted to increasing methane yield to meet energy demand. We can cite the three most popular types (plug-flow digesters, up-flow anaerobic sludge blankets, and covered lagoons) (Liu et al., 2021); (3) laboratory bioreactors are models of measurement and experimentation, such as the control of AD reactions, the methane concentration, and the microbial consortium's evolution (Habchi et al., 2023). Thus, in studying the efficiency of the AD process, we have several designs that depend on the researchers' objectives and the construction made using various materials packaged in bulk (El Asri & Afilal, 2018). Therefore, adopting AD bioreactors to produce biomethane meets user choice, requires effective sealing, and requires a controlled substrate supply.

The production of biomethane by AD is considered a socio-economically sustainable development tool because it intervenes in several areas of our daily lives; we can distinguish five main areas: (1) AD is used as a means of treatment and management of various types of organic waste, such as food waste, sewage sludge, agricultural waste, industrial effluent, and municipal waste by reducing their daily volume and stabilizing them before putting them in storage areas (Karouach et al., 2020). (2) It produces renewable energy (Elasri & El amin Afilal, 2016). (3) It is adopted in agriculture as much as a fertilizer-producing technology; the digestate produced at the end of the process is rich in nutrients and can be used as an organic amendment for agricultural soils either by counting or indirect application (Möller & Müller, 2012). (4) AD intervenes in protecting our planet through its contribution to the fight against climate change; AD reduces greenhouse gases because it allows the capture of methane produced by the anaerobic decomposition of waste (Zhao et al., 2020). (5) It is also adopted in water resources management; it allows water quality improvement by decomposing water effluents before their discharge into the ecosystem, thus helping to reduce water pollution (Silva et al., 2024). So, AD is critical in sustainable development and achieving SDGs.

8.4 Anaerobic digestion as a water solution

We have already discussed the scarcity of drinking water (3% of all global water reserves) and the difficulty of accessing this fraction (0.014%) on our planet. In addition to these problems, there is added pollution and drought. We are facing a water crisis, hence the need to protect drinking water and recycle water use. Among the current technologies that make it possible to reduce this crisis and contribute to the rationalization of water consumption is the production of biomethane by AD (Fig. 8.1).

8.4.1 Minimizing the impact of leachate on water resources

Globally, it has been estimated that 70% of the waste produced is disposed of in open dumpsites or landfills (Nabi et al., 2022). On the contrary, Mor & Ravindra (2023) stated that 70% to 90% of unsorted municipal solid waste is dumped either in low-lying areas or open dumps. These sites (landfills, open dumps, low-lying areas) generate liquid effluent called leachate from the waste moisture and rainwater infiltration into waste. China alone produces more than 50 million tonnes of landfill leachate annually (Song et al., 2023). This leachate is characterized by a large amount of ammonia, heavy metals, inorganic salts, pesticides, polycyclic aromatic compounds, microplastics, antibiotics, and xenobiotic organic compounds (Dagwar & Dutta, 2024; Karagozoglu & Asar, 2023). Leachate chemical compound types and their concentration wrongly depend on the site's age, waste type, and climate (Dagwar & Dutta, 2024; Nabi et al., 2022). Thus, all these substances described previously will percolate in the leachate to groundwater or runoff

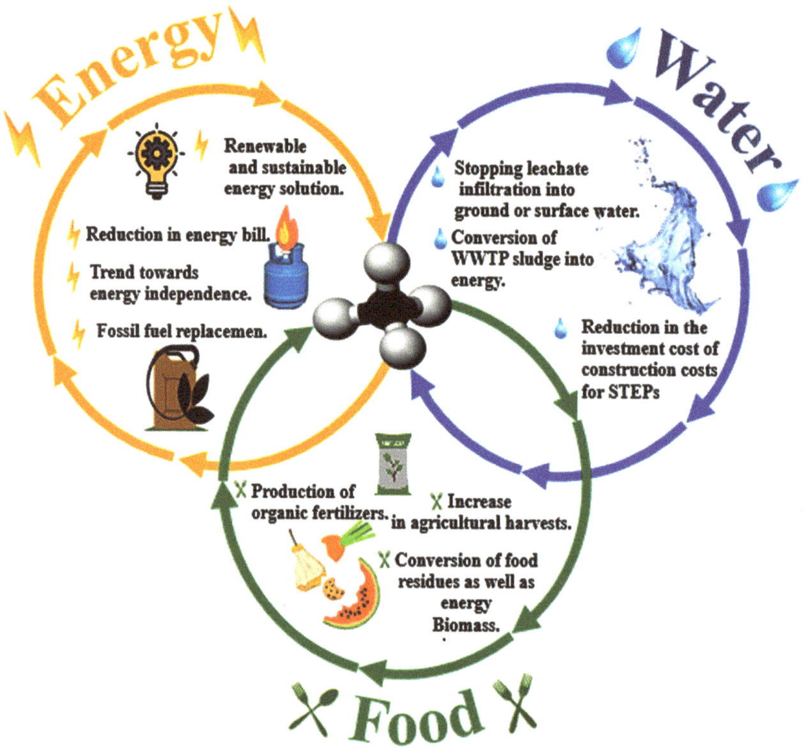

FIGURE 8.1 The role of biomethane in the water–energy–food nexus.

into the surface. For example, the Lebanon leachate produced by Tripoli's landfill is more than 210 tons per year, and 21 metals contaminate the Abou Ali River and the Mediterranean Sea. Gamar et al. (2018) and Moustafa et al. (2023) have raised a wake-up call for Moroccan policymakers and officials in the El Hajeb region to intervene to protect the water table from the leachate plume from the wild municipal landfill of Ait Bourzouine. Therefore, we must intervene in the face of this permanent leachate flow. Piadeh et al. (2024) recommended the adoption of biomethane production by AD to prevent the discharge of leachate from waste into rivers and groundwater; thus, the production of biomethane makes it possible to reduce the pollution of bodies of water and protect our water resources so we can achieve (SDG 6 and SDG 14). So, the recovery of organic waste and its leachate into biomethane makes it possible to protect water resources.

Several researchers have confirmed the recovery and recycling of leachate in digesters to increase biomethane production. Shahriari et al. (2012) recommended recycling leachate to produce biomethane from food waste.

This can reuse leachate to stimulate methanogenic activity and improve biogas production. Feng et al. (2019) confirmed that recirculating highly loaded organic matter in the around 6000 mg/L COD concentration leachate can improve the cumulative production of biomethane by up to 156%. This leachate concentration also promotes the proliferation of hydrogenotrophic *Methanobacterium*, increasing biomethane production. Leachate recycling also depends on the age of leachate used in this process. Nair et al. (2014) confirmed that the improvement in biomethane production varies from 19% to 41%, depending on the mixture ratio between old landfill leachate (more than 10 years) and young leachate. In another study, using leachate aged 3 months makes it possible to improve the cumulative production of biomethane to 43%. Liao et al. (2014) andSingh et al. (2023) showed that recycling leachate makes it possible to improve three primary parameters: (1) relief of acid inhibition caused by the accumulation of volatile fatty acids; (2) a buffer system will be installed to produce ammoniacal nitrogen; and (3) production of a biochemical balance between the different phyla of acidogenic and methanogenic microbes. These parameters made it possible to stabilize the biomethane yield to reach a range of 369 to 466 mL/gVS. Therefore, in addition to the fact that the recycling of leachate has reflected its improvement as a means of improving the production of biomethane, it also makes it possible to reduce the enormous quantities of leachate produced daily to avoid the costs of treatment, management, and evacuation. Recirculation of leachate also reduces the supply of potable water in the digesters. Therefore, this recycling of leachate must be adopted in all landfills and landfill centers rather than letting it accumulate and then having to treat it as a polluting waste; the recycling of leachate can help to minimize the impacts on water superficial and underground and reduce the supply of potable water.

8.4.2 Valorization of wastewater treatment residues

Faced with the minuscule quantity of drinking water accessible in our natural reservoirs, treating the residual water produced daily has become necessary to reuse it in human activities. The Earth's population produces more than 380 billion m^3 of wastewater yearly (Kundu et al., 2022). This enormous volume of wastewater will increase by 24% in 2030 compared withthe current level (Qadir et al., 2010). This water comes from different origins, such as residential, institutional, industrial, and commercial establishments. This increase is due to rapid urbanization in recent years. Thus, a large volume of water requires treatment and reuse to reduce water consumption per capita. After wastewater treatment, the result is sewage sludge, a combination of organic and inorganic substances, and is very rich in water, the main component. This sewage sludge as a biomass can be primary, which occurs in primary clarifiers, or secondary or activated sludge obtained after purification in secondary treatment bioreactors (Demirbas et al., 2017). In European countries, yearly sewage sludge generation is more than 50 million tonnes (Chen, Chaudhary, et al., 2020; Demirbas et al., 2017). The global quantity of sludge feedstock is around 462 billion tons Periyasamy et al. (2022) and Zhu et al. (2023) have characterized sludge production as a universal concern as it is increasing production due to the installation of several wastewater treatment plants from various industries. This leads to the production of a severe problem for the environment. Thus, we face a large amount of biomass that needs to be treated and valorized before it reaches nature.

There is a scientific competition to valorize the enormous quantity of sludge biomass produced. Sludge recovery is a significant parameter in the installation specifications for wastewater treatment plants (WWTP) because the cost of treating the sludge produced by the station represents 50% of the WWTP's total operating cost (Ripoll et al., 2020). So, several studies have confirmed the production of biomethane from wastewater treatment plant sludge to treat and produce renewable energy. In Poland, 1 ton of dry matter from WWTP's primary and secondary sewage sludge is estimated to produce more than 340 m^3 of biogas, that is, 820 kWh of electrical energy (Grosser & Neczaj, 2018). Periyasamy et al. (2022) described that biomethane from sludge treatment substitutes fossil fuels such as natural gas to produce heat and electricity. Di Capua et al. (2020) described the conversion of typical sludge to biomethane in conventional digesters between 0.19 and 0.24 m^3/kg VS. This amount of biomethane is considered that is considered as a low production. Among the typical problems of converting sludge into biomethane, that is, low biomethane conversion, is the low level of organic loading within this biomass. Azarmanesh et al. (2023) and Kwon et al. (2023) showed that the monodigestion of sewage sludge is mainly due to a low load of organic matter and a low C/N ratio. Among the solutions is the adoption of a co-digestion process; several researchers have confirmed that this is a reliable alternative option to resolve the disadvantages of single-substrate digestion linked to substrate characteristics (Bakraoui et al., 2019; Hagos et al., 2017; Karouach et al., 2021). Thus, to increase methane production and optimize sludge conversion, adopting co-digestion has become necessary to ensure sufficient energy production to supply the wastewater treatment plant and make it energy-independent.

Several works used other substrates with the sludge to achieve the previous objective (Table 8.1). Wickham et al. (2016) confirmed that the co-digestion of sludge with other substrates such as paper pulp reject, bakery waste, greasy residue, dairy waste, and dehydrated algae powder increases the biomethane yield by three to six times the monodigestion of sewage sludge. The co-digestion of water with microalgae increases the methane content to around 318 Ncm^3 CH_4/gVS. Ripoll et al. (2020) and Thorin et al. (2017) recommended the co-digestion of sewage sludge from water treatment with residual water from sherry-wine distilleries with a ratio of 50:50 to obtain a yield of 225 L CH_4/kgVS and microbial proliferation 5.5 times greater than

TABLE 8.1 Comparison of co-digestion of sewage sludge with different organic substrates.

Substrate with sewage sludge	Ration of mixture (%)	Digester type	Incubation temperature (°C)	HRT (d)	Biomethane production (mL/gVS)	References
Food waste	50:50	Batch	37	35	296.75	Zhang et al. (2016)
Orange peel waste	70:30	Semicontinuous	35	2	165	Serrano et al. (2014)
Fat, oil, and grease	35:65	Semicontinuous	55	15	421	Alqaralleh et al. (2018)
Restaurant grease trap	32:68	Semicontinuous	37	18.5	620	Liu and Buchanan (2011)
Sherry-wine distillery wastewater	50:50	Batch	35	20	250	Ripoll et al. (2020)
Fruit juice	70:30	Semicontinuous	35	20	205	Wickham et al. (2019)
Pig and bovine slaughterhouse	60:40	BMP	37	28	736.4	Salehiyoun et al. (2020)
Wheat straw	50:50	Batch	36	30	187.01	Li et al. (2020)
Microalgae	75:25	Batch	50	35	219	Caporgno et al. (2015)

monodigestion of sludge alone. Sillero et al. (2023) proposed a combined system of two digesters, the first (thermophilic), which feeds the second reactor (mesophilic) with the co-digestion of sludge with poultry manure and wine stillage. This system makes it possible to produce 1058 MWh/year of electricity and 4765 GJ/year of heat with a profitability indicator of 509,011 USD within a payback period of 4 years. In addition to biomethane production, this sludge waste has several advantages that can produce considerable hydrogen potential and reduce dependence on fossil fuels (Liu et al., 2019). This biomass sewage sludge conversion strategy could also reduce greenhouse gas emissions by 57% to 73% (Barghash et al., 2022). Eventually, wastewater sludge is a promising heterogeneous biomass for biomethane generation. It is a highly efficient approach that goes in the direction of environmental protection and sustainable development worldwide. Finally, producing biomethane by AD of sewage sludge is qualified as a promising new technology in biomass valorization because it achieves five objectives: (1) reduction in the volumes of waste produced, (2) reduction in production costs of their transport, (3) minimization of energy consumption, (4) avoiding waste placement in landfills and illegal dumps, (5) production of fertilizing potential from sewage sludge, (6) destruction of pathogenic germs, and (7) minimization of odor emissions.

8.5 Anaerobic digestion as an energy solution

Biomethane, a promising and sustainable energy solution, is poised to address several contemporary energy-related challenges. Biomethane produced by anaerobic digesters has long been a beacon of renewable and sustainable energy, playing a pivotal role in the global energy transition. By harnessing organic waste, reducing greenhouse gas emissions, and offering a versatile and stable energy source, biomethane paves the way for a cleaner and more resilient energy future, aligning with the global shift toward sustainable energy solutions.

8.5.1 Biomethane as a renewable energy source

Currently, energy is one of the main areas under scrutiny worldwide. The energy sector faces several challenges due to rising global energy demand. However, there is hope on the horizon. The ongoing depletion of fossil fuels and the problem of greenhouse gas emissions have complicated the environmental situation (Divya et al., 2015). Besides, the use of fossil fuels has several disadvantages, mainly economic, notably price fluctuations (Jame et al., 2021). All of this has drawn attention to renewable energy sources as solutions and alternative energy sources. It has become necessary to find energy sources that respect the environment and, at the same time, meet energy needs in the context of sustainable development. In this respect, biomethane technology by AD has attracted significant interest (Divya et al., 2015).

Biomethane, an odorless and easily flammable component of biogas, is a versatile resource. This gas, obtained from AD, shares the same properties as natural gas and can be used in many applications, sparking interest in its potential uses (Chinnici et al., 2018). Biomethane, an essential element varying from 45% to 75% by volume, is closely linked to the calorific value of biogas (Kabeyi & Olanrewaju, 2022). For instance, 1 m^3 of biogas can generate around 6 kWh of electricity, with the proportion of methane being the determining factor in the calorific value of biogas (Kabeyi & Olanrewaju, 2022). The energy value of biomethane is estimated at approximately 9.9 kWh per 1 m^3 of biomethane produced (Eswari et al., 2023). Purifying biogas to biomethane alone from other contaminating gases is crucial to meet the quality standards required for injection into natural gas networks or use as fuel for vehicles (El Asri et al., 2015). For example, polyetheretherketone membranes are applied to separate the biomethane from contaminating gases (Chinnici et al., 2018). Thus, purification is essential for increasing the calorific power of the biogas produced, protecting cogeneration engines, and ensuring a constant supply of renewable energy, further highlighting the versatility and importance of biomethane.

Biomethane is a natural product that can contribute to the energy transition using biomass as a raw material. So, biomethane is a renewable reservoir of energy that can be converted into various renewable products, such as fuel (Kabeyi & Olanrewaju, 2022). Therefore, biomethane generated from AD is a source of energy that can be used instead of fossil fuels, producing a clean, environmentally friendly fuel. Ahead of the race for sustainable development, using biomethane as an alternative to fossil fuels can alleviate current environmental problems by reducing the impact of global warming (Chynoweth et al., 2001). As part of this, around 35 Mtoe of biogas was produced in 2018 worldwide, while studies have estimated a much greater global potential for biomethane, which can satisfy 20% of global gas demand (Iglesias et al., 2021). Moreover, this approach is economically beneficial and can help us overcome significant economic concerns (Divya et al., 2015). For instance, the AD process is being undertaken in Europe by some countries for two reasons: the exploitation of waste as a raw material for the generation of clean energy and the integration of this technology into a framework that promotes the circular economy (Caposciutti et al., 2020). So, biomethane represents a renewable and sustainable energy solution that can play a vital role in the global energy transition.

8.5.2 Biomethane as energy security

Biomethane generation is economically feasible and can reduce the CO_2 footprint compared with other renewable energies such as wind and solar power. In contrast, other renewable energies require equipment, such as industrial manufacturing, which indirectly increases the CO_2 footprint

(Budzianowski & Postawa, 2017). Moreover, fossil fuels are subject to fluctuations in production and consumption, making them difficult to use despite their enormous potential (Bagi et al., 2017). So, unlike intermittent renewable energies such as solar, wind, and fossil fuels, biomethane can be produced and used continuously, providing a stable and reliable energy source.

Producing biomethane from local waste strengthens energy security by reducing dependence on imported fossil fuels and diversifying energy sources. Sulewski et al. (2023) have estimated that the global valorization of the biogas produced made it possible to valorize a high rate of 90% of global biomethane production in 2019. The biogas recovered by the plants undergoes purification. It is then used to produce heat and electricity for self-consumption by the plant itself and external use of the excess production via injection into the national electricity grid (Ardolino et al., 2018). In addition, the energy use of biomethane is not limited to a single type of use; it is possible to convert it into heat, electricity, electricity, and heat by cogeneration or fuel production, etc. However, biomethane is widely used to generate electricity in several countries worldwide. Indeed, electricity production from biomethane has considerable advantages, especially since production is independent of climatic conditions and is done according to demand under flexible conditions (Heiker et al., 2021). Biomethane is also used in transportation to supply vehicles (cars and trucks.) worldwide. In this sense, biomethane is transformed into biomethane by various technologies such as cryogenic separation (often applied for liquid fuel) and chemical absorption (Hoo et al., 2017). For example, Sweden presents 5% of its biomethane production for transport; it has exceeded 50,000 gas vehicles (Hoo et al., 2017). Therefore, the differentiated use of biomethane makes it possible to participate in energy security, such as biomass-producing countries. To understand the critical role of renewable energy in establishing energy security and encouraging energy independence, we will discuss some strategic programs based on biomethane as an energy source in the following axis.

8.5.3 Overview of strategic plans focused on biomethane

Biomethane production facilities are now widespread throughout the world. For example, 17,000 European biomethane plants have been installed, with Germany in first place, accounting for more than 50% of the total produced (Torrijos, 2016). Biomass from municipal solid waste feeds almost 1000 of these plants (Torrijos, 2016). Germany produced more than 33,000 GWh in 2018 and was, therefore, the first biomethane producer in the world (Calise et al., 2021). In 2018, China's production yearly exceeded 3200 GWh, making it one of the leading biomethane producers (Calise et al., 2021). So, several countries have adopted the biomethane technology in their strategic plans.

Sweden is one of the countries that has primarily encouraged using biomethane-based fuel; it uses 5% of its biogas production for transport.

From 2011 to 2015, the country exceeded 50,000 gas vehicles. The success of this country in the field of renewable energy is closely linked to the implementation of very effective environmental policies by the government and the introduction of subsidies and other strategies and efforts that all go in the direction of biogas development in the country (Hoo et al., 2017).

Malaysia is a country that exploits biomethane as a renewable energy source, which is of considerable importance to the country as part of its energy security program. The advantage of biomethane is that it represents a win–win strategy in this context while simultaneously making it possible to manage the problem of waste accumulation. Regarding raw materials, Malaysia has several sources, mainly oil mill rejects (palm oil), municipal solid waste, and livestock manure (Bong et al., 2017). Malaysia is a country that has taken a keen interest in renewable energies and has implemented several energy policies intending to ensure its energy security by reducing dependence on fossil fuels and protecting the environment. Biomethane is one of the energies that have aroused curiosity in this country as a strategy for treating organic waste and producing electrical energy. The policies have led to considerable improvements, with the country increasing its installed biogas capacity from 5.16 MW in 2012 to 6.58 MW in 2013, mainly from landfills and agricultural waste (Bong et al., 2017). In 2017, Malaysia installed 18 biomethane plants with a capacity of 31 MW, while 30 biomethane plants were set up in 2018 with a capacity of 56 MW thanks to the Feed-in Tariff projects. In addition, several biomethane companies exist in Malaysia, such as Cenergi SEA Sdn Bhd, which invests in renewable energy projects to find sustainable solutions to reduce carbon emissions. As a result, these projects have yielded positive results and have, among other things, provided 231 GWh of electricity to the country's national grid, and approximately 1,206,520 tonnes of carbon are saved (Amin et al., 2022). In 2019, the World Biogas Association reported that the country contains an installed biogas capacity of 68 MW, with palm oil mill effluent as the primary substrate of a 93% contribution and 5% landfill as the second source of raw material (Lim et al., 2021). So, Malaysia has made a great effort to invest in biomethane-based energy supplies.

In the United Kingdom, AD is of great interest for energy production. Indeed, this country has 661 active digesters, which offer biomethane, electricity, and heat for local heating. The various digesters are fed by substrates that come from agricultural, municipal, industrial, or sewage sludge. This country has seen a steady increase in biomethane production. For example, in 2010, the country managed to produce 117 GWh of electricity from biomethane, while in 2016, the production became 2052 GWh (Foster et al., 2021).

Brazil began to focus on biogas as an energy source in 2010. In 2015, this country installed 127 biomethane plants with a daily production of about 1.6 million m^3, equivalent to 3835 GWh of electricity. In 2016, Brazil reached a power generation capacity of 450 MW from biogas compared with 196 MW in

2015, showing a net increase in energy production via AD. This is due to this country's economic and technical viability (Freitas et al., 2019).

For the European Union, 17,376 biomethane plants produce 62.5 TWh of electrical energy. The EU's final energy consumption was 1084 Mtoe in 2015, contributing 16.7% from renewable energies. Using biomethane to produce energy can contribute at least 25% of bioenergy. The EU mainly uses energy crops to feed its digesters at 76% (w/w) of the total feed load, mainly maize capable of producing at least 200 m^3 per tonne of maize silage (Garcia et al., 2019). Europe is the world's largest biomethane producer, with 18,943 biogas plants in operation in 2019. Germany alone has around 10,000 plants in operation, making it Europe's leader, followed by Italy with around 1,700 plants, then France, Switzerland, and the United Kingdom, each with more than 500 units (Iglesias et al., 2021). By 2020, biomethane plants had reached around 900 installations, producing 3.3 billion m^3 of biomethane, equivalent to 32 TWh (Sulewski et al., 2023). While Denmark has installed 163 biogas plants, half of these use agricultural produce as the substrate for AD. So, 46% of the biogas produced in 2018 was used for electricity production and 3% for heat, with the remaining amount destined for the country's gas grid. Denmark has an 18.6% biogas contribution to gas consumption, the largest share in Europe (Korberg et al., 2021).

8.6 Anaerobic digestion as a food solution

Biomethane production through AD presents an integrated solution that addresses several contemporary food-related challenges. As we saw in the previous section, it not only makes it possible to recover food waste by producing a renewable energy source but also contributes to improving soil quality and promoting food security.

8.6.1 Why use anaerobic digestion process in food production?

The growth of the world's population implies an increase in food requirements, which poses a significant challenge for agricultural systems obliged to guarantee global food security (Calabi-Floody et al., 2018). Agriculture is the crucial sector underpinning the world's food supply, and its revolution took place in the last century. Improving the quantity and quality of agricultural production has become necessary in the face of this growing demand flow. Therefore, agriculture must adopt several technologies to achieve this objective.

The launch of mineral fertilizers is behind the exponential growth of the world's population. Mineral fertilizers seem of great importance for the fertilization of agricultural soils; in other words, they provide nutrients that will enable crops to develop optimally and reduce the difference between the plant's needs and the elements available to them in the soil. China, as an example, is the world's leading consumer of chemical fertilizers, accounting for over a third of global consumption and 2.8 times higher than the United States due to the scale of its agricultural activities and large cultivated area, that is,164 million hectares (Martínez-Dalmau et al., 2021). Wilson and Shakya (2023) described that chemical fertilization increased sharply during the 20th century, from 20 million tonnes in 1950 to 190 million tonnes in 2020. However, the intensification of the use of these products is not advantageous, as it has led to several ecosystemic and environmental disturbances, in particular, the release of greenhouse gases, pollution of environmental components (air, water, and soil), and destruction of soil biodiversity. Unfortunately, the problem with nitrogen fertilizers is that almost half of what is applied is lost, that is, not used by plants, but either volatilized into the atmosphere, runoff, or leached (Martínez-Dalmau et al., 2021; Kovačić et al., 2022). However, 1275 gigagrams of N_2O were released into the atmosphere in China country in 2017 (Martínez-Dalmau et al., 2021). In addition to their environmental danger, the price of these chemical fertilizers is gradually increasing. Several researchers think that it is a growing market that is flourishing. Wilson and Shakya (2023) described an increase in the global fertilizer market of 3.8% to $155 billion in 2025. Other researchers estimated another 14.2% increase from $1.1 million in 2016 to $3.12 million in 2024 (Joshi & Gauraha, 2022). Thus, we must seek a technology that produces fertilizer without any danger to the environment and reduces the importation of these chemicals. Among several technologies, the technology of biomethane production is a current trend that can be exploited to produce biological fertilizers from various types of waste, including food industry waste (Czekała et al., 2022).

Once the biomethane from the AD of organic waste has been obtained, the digested substrate, known as digestate, is recovered. It is a valuable source of nutrients, such as nitrogen (N), phosphorus (P), potassium (K), and sulfur (S), which are easily assimilated by plants, organic matter, and several microelements that are highly beneficial to the soil ecosystem. An AD plant is estimated to produce more than 10,000 tonnes of digestate yearly (Kaur et al., 2020). Furthermore, the digestate is very favorable for crops, as it increases the nutrient content available to the plant, which can replace chemical fertilizers in the field (Lamolinara et al., 2022). So, the biomethane process generates an enormous quantity of digestate that can used as a fertilizer in food production, thus ensuring the economic viability of the AD process and protecting the environment from any harmful effects.

8.6.2 What is anaerobic digestate?

The biomethane digestate represents 90%–95% of the raw material outgoing into the digester. However, organic matter decomposition varies from 20% to 95%, depending on the biomass and the design of the AD system

(Lamolinara et al., 2022). Digestate is a biomethane residue made up of two different fractions: The solid component is very rich in organic matter, which presents a very significant impact on soil composition by improving the physicochemical properties of the treated soil, while the liquid component is concentrated mainly in N and K, which is why this fraction can play the role of a fertilizer (Koszel & Lorencowicz, 2015; Kovačić et al., 2022). Digestate also contains several compounds: trace elements, water and microorganisms, and remnants of organic matter that are not degraded during biomethane production. The composition varies from one digestate to another, depending on the type of substrate (food waste, municipal sludge, agricultural waste, municipal solid waste, etc.), digester operating conditions, and other factors (Guan et al., 2024). We can conclude that their composition varies when comparing the digestate composition of different biomethane-producing substrates (Table 8.2).

The EU, for example, uses a significant proportion exceeding 95% of its digestate production as liquid fertilizer applied to agriculture, while less than 3% is valorized into other products (Lu & Xu, 2021). To improve soil fertility, we use biomethane digestate as fertilizer to increase plant quality and resistance to all biotic or abiotic factors (Koszel & Lorencowicz, 2015). So, local availability, economic feasibility, and sustainability are three vital points of using digestate to produce mineral fertilizers. It is, therefore, a promising approach that can contribute to the circular economy. In this sense, the production of mineral fertilizers from digestate is based on extracting nutrients and reusing them to create various end products according to needs and respecting quality standards (Chojnacka & Moustakas, 2024). On the contrary, it is essential to note that recently, we have been able to manufacture products derived from digestate in parallel with the application of raw digestate, such as compost production, biochar production, and the manufacture of highly concentrated fertilizer pellets in nutrients and organic matter (Czekała et al., 2022).

In general, we can conclude that busing digestate from biomethane production to fertilize fields offers many advantages: (1) minimizing unpleasant odors, (2) eliminating potential pathogens, and reducing dependence on plant protection products (Koszel & Lorencowicz, 2015), (3) the organic carbon substances contained in digestate favor the retention capacity of mobile forms in the soil, resulting in a lower nitrate leaching potential than with mineral fertilizers (Jurgutis et al., 2021), and (4) digestate has other properties that encourage its direct application for fertilization, including a slightly acidic pH and low dry mass; it does not require additional treatment (Czekała et al., 2022).

8.6.3 Anaerobic digestate as a stimulator of food production

The biomethane digestate improves the chemical characteristics of the soil and can be adopted as an alternative to chemical fertilizers; the difference between the two is that the fertilizing effect of digestate is not immediate, as the

TABLE 8.2 The main properties of digestates from different organism substrates.

Substrate/origin	Fraction	pH	TS (%)	VS (%TS)	Total N	C/N	N-NH_3 (g/L)	References
Municipal solid waste (organic fraction)	Solid	8.80	7.23–94.78	68.0–71.0	–	12.1–20.9	–	Logan and Visvanathan (2019)
	Liquid	8.34–8.80	2.0–19.20	66.4	–	2.7	3.84	
Food waste	Solid	7.97	9.00	80.4	–	–	5.7	Peng and Pivato (2019)
	Liquid	7.9	3.23	68.4	–	–	2.7	
Co-digestion of agricultural residues	Whole	7.5–8.1	–	62.1–77 (% DM)	44–120	–	–	Monlau et al. (2015)
Typical anaerobic digestates	Whole	7.3–9.0	–	–	0.12–1.5 (% FM)	2.0–24.8	–	Nkoa (2014)
Slaughterhouse waste	Whole	7.9–8.3	2.2–4.9	1.6–3.9	6.4–8.1 (g/kg)	–	–	Bauer et al. (2021)
Municipal sludge waste	Whole	7.6–7.8	22.2–43.5 (g/kg)	13.8–32.78 (g/kg)	2.2–3.3 (g/kg)	6.14–8.36		Guan et al. (2024)
Co-digestion of cow manure, triticale, and hybrid type of corn of wheat and rye	Whole	8.2	–	–	2.9 (% dm)	13	–	Pivato et al. (2016)

(*Continued*)

TABLE 8.2 (Continued)

Substrate/origin	Fraction	pH	TS (%)	VS (%TS)	Total N	C/N	N-NH_3 (g/L)	References
Biogas plant	Whole	7.6–8.3	19.9–78.8 (g/kg FM)	12.3–63.7 (g/ kg FM)	2.2–8.7 (g/kg FM)	1.5–6.1	–	Chojnacka and Moustakas (2024)

quantity of nitrogen directly assimilable by plants contained in organic waste is lower than that of mineral fertilizers (Odlare et al., 2011). However, the use of digestate, in its initial state or just its liquid fraction after dilution, for hydroponic lettuce production has proved beneficial. It gives results similar to the effect of commercial fertilizer application. In this respect, digestate is well suited to vegetables that grow for short periods and require high nitrogen concentrations (Möller et Müller, 2012). So, digestate offers a valuable advantage for crops, particularly organic vegetables, in terms of short-term nitrogen release: The process works at the same speed as organic fertilizers and even better under low-temperature conditions, where release becomes more optimal.

Biomethane digestate has undoubtedly been widely studied as a fertilizer and soil improver. Karimi et al. (2022) have demonstrated that the optimum concentration of digestate to be applied is not constant but is influenced by several parameters, notably the type of raw material used, the type of soil concerned, and its fracturing. In addition, they conclude that many soil microbiological responses can be obtained due to the diversity of digestate composition. Chiew et al. (2015) declared that digestate benefits soil and plants, enriching them with macro- and micronutrients. Zhao et al. (2014) have shown that digestate can improve seed germination and seedling growth, but high concentrations can be harmful. Soils of low fertility treated with digestate continuously yielded three times more semi-natural grass (Jurgutis et al., 2021). Digestate is one of the most potassium-releasing organic fertilizers and helps reduce the fixation of this element in soils (Bader et al., 2021). So, the spreading digestate on agricultural land is highly beneficial but farms with intensive livestock production systems generate vast quantities of digestate, resulting in excess available land. In this context, Shi et al. (2018) have reported numerous management approaches that can exploit digestate for nutrient recovery, including chemical precipitation, membrane separation techniques, ammonia stripping, ion exchange, and adsorption, as well as thermal treatments such as incineration, pyrolysis, and hydrothermal carbonization. Another study conducted by Doyeni et al. (2021) evaluated the application of several digestates from different feedstocks (chicken, cow, and pig manure) on spring wheat (*Triticum aestivum*). The digestates had similar effects on biomass yield, which was increased by the digestate compared with the control. Analysis of their effects revealed that the digestates improved plant performance regarding their content of various phytochemicals, notably phenols, flavonoids, nehoesperidin, hesperitin, naringin, and narirutin. These compounds have antioxidant, anti-inflammatory, and anti-carcinogenic properties. The digestates have, therefore, improved the nutraceutical qualities of cucumbers, which means beneficial effects on consumer health. Digestate from the AD of animal manure was the subject of another study on the tomato plant, and the results were exciting. The digestates improved the content of tomato fruit in various essential compounds, including concentrations of total nitrogen and total P, soluble sugars, vitamin C, proteins, and others, consequently obtaining fruit of the highest nutritional quality (Yu et al., 2010). Similarly, another study on tomato plants showed the effect of two different digestates on the growth and quality of the plant's fruit, with results varying according to the type of digestate and the fraction used (liquid or solid). In general, both digestates positively affected tomatoes' growth and nutritional value by increasing their content of various compounds (phenolic compounds, vitamin C, and flavonoids) (Panuccio et al., 2021). So, the digestate issued by biomethane production is a compound that can go a long way in producing foods of high nutritional value and improving crop growth. These works show that digestate can offer significant economic benefits through improved feed quality and environmental benefits through the sustainable management of digestate after AD.

8.7 Conclusion

In this chapter, we saw that biomethane produced by AD plays a central role in establishing the water–energy–food nexus. Biomethane helps us with the water level; it allows the establishment of sustainable management of water resources. By treating wastewater and reducing leachate pollution, biomethane production protects aquatic ecosystems and ensures more efficient and responsible use of water resources. On the energy level, recovering the organic waste produced daily in our biomethane ecosystem offers a permanent supply of renewable and sustainable energy, contributing to reducing greenhouse gas emissions, transitioning to a low-carbon economy, and establishing energy independence. The efficient management of food waste, agricultural residues, and sewage sludge through AD does not just produce energy in the form of biomethane. However, it improves soil fertility through digestate as an organic amendment. This practice supports more sustainable and productive agriculture, reducing dependence on chemical fertilizers and thus contributing to food security on our planet.

Therefore, adopting biomethane in the water–energy–food nexus demonstrates the power of holistic and interconnected approaches to solving the environmental, energy, and food crises that threaten our existence. They make it possible to connect these areas and maximize synergies between them. Thus, biomethane not only allows better resource management but also increases resilience to the challenges of climate change and population growth. Thus, adopting biomethane in the water–energy–food nexus is a path toward a more sustainable and resilient future.

References

Ahmad, T., & Zhang, D. (2020). A critical review of comparative global historical energy consumption and future demand: The story told so far. *Energy Reports, 6*, 1973–1991. https://210.1080/10.1016/j.egyr.2020.07.020.

Ajay, C. M., Mohan, S., & Dinesha, P. (2021). Decentralized energy from portable biogas digesters using domestic kitchen waste: A review. *Waste Management (New York, N.Y.), 125*, 10–26. https://210.1080/10.1016/j.wasman.2021.02.031.

Al-Obadi, M., Ayad, H., Pokharel, S., & Ayari, M. A. (2022). Perspectives on Food Waste Management: Prevention and Social Innovations. *Sustainable Production and Consumption, 31*, 190–208. https://210.1080/10.1016/j.spc.2022.02.012.

Alqaralleh, R. M., Kennedy, K., & Delatolla, R. (2018). Improving biogas production from anaerobic co-digestion of thickened waste activated sludge (TWAS) and fat, oil and grease (FOG) using a dual-stage hyper-thermophilic/thermophilic semi-continuous reactor. *Journal of Environmental Management, 217*, 416–428. https://210.1080/10.1016/j.jenvman.2018.03.123.

Amin, M. A., Shukor, H., Yin, L. S., Kasim, F. H., Shoparwe, N. F., Makhtar, M. M. Z., & Yaser, A. Z. (2022). Methane biogas production in Malaysia: Challenge and future plan. *International Journal of Chemical Engineering, 2022*. https://210.1080/10.1155/2022/2278211.

Ardolino, F., Parrillo, F., & Arena, U. (2018). Biowaste-to-biomethane or biowaste-to-energy? An LCA study on anaerobic digestion of organic waste. *Journal of Cleaner Production, 174*, 462–476. https://210.1080/10.1016/j.jclepro.2017.10.320.

Azarmanesh, R., Qaretapeh, M. Z., Zonoozi, M. H., Ghiasinejad, H., & Zhang, Y. (2023). Anaerobic co-digestion of sewage sludge with other organic wastes: A comprehensive review focusing on selection criteria, operational conditions, and microbiology. *Chemical Engineering Journal Advances, 14*, 100453. https://210.1080/10.1016/j.ceja.2023.100453.

Bader, B. R., Taban, S. K., Fahmi, A. H., Abood, M. A., & Hamdi, G. J. (2021). Potassium Availability in Soil Amended with Organic Matter and Phosphorous Fertiliser under Water Stress during Maize (Zea Mays L) Growth. *Journal of the Saudi Society of Agricultural Sciences, 20*(6), 390–394. https://210.1080/10.1016/j.jssas.2021.04.006.

Bagi, Z., Ács, N., Böjti, T., Kakuk, B., Rákhely, G., Strang, O., Szuhaj, M., Wirth, R., & Kovács, K. L. (2017). Biomethane: The energy storage, platform chemical and greenhouse gas mitigation target. *Hungary Anaerobe, 46*, 13–22. https://210.1080/10.1016/j.anaerobe.2017.03.001.

Bakraoui, M., Karouach, F., Ouhammou, B., Aggour, M., Essamri, A., & El Bari, H. (2019). Kinetics study of the methane production from experimental recycled pulp and paper sludge by CSTR technology. *Journal of Material Cycles and Waste Management, 21*(6), 1426–1436. https://210.1080/10.1007/s10163-019-00894-6.

Barghash, H., AlRashdi, Z., Okedu, K., & Desmond, P. (2022). Life-cycle assessment study for bio-hydrogen gas production from sewage treatment plants using solar PVs. *Energies, 15*(21), 8056. https://210.1080/10.3390/en15218056.

Bauer, L., Ranglová, K., Masojídek, J., Drosg, B., Meixner, K. (2021). Digestate as sustainable nutrient source for microalgae—challenges and prospects. *Applied Sciences, 11*, 1056. https://doi.org/10.3390/app11031056.

Biswas, S., Kulkarni, A. P., Giddey, S., & Bhattacharya, S. (2020). A review on synthesis of methane as a pathway for renewable energy storage with a focus on solid oxide electrolytic cell-based processes. *Frontiers in Energy Research, 8*, 570112. https://210.1080/10.3389/fenrg.2020.570112.

Blair, K. J., Moran, D., & Alexander, P. (2023). Worldviews, values and perspectives towards the future of the livestock sector. *Agriculture and Human Values, 41*, 91–108. https://210.1080/10.1007/s10460-023-10469-9.

Bong, C. P. C, Ho, W. S., Hashim, H., Lim, J. S., Ho, C. S., Peng Tan, W. S., & Lee, C. T. (2017). Review on the renewable energy and solid waste management policies towards biogas development in Malaysia. *Renewable and Sustainable Energy Reviews, 70*, 988–998. https://210.1080/10.1016/j.rser.2016.12.004.

Brouwer, F., Avgerinopoulos, G., Fazekas, D., Laspidou, C., Mercure, J. F., Pollitt, H., Ramos, E. P., & Howells, M. (2018). Energy modelling and the nexus concept. *Energy Strategy Reviews, 19*, 1–6. https://210.1080/10.1016/j.esr.2017.10.005.

Budzianowski, W. M., & Postawa, K. (2017). Renewable energy from biogas with reduced carbon dioxide footprint: Implications of applying different plant configurations and operating pressures. *Renewable and Sustainable Energy Reviews, 68*, 852–868. https://210.1080/10.1016/j.rser.2016.05.076.

Calabi-Floody, M., Medina, J., Rumpel, C., Condron, L. M., Hernandez, M., Dumont, M., & Mora, M. D. L. L. (2018). Smart Fertilizers as a Strategy for Sustainable Agriculture. In *Advances in Agronomy, 147*, Elsevier, 119–157.

Calbry-Muzyka, A., Madi, H., Rüsch-Pfund, F., Gandiglio, M., & Biollaz, S. (2022). Biogas composition from agricultural sources and organic fraction of municipal solid waste. *Renewable Energy, 181*, 1000–1007. https://210.1080/10.1016/j.renene.2021.09.100.

Calise, F., Cappiello, F. L., Cimmino, L., D'accadia, M. D., & Vicidomini, M. (2021). A review of the state of the art of biomethane production: Recent advancements and integration of renewable energies. *Energies, 14*(16), 4895. https://210.1080/10.3390/en14164895.

Caporgno, M. P., Trobajo, R., Caiola, N., Ibáñez, C., Fabregat, A., & Bengoa, C. (2015). Biogas production from sewage sludge and microalgae co-digestion under mesophilic and thermophilic conditions. *Renewable Energy, 75*, 374–380. https://210.1080/10.1016/j.renene.2014.10.019.

Caposciutti, G., Baccioli, A., Ferrari, L., & Desideri, U. (2020). Biogas from anaerobic digestion: Power generation or biomethane production? *Energies, 13*(3), 743. https://210.1080/10.3390/en13030743.

Chanthakett, A., Arif, M. T., Khan, M. M. K., & Subhani, M. (2024). Hydrogen production from municipal solid waste using gasification method. In *Hydrogen Energy Conversion and Management*. Elsevier, 103–131. https://210.1080/10.1016/B978-0-443-15329-7.00012-0.

Chen, C., Chaudhary, A., & Mathys, A. (2020). Nutritional and environmental losses embedded in global food waste. *Resources, Conservation and Recycling, 160*, 104912. https://210.1080/10.1016/j.resconrec.2020.104912.

Chen, S., Zhao, Z., Soomro, A., Ma, S., Wu, M., Sun, Z., & Xiang, W. (2020). Hydrogen-rich syngas production via sorption-enhanced steam gasification of sewage sludge. *Biomass and Bioenergy, 138*, 105607. https://210.1080/10.1016/j.biombioe.2020.105607.

Chinnici, G., Selvaggi, R., D'Amico, M., & Pecorino, B. (2018). Assessment of the potential energy supply and biomethane from the anaerobic digestion of agro-food feedstocks in Sicily. *Renewable and Sustainable Energy Reviews, 82*, 6–13. https://210.1080/10.1016/j.rser.2017.09.018.

Chojnacka, K., & Moustakas, K. (2024). Anaerobic Digestate Management for Carbon Neutrality and Fertilizer Use: A Review of Current Practices and Future Opportunities. *Biomass and Bioenergy, 180*, 106991. https://210.1080/10.1016/j.biombioe.2023.106991.

Chynoweth, D. P., Owens, J. M., & Legrand, R. (2001). Renewable methane from anaerobic digestion of biomass. *Renewable Energy, 22*, 1–8. https://210.1080/10.1016/S0960-1481(00)00019-7.

Cristina, G., Camelin, E., Tommasi, T., Fino, D., & Pugliese, M. (2020). Anaerobic digestates from sewage sludge used as fertilizer on a poor alkaline sandy soil and on a peat substrate:

Effects on tomato plants growth and on soil properties. *Journal of Environmental Management, 269*, 110767. https://210.1080/10.1016/j.jenvman.2020.110767.
Czekała, W., Jasiński, T., Grzelak, M., Witaszek, K., & Dach, J. (2022). Biogas Plant Operation: Digestate as the Valuable Product. *Energies, 15*(21), 8275. https://210.1080/10.3390/en15218275.
Dagwar, P. P., & Dutta, D. (2024). Landfill leachate a potential challenge towards sustainable environmental management. *Science of the Total Environment, 926*, 171668. https://210.1080/10.1016/j.scitotenv.2024.171668.
Demirbas, A., Edris, G., & Alalayah, W. M. (2017). Sludge production from municipal wastewater treatment in sewage treatment plant. *Energy Sources Part A: Recovery, Utilization, and Environmental Effects, 39*, 999–1006. https://210.1080/10.1080/15567036.2017.1283551.
Di Capua, F., Spasiano, D., Giordano, A., Adani, F., Fratino, U., Pirozzi, F., & Esposito, G. (2020). High-solid anaerobic digestion of sewage sludge: Challenges and opportunities. *Applied Energy, 278*, 115608. https://210.1080/10.1016/j.apenergy.2020.115608.
Divya, D., Gopinath, L. R., & Christy, P. M. (2015). A review on current aspects and diverse prospects for enhancing biogas production in sustainable means. *Renewable and Sustainable Energy Reviews, 42*, 690–699. https://210.1080/10.1016/j.rser.2014.10.055.
Dong, K., Dong, X., & Jiang, Q. (2020). How renewable energy consumption lower global CO_2 emissions? Evidence from countries with different income levels. *World Economics, 43*, 1665–1698. https://210.1080/10.1111/twec.12898.
Doyeni, M. O., Baksinskaite, A., Suproniene, S., & Tilvikiene, V. (2021). Effect of Animal Waste Based Digestate Fertilization on Soil Microbial Activities, Greenhouse Gas Emissions and Spring Wheat Productivity in Loam and Sandy Loam Soil. *Agronomy, 11*(7), 1281. https://210.1080/10.3390/agronomy11071281.
El Asri, O., & Afilal, M. E. (2018). Comparison of the experimental and theoretical production of biogas by monosaccharides, disaccharides, and amino acids. *International Journal of Environmental Science and Technology, 15*(9), 1957–1966. https://210.1080/10.1007/s13762-017-1570-1.
El Asri, O., Hafidi, I., & Afilal, M. E. (2015). Comparison of biogas purification by different substrates and construction of a biogas purification system. *Waste Biomass Valorization, 6*, 459–464. https://210.1080/10.1007/s12649-015-9378-z.
El Asri, O., Afilal, M. E., Laiche, H., & Elfarh, L. (2020). Evaluation of physicochemical, microbiological, and energetic characteristics of four agricultural wastes for use in the production of green energy in Moroccan farms. *Morocco Chemical and Biological Technologies in Agriculture, 7*(1). https://210.1080/10.1186/s40538-020-00187-3.
El Asri, O., Fadlaoui, S., & Afilal, M. E. (2022a). Applications of microbes in municipal solid waste treatment. In Inamuddin, M. I. Ahamed, & R. Prasad (Eds.). *Application of Microbes in Environmental and Microbial Biotechnology, Environmental and Microbial Biotechnology* (pp. 587–607). Singapore: Springer Nature Singapore. https://210.1080/10.1007/978-981-16-2225-0_21.
El Asri, O., Fadlaoui, S., & Ramdani, M. (2022b). Anaerobic digestion technology of solid and liquid forms of olive wastes in the mediterranean region. In M. F. Ramadan, & M. A. Farag (Eds.). *Mediterranean Fruits Bio-Wastes* (pp. 67–89). Cham: Springer International Publishing. https://210.1080/10.1007/978-3-030-84436-3_4.
El Gnaoui, Y., Sounni, F., Bakraoui, M., Karouach, F., Benlemlih, M., Barz, M., & El Bari, H. (2020). Anaerobic co-digestion assessment of olive mill wastewater and food waste: Effect of mixture ratio on methane production and process stability. *Journal of Environmental Chemical Engineering, 8*(4), 103874. https://210.1080/10.1016/j.jece.2020.103874.

Elasri, O., & El amin Afilal, M. (2016). Potential for biogas production from the anaerobic digestion of chicken droppings in Morocco. *International Journal of Recycling of Organic Waste in Agriculture, 5*(3), 195–204. https://210.1080/10.1007/s40093-016-0128-4.

Elasri, O., Salem, M., Ramdani, M., Zaraali, O., & Lahbib, L. (2018). Effect of increasing inoculum ratio on energy recovery from chicken manure for better use in Egyptian agricultural farms. *Chemical and Biological Technologies in Agriculture, 5*(1). https://210.1080/10.1186/s40538-018-0129-9.

Erenstein, O., Jaleta, M., Mottaleb, K. A., Sonder, K., Donovan, J., & Braun, H. J. (2022). *Global Trends in Wheat Production, Consumption and Trade Wheat Improvement: Food Security in a Changing Climate.* Mexico: Springer International Publishing, 47–66. 10.1007/978-3-030-90673-3_4.

Eswari, A. P., Ravi, Y. K., Kavitha, S., & Rajesh Banu, J. (2023). Recent insight into anaerobic digestion of lignocellulosic biomass for cost effective bioenergy generation. *e-Prime - Advances in Electrical Engineering, Electronics and Energy, 3*, 100119. https://210.1080/10.1016/j.prime.2023.100119.

Faverdin, P., Guyomard, H., Puillet, L., & Forslund, A. (2022). Animal board invited review: Specialising and intensifying cattle production for better efficiency and less global warming: Contrasting results for milk and meat co-production at different scales. *Animal, 16*(1), 100431. https://210.1080/10.1016/j.animal.2021.100431.

Feng, S., Hong, X., Wang, T., Huang, X., Tong, Y., & Yang, H. (2019). Reutilization of high COD leachate via recirculation strategy for methane production in anaerobic digestion of municipal solid waste: Performance and dynamic of methanogen community. *Bioresource Technology, 288*, 121509. https://210.1080/10.1016/j.biortech.2019.121509.

Foster, W., Azimov, U., Gauthier-Maradei, P., Molano, L. C., Combrinck, M., Munoz, J., Esteves, J. J., & Patino, L. (2021). Waste-to-energy conversion technologies in the UK: Processes and barriers—A review. *Renewable and Sustainable Energy Reviews, 135*. https://210.1080/10.1016/j.rser.2020.110226 .

Freitas, F. F., De Souza, S. S., Ferreira, L. R. A., Otto, R. B., Alessio, F. J., De Souza, S. N. M., Venturini, O. J., & Junior, O. H. A. (2019). The Brazilian market of distributed biogas generation: Overview, technological development and case study. *Renewable and Sustainable Energy Reviews, 101*, 146–157. https://210.1080/10.1016/j.rser.2018.11.007.

Gamar, A., Zair, T., El Kabriti, M., & El Hilali, F. (2018). Study of the impact of the wild dump leachates of the region of El Hajeb (Morocco) on the physicochemical quality of the adjacent water table. *Karbala International Journal of Modern Science, 4*(4), 382–392. https://210.1080/10.1016/j.kijoms.2018.10.002.

Garcia, N. H., Mattioli, A., Gil, A., Frison, N., Battista, F., & Bolzonella, D. (2019). Evaluation of the methane potential of different agricultural and food processing substrates for improved biogas production in rural areas. *Renewable and Sustainable Energy Reviews, 112*, 1–10. https://210.1080/10.1016/j.rser.2019.05.040.

Gnaoui, Y. E., Karouach, F., Bakraoui, M., Barz, M., & Bari, H. E. (2020). Mesophilic anaerobic digestion of food waste: Effect of thermal pretreatment on improvement of anaerobic digestion process. *Energy Reports, 6*, 417–422. https://210.1080/10.1016/j.egyr.2019.11.096.

Guan, D., Zhao, J., Wang, Y., Fu, Z., Zhang, D., Zhang, H., Xie, J., Sun, Y., Zhu, J., & Wang, D. (2024). A Critical Review on Sustainable Management and Resource Utilization of Digestate. *Process Safety and Environmental Protection, 183*, 339–354. https://210.1080/0.1016/j.psep.2024.01.029.

González-Rosell, A., Arfa, I., & Blanco, M. (2023). Introducing GoNEXUS SEF: A solutions evaluation framework for the joint governance of water, energy, and food resources. *Sustainability Science, 18*, 1683–1703. https://210.1080/10.1007/s11625-023-01324-1.

Grosser, A., & Neczaj, E. (2018). Sewage sludge and fat rich materials co-digestion—Performance and energy potential. *Journal of Cleaner Production, 198*, 1076–1089. https://210.1080/10.1016/j.jclepro.2018.07.124.

Habchi, S., Lahboubi, N., Sallek, B., & El Bari, H. (2023). Response surface methodology for anaerobic digestion of waste from poultry slaughterhouse: Optimization of load and hydraulic retention time. *Results in Engineering, 18*, 101215. https://210.1080/10.1016/j.rineng.2023.101215.

Hagos, K., Zong, J., Li, D., Liu, C., & Lu, X. (2017). Anaerobic co-digestion process for biogas production: Progress, challenges and perspectives. *Renewable and Sustainable Energy Reviews, 76*, 1485–1496. https://210.1080/10.1016/j.rser.2016.11.184.

Hamraoui, K., Gil, A., El Bari, H., Siles, J. A., Chica, A. F., & Martín, M. A. (2020). Evaluation of hydrothermal pretreatment for biological treatment of lignocellulosic feedstock (pepper plant and eggplant). *Waste Management, 102*, 76–84. https://210.1080/10.1016/j.wasman.2019.10.020.

Heiker, M., Kraume, M., Mertins, A., Wawer, T., & Rosenberger, S. (2021). Biogas plants in renewable energy systems—A systematic review of modeling approaches of biogas production. *Applied Sciences, 11*(8), 3361. https://210.1080/10.3390/app11083361.

Hoo, P. Y., Hashim, H., Ho, W. S., & Tan, S. T. (2017). Successful biogas implementation—A mini-review on biogas utilization, energy policies and economic incentives. *Chemical Engineering Transactions, 61*, 1495–1500. https://210.1080/10.3303/CET1761247.

Huckleberry, J. K., & Potts, M. D. (2019). Constraints to implementing the food-energy-water nexus concept: Governance in the Lower Colorado River Basin. *Environmental Science and Policy, 92*, 289–298. https://210.1080/10.1016/j.envsci.2018.11.027.

Hussien, W. A., Memon, F. A., & Savic, D. A. (2016). Assessing and modelling the influence of household characteristics on per capita water consumption. *Water Resources Management, 30*, 2931–2955. https://210.1080/10.1007/s11269-016-1314-x.

Iglesias, R., Muñoz, R., Polanco, M., Díaz, I., Susmozas, A., Moreno, A. D., Guirado, M., Carreras, N., & Ballesteros, M. (2021). Biogas from anaerobic digestion as an energy vector: Current upgrading development. *Energies, 14*(10), 2742. https://210.1080/10.3390/en14102742.

Jame, R., Alaysuy, O., Omer, N., Alatawi, N. M., & Elsaim, M. H. (2021). Production of biomethane gas as renewable energy source—A review. *Journal of Energy, Environmental & Chemical Engineering, 6*(4), 124. https://210.1080/10.11648/j.jeece.20210604.14.

Jamshidi, M., & Askarzadeh, A. (2019). Techno-economic analysis and size optimization of an off-grid hybrid photovoltaic, fuel cell and diesel generator system. *Sustainable Cities and Society, 44*, 310–320. https://210.1080/10.1016/j.scs.2018.10.021.

Joshi, S. K., & Gauraha, A. K. (2022). Global biofertilizer market: Emerging trends and opportunities. In *Trends of Applied Microbiology for Sustainable Economy*. India: Elsevier, 689–697. 10.1016/B978-0-323-91595-3.00024-0.

Jurgutis, L., Šlepetienė, A., Šlepetys, J., & Cesevičienė, J. (2021). Towards a Full Circular Economy in Biogas Plants: Sustainable Management of Digestate for Growing Biomass Feedstocks and Use as Biofertilizer. *Energies, 14*(14), 4272. https://210.1080/10.3390/en14144272.

Kabeyi, M. J. B., & Olanrewaju, O. A. (2022). Biogas production and applications in the sustainable energy transition. *Journal of Energy, 2022*, 1–43. https://210.1080/10.1155/2022/8750221.

Karagozoglu, M. B., & Asar, M. (2023). Physico-chemical characteristics of leachate from landfills and its impact on surface water quality: case study in Osmaniye/Turkey. *Sustainable Water Resources Management, 9*(4), 125. https://210.1080/10.1007/s40899-023-00909-z.

Karouach, F., Bakraoui, M., El Gnaoui, Y., Lahboubi, N., & El Bari, H. (2020). Effect of combined mechanical–ultrasonic pretreatment on mesophilic anaerobic digestion of household organic waste fraction in Morocco. *Energy Reports, 6*, 310–314. https://210.1080/10.1016/j.egyr.2019.11.081.

Karouach, F., Bakraoui, M., Zguani, A., Hammadi, A., & El Bari, H. (2021). Co-digestion of industrial recycled pulp and paper sludge with vinasse wastewater: experimental and theoretical study. *International Journal of Environmental Science and Technology, 18*(11), 3651–3664. https://210.1080/10.1007/s13762-020-03111-2.

Kaur, G., Wong, J. W. C., Kumar, R., Patria, R. D., Bhardwaj, A., Uisan, K., & Johnravindar, D. (2020). Value Addition of Anaerobic Digestate From Biowaste: Thinking Beyond Agriculture. *Current Sustainable/Renewable Energy Reports, 7*(2), 48–55. https://210.1080/10.1007/s40518-020-00148-2.

Kohli, V., Chakravarty, S., Chamola, V., Sangwan, K. S., & Zeadally, S. (2023). An analysis of energy consumption and carbon footprints of cryptocurrencies and possible solutions. *Digital Communications and Networks, 9*(1), 79–89. https://210.1080/10.1016/j.dcan.2022.06.017.

Korberg, A. D., Mathiesen, B. V., Clausen, L. R., & Skov, I. R. (2021). The role of biomass gasification in low-carbon energy and transport systems. *Smart Energy, 1*, 100006. https://210.1080/10.1016/j.segy.2021.100006.

Koszel, M., & Lorencowicz, E. (2015). Agricultural Use of Biogas Digestate as a Replacement Fertilizers. *Agriculture and Agricultural Science Procedia, 7*, 119–124. https://210.1080/10.1016/j.aaspro.2015.12.004.

Kovačić, Đ., Lončarić, Z., Jović, J., Samac, D., Popović, B., & Tišma, M. (2022). Digestate management and processing practices: a review. *Applied Sciences, 12*(18), 9216. https://210.1080/9210.3390/app12189216.

Kundu, D., Dutta, D., Samanta, P., Dey, S., Sherpa, K. C., Kumar, S., & Dubey, B. K. (2022). Valorization of wastewater: A paradigm shift towards circular bioeconomy and sustainability. *Science of the Total Environment, 848*, 157709. https://210.1080/10.1016/j.scitotenv.2022.157709.

Kwon, Y., Park, J., Kim, G. B., Jo, Y., Park, S., & Kim, S. H. (2023). Anaerobic digestion of sewage sludge using anaerobic dynamic membrane bioreactor under various sludge composition and organic loading rates. *Bioresource Technology, 384*, 129275. https://210.1080/10.1016/j.biortech.2023.129275.

Lay, J.-J., Li, Y.-Y., & Noike, T. (1997). Influences of pH and moisture content on the methane production in high-solids sludge digestion. *Water Research, 31*, 1518–1524. https://210.1080/10.1016/S0043-1354(96)00413-7.

Lamolinara, B., Pérez-Martínez, A., Guardado-Yordi, E., Fiallos, C. G., Diéguez-Santana, K., & Ruiz-Mercado, G. J. (2022). Anaerobic digestate management, environmental impacts, and techno-economic challenges. *Waste Management, 140*, 14–30. https://210.1080/10.1016/j.wasman.2021.1012.1035.

Li, P., He, C., Yu, R., Shen, D., & Jiao, Y. (2020). Anaerobic co-digestion of urban sewage sludge with agricultural biomass. *Waste and Biomass Valorization, 11*(11), 6199–6209. https://210.1080/10.1007/s12649-019-00870-z.

Liao, X., Zhu, S., Zhong, D., Zhu, J., & Liao, L. (2014). Anaerobic co-digestion of food waste and landfill leachate in single-phase batch reactors. *Waste Management, 34*(11), 2278–2284. https://210.1080/10.1016/j.wasman.2014.06.014.

Lim, Y. F., Chan, Y. J., Abakr, Y. A., Sethu, V., Selvarajoo, A., Singh, A., Lee, J., & Gareth, M. (2021). Review of biowastes to energy in Malaysia: Current technology, scalability and socioeconomic analysis. *Cleaner Engineering and Technology, 4*, 100257. https://210.1080/10.1016/j.clet.2021.100257.

Liu, S., Cao, L., Xu, F., Yang, L., Li, Y., & Inalegwu, O. S. (2021). Integration of algae cultivation to anaerobic digestion for biofuel and bioenergy production. *Advances in Bioenergy, 6*, 199–300. https://210.1080/10.1016/bs.aibe.2021.06.002.

Liu, Y., Lin, R., Man, Y., & Ren, J. (2019). Recent developments of hydrogen production from sewage sludge by biological and thermochemical process. *International Journal of Hydrogen Energy, 44*(36), 19676–19697. https://210.1080/10.1016/j.ijhydene.2019.06.044.

Liu, Z., & Buchanan, I. D. (2011). Anaerobic co-digestion of municipal wastewater sludge and restaurant grease. *Water Quality Research Journal of Canada, 46*(4), 290–299. https://210.1080/10.2166/wqrjc.2011.122Canada.

Logan, M., & Visvanathan, C. (2019). Management strategies for anaerobic digestate of organic fraction of municipal solid waste: Current status and future prospects. *Waste Management & Research, 37*(1_suppl), 27–39. https://210.1080/10.1177/0734242X18816793.

Lu, J., & Xu, S. (2021). Post-treatment of food waste digestate towards land application: A review. *Journal of Cleaner Production, 303*, 127033. https://210.1080/127010.121016/j.jclepro.122021.127033.

Martínez-Dalmau, J., Berbel, J., & Ordóñez-Fernández, R. (2021). Nitrogen fertilization. A review of the risks associated with the inefficiency of its use and policy responses. *Sustainability. 13*(10), 5625. https://210.1080/5610.3390/su13105625.

Meng, Y., Li, S., Yuan, H., Zou, D., Liu, Y., Zhu, B., & Li, X. (2015). Effect of lipase addition on hydrolysis and biomethane production of Chinese food waste. *Bioresource Technology, 179*, 452–459. https://210.1080/10.1016/j.biortech.2014.12.015.

Moodley, P., & Trois, C. (2021). *Lignocellulosic Biorefineries: The Path Forward: Sustainable Biofuels: Opportunities and Challenges*. Elsevier, 21–42. https://10.1016/B978-0-12-820297-5.00010-4.

Mor, S., & Ravindra, K. (2023). Municipal solid waste landfills in lower- and middle-income countries: Environmental impacts, challenges and sustainable management practices. *Process Safety and Environmental Protection, 174*, 510–530. https://210.1080/10.1016/j.psep.2023.04.014.

Morita (2021). *Past Trends in Water Productivity at the Global and Regional Scale*. Elsevier BV, 99–118. https://10.1016/b978-0-323-91277-8.00003-4.

Morone, P., Koutinas, A., Gathergood, N., Arshadi, M., & Matharu, A. (2019). Food waste: Challenges and opportunities for enhancing the emerging bio-economy. *Journal of Cleaner Production, 221*, 10–16. https://210.1080/10.1016/j.jclepro.2019.02.258.

Monlau, F., Sambusiti, C., Ficara, E., Aboulkas, A., Barakat, A., & Carrère, H. (2015). New opportunities for agricultural digestate valorization: current situation and perspectives. *Energy & Environmental Science, 8*(9), 2600–2621. https://210.1080/2610.1039/C2605EE01633A.

Moustafa, A., Hamzeh, M., Net, S., Baroudi, M., & Ouddane, B. (2023). Seasonal variation of leachate from municipal solid waste landfill of Tripoli–Lebanon (case study). *International Journal of Environmental Science and Technology, 20*(11), 12545–12558. https://210.1080/10.1007/s13762-023-04834-8.

Music, W., & Gonfa, G. (2023). Fresh water resource, scarcity, water salinity challenges and possible remedies: A review. *Heliyon, 9*(8), e18685. https://210.1080/10.1016/j.heliyon.2023.e18685.

Möller, K., & Müller, T. (2012). Effects of anaerobic digestion on digestate nutrient availability and crop growth: A review. *Engineering in Life Sciences, 12*(3), 242–257. https://210.1080/10.1002/elsc.201100085.

Nabi, M., Liang, H., Cheng, L., Yang, W., & Gao, D. (2022). A comprehensive review on the use of conductive materials to improve anaerobic digestion: Focusing on landfill leachate treatment. *Journal of Environmental Management, 309*, 114540. https://210.1080/10.1016/j.jenvman.2022.114540.

Nair, A., Sartaj, M., Kennedy, K., & Coelho, N. M. G. (2014). Enhancing biogas production from anaerobic biodegradation of the organic fraction of municipal solid waste through leachate blending and recirculation. *Waste Management and Research, 32*(10), 939–946. https://210.1080/10.1177/0734242X14546036.

Niccolucci, V., Botto, S., Rugani, B., Nicolardi, V., Bastianoni, S., & Gaggi, C. (2011). The real water consumption behind drinking water: The case of Italy. *Journal of Environmental Management, 92*(10), 2611–2618. https://210.1080/10.1016/j.jenvman.2011.05.033.

Nkoa, R. (2014). Agricultural benefits and environmental risks of soil fertilization with anaerobic digestates: a review. *Agronomy for Sustainable Development, 34*, 473–492. https://210.1080/410.1007/s13593-13013-10196-z.

Noussan, M., Negro, V., Prussi, M., & Chiaramonti, D. (2024). The potential role of biomethane for the decarbonization of transport: An analysis of 2030 scenarios in Italy. *Applied Energy, 355*, 122322. https://210.1080/10.1016/j.apenergy.2023.122322.

Otles, S., Despoudi, S., Bucatariu, C., & Kartal, C. (2015). *Food waste management, valorization, and sustainability in the food industry*. Elsevier BV, 3–23. 10.1016/b978-0-12-800351-0.00001-8.

Ouahid, E. A., Mohamed, R., Meryam, O., Yahya, H. S. Y., Latrach, L., Soufiane, F., & Abdelhafid, C. (2019). Evaluation of the energetic valorization of the lagoon and Mediterranean algae (Caulerpa prolifera & Corallina elongata) by anaerobic digestion. *Scientific African, 5*, e00111. https://210.1080/10.1016/j.sciaf.2019.e00111.

Ourradi, H., Lahboubi, N., Habchi, S., Hanine, H., & El bari, H. (2022). Methane production from date seed cake (Phoenix dactylifera L.) using mesophilic fed-batch anaerobic digestion. *Cleaner Waste Systems, 2*, 100009. https://210.1080/10.1016/j.clwas.2022.100009.

Panuccio, M. R., Mallamaci, C., Attinà, E., & Muscolo, A. (2021). Using digestate as fertilizer for a sustainable tomato cultivation. *Sustainability, 13*(3), 1574. https://210.1080/1510.3390/su13031574.

Periyasamy, S., Kavitha, S., Isabel, B. J., Temesgen, T., Banu, R. J., & Sivashanmugam, P. (2022). Biogas recovery from sludge. In *Clean Energy and Resource Recovery*. Elsevier, 381–394. https://210.1080/10.1016/B978-0-323-90178-9.00008-1.

Peng, W., & Pivato, A. (2019). Sustainable management of digestate from the organic fraction of municipal solid waste and food waste under the concepts of back to earth alternatives and circular economy. *Waste and Biomass Valorization, 10*(2), 465–481. https://210.1080/410.1007/s12649-12017-10071-12642.

Pivato, A., Vanin, S., Raga, R., Lavagnolo, M. C., Barausse, A., Rieple, A., Laurent, A., & Cossu, R. (2016). Use of digestate from a decentralized on-farm biogas plant as fertilizer in soils: An ecotoxicological study for future indicators in risk and life cycle assessment. *Waste management, 49*, 378–389. https://210.1080/310.1016/j.wasman.2015.1012.100.

Qadir, M., Bahri, A., Sato, T., & Al-Karadsheh, E. (2010). Wastewater Production, Treatment, and Irrigation in Middle East and North Africa. *Irrigation and Drainage Systems, 24*(1-2), 37–51. https://doi.org/10.1007/s10795-009-9081-y.

Ravindran, B., Karmegam, N., Awasthi, M. K., Chang, S. W., Selvi, P. K., Balachandar, R., Chinnappan, S., Azelee, N. I. W., & Munuswamy-Ramanujam, G. (2022). Valorization of food waste and poultry manure through co-composting amending saw dust, biochar and mineral salts for value-added compost production. *Bioresource Technology, 346*, 126442. https://210.1080/10.1016/j.biortech.2021.126442.

Rempel, A., & Gupta, J. (2022). Equitable, effective, and feasible approaches for a prospective fossil fuel transition. *WIREs Climate Change, 13*(2), e756. https://210.1080/10.1002/wcc.756.

Ripoll, V., Agabo-García, C., Perez, M., & Solera, R. (2020). Improvement of biomethane potential of sewage sludge anaerobic co-digestion by addition of "sherry-wine" distillery wastewater. *Journal of Cleaner Production, 251*, 119667. https://210.1080/10.1016/j.jclepro.2019.119667.

Sahoo, A., Dwivedi, A., Madheshiya, P., Kumar, U., Sharma, R. K., & Tiwari, S. (2023). Insights into the management of food waste in developing countries: With special reference to India. *Environmental Science and Pollution Research, 31*(12), 17887–17913. https://210.1080/10.1007/s11356-023-27901-6.

Salehiyoun, A. R., Di Maria, F., Sharifi, M., Norouzi, O., Zilouei, H., & Aghbashlo, M. (2020). Anaerobic co-digestion of sewage sludge and slaughterhouse waste in existing wastewater digesters. *Renewable Energy, 145*, 2503–2509. https://210.1080/10.1016/j.renene.2019.08.001.

Semrau, R., & Engell, S. (2023). Process as a battery: Robust dynamic optimal operation of zeolite crystallization in a COBR with respect to the carbon footprint of electric power. *Computer Aided Chemical Engineering, 52*, 1129–1134. https://210.1080/10.1016/B978-0-443-15274-0.50180-3.

Serrano, A., López, J. A. S., Chica, A. F., Martin, M. A., Karouach, F., Mesfioui, A., & El Bari, H. (2014). Mesophilic anaerobic co-digestion of sewage sludge and orange peel waste. *Environmental Technology (United Kingdom), 35*(7), 898–906. https://210.1080/10.1080/09593330.2013.855822.

Shahriari, H., Warith, M., Hamoda, M., & Kennedy, K. J. (2012). Effect of leachate recirculation on mesophilic anaerobic digestion of food waste. *Waste Management, 32*(3), 400–403. https://210.1080/10.1016/j.wasman.2011.10.022.

Shatat, M., Worall, M., & Riffat, S. (2013). Opportunities for solar water desalination worldwide: Review. *Sustainable Cities and Socities, 9*, 67–80. https://210.1080/10.1016/j.scs.2013.03.004.

Shi, L., Simplicio, W. S., Wu, G., Hu, Z., Hu, H., & Zhan, X. (2018). Nutrient recovery from digestate of anaerobic digestion of livestock manure: a review. *Current Pollution Reports, 4*, 74–83. https://210.1080/10.1007/s40726-40018-40082-z.

Sillero, L., Sganzerla, W. G., Carneiro, T. F., Solera, R., & Perez, M. (2023). Techno-economic analysis of single-stage and temperature-phase anaerobic co-digestion of sewage sludge, wine vinasse, and poultry manure. *Journal of Environmental Management, 325*, 116419. https://210.1080/10.1016/j.jenvman.2022.116419.

Silva, J., Gonçalves, J. C., Rocha, C., Vilaça, J., & Madeira, L. M. (2024). Biomethane production from biogas obtained in wastewater treatment plants: Process optimization and economic analysis. *Renewable Energy, 220*, 119469. https://210.1080/10.1016/j.renene.2023.119469.

Singh, D., Tembhare, M., Machhirake, N., & Kumar, S. (2023). Impact of municipal solid waste landfill leachate on biogas production rate. *Journal of Environmental Management, 336*, 117643. https://210.1080/10.1016/j.jenvman.2023.117643.

Sinha, S., & Tripathi, P. (2021). Trends and challenges in valorisation of food waste in developing economies: A case study of India. *Case Studies in Chemical and Environmental Engineering, 4*, 100162. https://210.1080/10.1016/j.cscee.2021.100162.

Skuse, C., Gallego-Schmid, A., Azapagic, A., & Gorgojo, P. (2021). Can emerging membrane-based desalination technologies replace reverse osmosis? *Desalination, 500*, 114844. https://210.1080/10.1016/j.desal.2020.114844.

Song, J., Li, M., Wang, C., Fan, Y., Li, Y., Wang, Y., Zhang, W., Li, H., & Wang, H. (2023). Enhanced treatment of landfill leachate by biochar-based aerobic denitrifying bacteria functional microbial materials: Preparation and performance. *Frontiers in Microbiology, 14*, 1139650. https://210.1080/10.3389/fmicb.2023.1139650.

Sorrell, S., Speirs, J., Bentley, R., Brandt, A., & Miller, R. (2010). Global oil depletion: A review of the evidence. *Energy Policy, 38*(9), 5290–5295. https://210.1080/10.1016/j.enpol.2010.04.046.

Sulewski, P., Ignaciuk, W., Szymańska, M., & Wąs, A. (2023). Development of the biomethane market in Europe. *Energies, 16*(4), 2001. https://210.1080/10.3390/en16042001.

Taguta, C., Senzanje, A., Kiala, Z., Malota, M., & Mabhaudhi, T. (2022). Water-energy-food nexus tools in theory and practice: A systematic review. *Frontiers in Water, 4*, 837316. https://210.1080/10.3389/frwa.2022.837316.

Tashtoush, F. M., Al-Zubari, W. K., & Shah, A. (2019). A review of the water–energy–food nexus measurement and management approach. *International Journal of Energy and Water Resources, 3*, 361–374. https://210.1080/10.1007/s42108-019-00042-8.

Thorin, E., Olsson, J., Schwede, S., & Nehrenheim, E. (2017). Biogas from co-digestion of sewage sludge and microalgae. *Energy Procedia, 105*, 1037–1042. https://210.1080/10.1016/j.egypro.2017.03.449.

Torrijos, M. (2016). State of development of biogas production in Europe. *Procedia Environmental Sciences, 35*, 881–889. https://210.1080/10.1016/j.proenv.2016.07.043.

Wickham, R., Galway, B., Bustamante, H., & Nghiem, L. D. (2016). Biomethane potential evaluation of co-digestion of sewage sludge and organic wastes. *International Biodeterioration and Biodegradation, 113*, 3–8. https://210.1080/10.1016/j.ibiod.2016.03.018.

Wickham, R., Xie, S., Galway, B., Bustamante, H., & Nghiem, L. D. (2019). Pilot-scale operation experience of anaerobic co-digestion for possible full scale implementation. *International Biodeterioration and Biodegradation, 142*, 137–142. https://210.1080/10.1016/j.ibiod.2019.05.020.

Wilson, W. W., & Shakya, S. (2023). Quantifying impacts of competition and demand on the risk for fertilizer plant locations. *Journal of Commodity Markets, 30*, 100326. https://210.1080/10.1016/j.jcomm.2023.100326.

Yu, F.-B., Luo, X.-P., Song, C.-F., Zhang, M.-X., & Shan, S.-D. (2010). Concentrated biogas slurry enhanced soil fertility and tomato quality. *Acta Agriculturae Scandinavica Section B–Soil and Plant Science, 60*(3), 262–268. https://210.1080/210.1080/09064710902893385.

Zhang, J., Lv, C., Tong, J., Liu, J., Liu, J., Yu, D., Wang, Y., Chen, M., & Wei, Y. (2016). Optimization and microbial community analysis of anaerobic co-digestion of food waste and sewage sludge based on microwave pretreatment. *Bioresource Technology, 200*, 253–261. https://210.1080/10.1016/j.biortech.2015.10.037.

Zhao, Z., Bian, R., Zhao, F., & Chai, X. (2020). Implications of municipal solid waste disposal methods in China on greenhouse gas emissions. *Environmental Progress & Sustainable Energy, 39*(3), e13372. https://210.1080/10.1002/ep.13372.

Zhu, X., Xu, Y., Zhen, G., Lu, X., Xu, S., Zhang, J., Gu, L., Wen, H., Liu, H., Zhang, X., & Wu, Z. (2023). Effective multipurpose sewage sludge and food waste reduction strategies: A focus on recent advances and future perspectives. *Chemosphere, 311*, 136670. https://210.1080/10.1016/j.chemosphere.2022.136670.

Chapter 9

Biohydrogen production for power and transport

Saida Tayibi[1], Doha Elalami[2], Santi Chuetor[3,4], and Abdellatif Barakat[5,6]
[1]*AgroBiosciences, College for Sustainable Agriculture and Environmental Sciences, Mohammed VI Polytechnic University, Hay Moulay Rachid, Ben Guerir, Rehamna Province, Morocco,* [2]*Department of Process Engineering, Mohammadia School of Engineering, Mohammed V University, Rabat, Morocco,* [3]*Department of Chemical Engineering, Faculty of Engineering, King Mongkut's University of Technology North Bangkok, Bangkok, Thailand,* [4]*Center of Eco-Materials and Cleaner Technology, King Mongkut's University of Technology, North Bangkok, Bangkok, Thailand,* [5]*UMR IATE, INRAE, Montpellier University, Institute Agro of Montpellier, Montpellier, France,* [6]*CAES, University Mohammed VI Polytechnic (UM6P), Benguérir, Morocco*

9.1 Introduction

Petroleum fuel, alongside natural gas and coal, has been integral to global energy consumption for decades, significantly influencing the environment and climate through the release of greenhouse gases such as carbon dioxide (CO_2), methane (CH_4), nitrogen dioxide (NO_2), and ozone (O_3), all of which are primary contributors to global warming (Faye et al., 2022). Specifically, the anticipated trajectory of global energy demand is projected to escalate to 812 quadrillion kilojoules by the year 2035 (Kaur et al., 2019). Concurrently, the demand for liquid fossil fuels is forecasted to surge to 112.2 million barrels per day by the same year (Kaur et al., 2019). This escalating demand underscores the urgent need for sustainable energy solutions to both mitigate the adverse impacts of climate change and diminish dependence on finite fossil fuel resources. Consequently, there has been a marked surge in intensive research efforts aimed at exploring and harnessing alternative energy sources (Hassan et al., 2024; Saravanan et al., 2021; Srivastava et al., 2020). Among these, biohydrogen emerges as a frontrunner, offering a renewable and environmentally friendly option for both power generation and transportation (Al- Janabi et al., 2024; Bastola & Hiloidhari, 2024; Sahrin et al., 2022; Xu et al., 2022). Biohydrogen, derived from biomass through various biological and thermochemical processes, presents a promising route toward a cleaner and more sustainable energy

Biogas in the Circular Economy. DOI: https://doi.org/10.1016/B978-0-443-29230-9.00012-7

future (Ali et al., 2024). The feedstock for biohydrogen production could be provided from diverse sustainable and natural resources, including energy crops, industrial and municipal waste, and agricultural and forestry residues. It was reported that these resources could be classified into different classes of raw materials as follows: for example, (1) plant straw materials (corn straw, rice straw, etc.), (2) animal manures (cow, chicken, sheep, etc.), (3) food industry by-products (olive cake, apple pomace, argan cake, rapeseed cake, etc.), (4) garbage (landfill leachate, household waste, etc.), and (5) other materials including sewage sludge, sediment, and microalgae. Contrary to the production of hydrogen using fossil fuel, the process of the biohydrogen production from biomass could be considered as carbon-neutral emission process as the carbon in biomass come from the capture of the atmospheric CO_2 by plants; therefore, the carbon dioxide that is generated during the production of biohydrogen from biomass could be considered as the recycling of carbon elements (Rivera-Tinoco & Bouallou, 2010).

In this book chapter, we discuss the multifaceted realm of biohydrogen production, transportation, storage, and application. Our exploration begins with an overview of the diverse technologies employed in biohydrogen production, ranging from biologic processes such as fermentation and biophotolysis to electrochemical and thermochemical pathways. Understanding these methodologies is crucial for unlocking the full potential of biohydrogen as a viable energy source. Subsequently, we examine the intricacies of biohydrogen transportation and storage technologies, essential for bridging the gap between production and utilization. From gaseous and liquid transportation to innovative biohydrogen carriers and storage solutions, each avenue presents unique challenges and opportunities in facilitating the widespread adoption of biohydrogen. Moving forward, we delve into the myriad applications of biohydrogen, particularly in power generation and transportation. Fuel cells stand out as a cornerstone technology, enabling the conversion of biohydrogen into electricity with high efficiency and minimal environmental impact. Moreover, the utilization of biohydrogen in various industrial sectors, including the chemical industry, plant power generation, and fertilizer production, underscores its versatility and potential to revolutionize diverse sectors. Despite its promise, the journey toward mainstream adoption of biohydrogen is fraught with challenges. Technical and operational hurdles, economic constraints, storage limitations, and environmental concerns pose significant obstacles that must be addressed to realize the full potential of biohydrogen as a sustainable energy solution. Therefore, this book chapter aims to provide a comprehensive overview of the advancements, challenges, and prospects in biohydrogen production for power generation and transportation. By elucidating the intricacies of biohydrogen technology and highlighting areas for further research and development, we hope to contribute to the collective effort toward a cleaner, greener, and more sustainable energy future (Fig. 9.1).

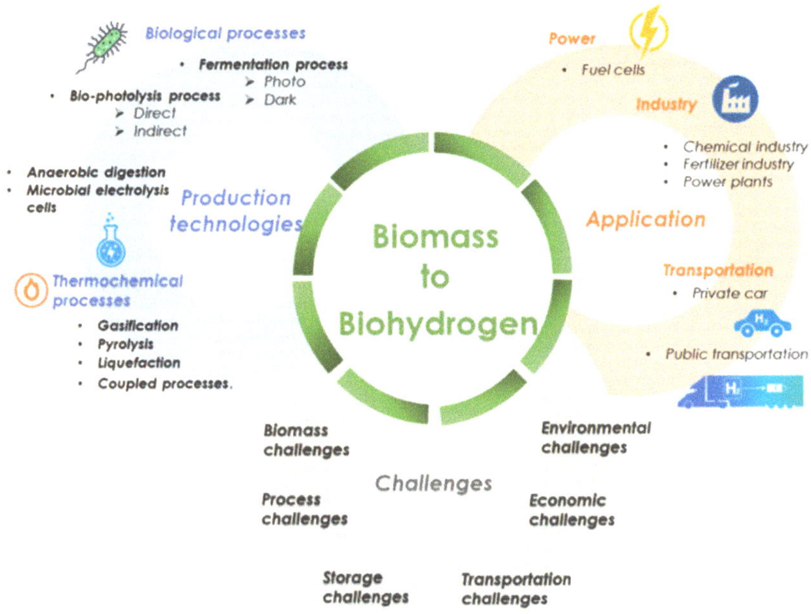

FIGURE 9.1 Overview of the production technology biohydrogen from biomass, its main applications, and main challenges.

9.2 Biohydrogen production technologies

9.2.1 Biologic processes

Biohydrogen production is more intriguing than conventional hydrogen due to its renewable sources, environmental benefits, and energy efficiency. It utilizes biomass and organic waste, reducing carbon footprints and promoting sustainable waste management. Biohydrogen can be economically advantageous by converting waste into fuel and fostering rural development. Additionally, biohydrogen can be integrated into biorefineries, producing multiple products from biomass, thereby increasing overall sustainability and efficiency. These factors collectively make biohydrogen a promising and sustainable alternative for hydrogen production. Fig. 9.2 presents the main biological processes used to produce biohydrogen (Fig. 9.2 and Table 9.1).

9.2.1.1 Biophotolysis

Bio-photolysis, also known as biological photolysis, is a process that uses microorganisms such as algae or photosynthetic bacteria to produce hydrogen gas through the photolysis of water. This process harnesses the natural

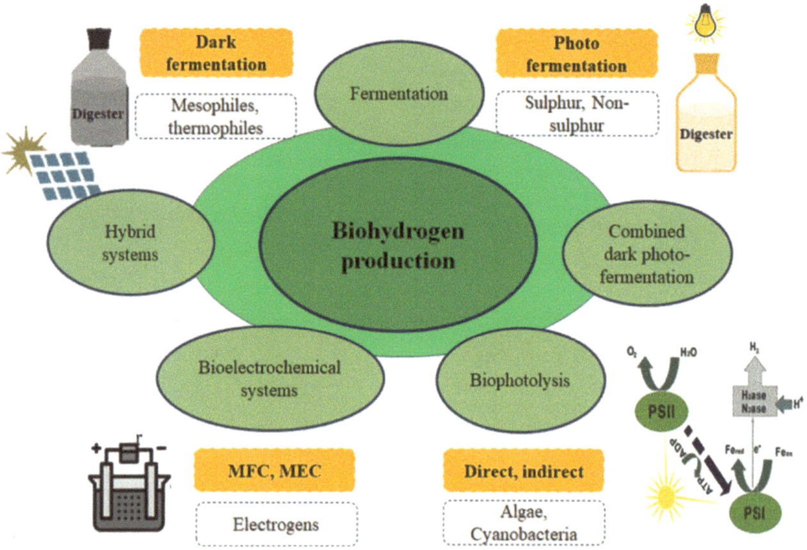

FIGURE 9.2 Bioprocesses used to produce biohydrogen.

photosynthetic abilities of these organisms to split water molecules into hydrogen and oxygen using sunlight as an energy source. Some types of photoautrophic microorganisms, such as green algae (e.g., *Chlamydomonas reinhardtii*) or cyanobacteria (e.g., *Nostoc, Anabaena, Calothrix, and Oscillatoria*), possess the ability to perform photosynthesis. These organisms contain chlorophyll and other pigments that can capture sunlight and convert it into chemical energy through photosynthetic reactions. They generate enzymes through nitrogen fixation that catalyze hydrogen production. The selected photosynthetic organisms are cultivated in a suitable growth medium under controlled environmental conditions, such as temperature, light intensity, and nutrient availability. This cultivation process allows the organisms to grow and multiply while photosynthetically converting carbon dioxide and water into organic compounds and oxygen. In fact, under certain conditions, such as nitrogen deprivation or sulfur depletion, some photosynthetic organisms undergo a metabolic shift and produce hydrogen gas as a by-product of photosynthesis. However, bio-photolysis also faces challenges such as low hydrogen production rates, sensitivity to environmental conditions, and the need for further research and optimization to improve efficiency and scalability. Nonetheless, ongoing research efforts are focused on developing and optimizing bio-photolysis systems for practical applications in sustainable hydrogen production.

TABLE 9.1 Biohydrogen production through biological processes and bioelectrochemical systems in literature

Process type	Substrate	Hydrogen yield	Conditions	References
Anaerobic digestion	Food waste, sewage sludge (1:1)	75 mL H_2/g VS	pH 5.5, 35°C, batch	Silva et al. (2018)
	Molasses	2.5 mol H_2/ mol sucrose	pH 5.5, 37°C, CSTR	Aceves-Lara et al. (2008)
	Corn straw	1 mL H_2/g VS	pH 5.5, 35°C, batch	Li and Chen (2007)
		68 mL H_2/g VS	pH 5.5, 35°C, batch, steam explosion (1.2 MPa for 10 min)	
	Microalgae (*Scenedesmus obtusiusculus*)	28.5 mL H_2/g VS	pH 7.5, 37°C, Batch	Rincón-Pérez et al. (2020)
Dark fermentation	Glucose	2.5 mol H_2/mol glucose	pH 6.0, 35°C, Batch reactor	Kapdan and Kargi (2006)
	Food waste	9 mL H_2/gTS	pH 6.0, 30°C, Batch reactor	Hovorukha et al. (2021)
	Corn stover	35 mL H_2/gTS	pH 6.5, 35°C, Batch	Zhang et al. (2020)
	Rice straw	5.3 mL H_2/gTS	pH 6.5, 55°C, Batch	Chen et al. (2012)
	Cheese whey	90 mL/ gTOC	pH 7.5, 39°C, Batch reactor	De Gioannis et al. (2014)

(*Continued*)

TABLE 9.1 (Continued)

Process type	Substrate	Hydrogen yield	Conditions	References
Photo-fermentation	Glucose	110 mL H_2/g glucose	135 W/ m^2, pH 7.8, 30°C, *Rhodobacter sphaeroides*, Batch	Fang et al. (2006)
	Cyanobacteria	13 mL H_2/L.h	140 μE/m^2.S, pH 7.5, 30°C, Batch	Azwar et al. (2014)
	Corn stover	140 m H_2/gTS	pH of 6.5, light intensity of 3500 Lux, and temperature of 30°C	Zhang et al. (2020)
	Acetate	0.65 mole /mole of substrate	pH of 7.0, light intensity of 10.25 W/m^2 and temperature of 30°C	Loss et al. (2013)
Sequential dark and photo-fermentation	Cassava starch	DF: 2.53 mol/mol DF + PF: 3.54 mol/mol	pH 7.0, 30°C, under 4000-lux	Sun et al. (2010)
	Corn stover	80 mL H_2/gTS	pH of 6.5, light intensity of 3500 Lux, and temperature of 30°C	Zhang et al. (2020)
	Water hyacinth	112.3 mL/g TVS to 751.5 mL/g TVS	pH of 6.0, light intensity of 6000 lux and temperature of 35°C	Cheng et al. (2011)
Bioelectrochemical system	Synthetic organic solution	38 m H_2	0.8 V applied voltage,	Yuan et al. (2015)
	Crude glycerol	106 mL H_2/g COD	0.6 V applied voltage,	Chookaew et al. (2014)

9.2.1.2 Anaerobic digestion for biohydrogen

Anaerobic digestion (AD) for biohydrogen production involves the bioconversion of organic substrates into hydrogen gas under oxygen-free conditions, primarily through the metabolic activities of specific bacterial communities. The process emphasizes the hydrolysis and acidogenesis phases of AD while inhibiting methanogenesis to prevent the conversion of hydrogen to methane. During hydrolysis, complex organic molecules like carbohydrates, proteins, and lipids are enzymatically broken down into simpler monomers such as sugars, amino acids, and fatty acids. These monomers are subsequently metabolized by acidogenic bacteria (e.g., species of *Clostridium* and *Enterobacter*), producing volatile fatty acids, alcohols, carbon dioxide, and hydrogen. Optimal biohydrogen production requires maintaining an acidic pH around 5.5 to inhibit methanogens, utilizing carbohydrate-rich substrates, and operating under mesophilic (30°C–40°C) or thermophilic (50°C–60°C) conditions to favor hydrogen-producing bacteria (Khan et al., 2018). This modified AD approach not only offers a sustainable method for biohydrogen production but also contributes to waste management and renewable energy generation (Chen et al., 2012). Recently, researchers in AD for biohydrogen production are focusing on several innovative strategies and technologies to enhance hydrogen yields and process efficiency. Advanced pretreatment technologies, such as hydrothermal and enzymatic treatments, improve substrate biodegradability (Li & Chen, 2007). In addition, the integration of AD with other renewable energy systems, like solar and wind, aims to create hybrid systems that enhance energy efficiency (Ghimire et al., 2015). Engineering microbial communities to promote synergistic interactions and applying nanotechnology to enhance microbial activity are also key trends. Artificial intelligence and machine learning are being used to optimize processes and predict outcomes (Gupta et al., 2023), leading to more efficient hydrogen production. These trends collectively aim to make AD a more viable and sustainable method for biohydrogen production (Zhang & Xue, Wang, et al., 2023).

9.2.1.3 Photo-fermentation process

Photo-fermentation (PF) is a biological process that utilizes some types of microorganisms (e.g., Gram-negative prokaryotes) to produce hydrogen gas from organic compounds under light conditions. Unlike bio-photolysis, which directly utilizes photosynthetic organisms to dissociate water into hydrogen and oxygen, PF involves the fermentation of organic substrates by photosynthetic bacteria or algae to generate organic acids, alcohols, and hydrogen gas as a metabolic by-product under anaerobic conditions. Besides the high hydrogen yield that can be obtained using some types of bacteria, thy are sensitive to environmental conditions (e.g., light intensity, temperature, and pH) and competition from other microbial populations, which requires further research and optimization to improve efficiency and scalability. Most industrial hydrogen production today relies on other methods, such as steam methane reforming and electrolysis (Akhlaghi & Najafpour-Darzi, 2020). Currently, there are no industrial-scale PF plants dedicated to producing hydrogen. In a previous study, a photo-bioreactor pilot scale was operated in outdoor condition using sunlight to reduce the energy supply to the system. It was found that the unstability of light intensity and the contamination reduced highly the performance of the photo-bioreactor (Chen et al., 2008). Thus, significant technological and economic challenges remain to be addressed before scaling up to industrial levels can be viable (Ahmed et al., 2021; Gupta et al., 2024). Over the past decade, significant progress has been made in optimizing and understanding biohydrogen production, paving the way for its potential commercialization and wider application in renewable energy systems. Researchers have focused on improving the hydrogen production efficiency of purple nonsulfur bacteria such as *Rhodobacter sphaeroides* and *Rhodopseudomonas palustris*. Genetic engineering techniques have been used to eliminate competing metabolic pathways and to overexpress hydrogenase enzymes that are critical for hydrogen production. These modifications have resulted in bacterial strains with significantly higher hydrogen yields (Khetkorn et al., 2013). Effective utilization of light energy is paramount for the efficiency of PF processes. Advances have been made in the development of photobioreactors that optimize light distribution and intensity. The use of LED lights with specific wavelengths tailored to the absorption spectra of photosynthetic bacteria has been shown to enhance hydrogen production rates (Chen et al., 2016). Additionally, the design of photobioreactors has evolved to ensure better light penetration and reduce shading effects, thus improving the overall efficiency of the process (Mirza et al., 2019). The type of organic substrates used in PF significantly affects hydrogen production. Recent studies have explored a variety of waste materials, including agricultural residues, food waste, and industrial effluents, as potential feedstocks, as shown in Table 9.1. Using these waste materials not only provides a cost-effective substrate but also addresses waste management issues, making the process more sustainable. Advances in pretreatment methods to break down complex organic materials into simpler forms that are more readily metabolized by bacteria have also contributed to improved hydrogen yields (Monlau, Barakat, et al., 2013). Moreover, coupling PF with wastewater treatment processes can provide a dual benefit of hydrogen production and wastewater remediation, making the technology more attractive for practical applications (Elsayad et al., 2024).

9.2.1.4 Dark fermentation

Dark fermentation (DF) and PF are both biological processes used to produce hydrogen, but they differ significantly in their mechanisms, conditions, and efficiency. DF involves the anaerobic breakdown of organic substrates by microorganisms in the absence of light. The process uses heterotrophic bacteria or microalgae that do not require light for growth such as *Clostridium* and *Enterobacter* species. It involves stages such as hydrolysis, acidogenesis, and acetogenesis and operates under simpler conditions and consumes less energy compared with PF, as the latter requires light management which can affect the bioreactor design. However, lower hydrogen yield can be obtained in DF compared with PF (Brentner et al., 2010). Combining DF and PF can enhance overall hydrogen yields. In fact, DF produces organic acids and other intermediates that serve as substrates for PF (Chen et al., 2008). Additionally, integrating DF with AD can facilitate the complete utilization of organic waste, producing hydrogen in the first stage and methane in the second stage, thus improving the overall energy recovery from waste (Bertasini et al., 2024). Advancements in reactor design have significantly contributed to the efficiency of DF processes. Researchers have developed various reactor configurations, such as continuously stirred tank reactors (CSTR), anaerobic sequencing batch reactors, and biofilm reactors, to optimize hydrogen production (Show et al., 2011). Innovative reactor designs, including those incorporating biochar and other support materials, have improved microbial adhesion and stability, thereby enhancing hydrogen production rates.

9.2.2 Bioelectrochemical systems

Bioelectrochemical systems (BES) represent a promising technology for the sustainable production of biohydrogen, integrating microbial metabolism with electrochemical processes (De Gioannis et al., 2023). Microbial fuel cells (MFCs) and microbial electrolysis cells (MECs) are the two primary types of bioelectrochemical systems used for biohydrogen production. MFCs generate electricity through the metabolic activities of microorganisms, whereas MECs use an external power source to drive hydrogen production from organic substrates. Recent research has focused on improving the performance and efficiency of these systems through various innovations in microbial selection, electrode materials, and reactor design (Arun et al., 2024). As with fermentation, researchers have identified and designed microbial strains with enhanced electrochemical activity and hydrogen production potentials. For example, electrogenic bacteria such as *Geobacter sulfurreducens* and *Shewanella oneidensis* have been extensively studied for their ability to efficiently transfer electrons to electrodes. Genetic modifications have been used to enhance the expression of key enzymes involved in electron transfer and hydrogen production, thereby improving the overall performance of the system (Surti et al., 2021). In addition, the development of advanced electrode materials has significantly contributed to the efficiency of bioelectrochemical systems. Innovations in electrode design, such as the use of nanomaterials, conductive polymers, and three-dimensional structures, have improved the surface area and conductivity of electrodes, facilitating better microbial adhesion and electron transfer. In addition, the incorporation of catalysts, such as platinum and nickel, has enhanced the hydrogen evolution reaction, leading to higher hydrogen yields (Kumar et al., 2021). Moreover, innovations such as the use of membrane-less reactors and the integration of bioelectrochemical systems with other processes, like AD, have been efficient in enhancing overall performance (Rosenbaum & Franks, 2014). In fact, integrating bioelectrochemical systems with other renewable energy technologies has been a key focus of recent research. For example, coupling BESs with solar panels or wind turbines can provide a sustainable power source for the electrolysis process in MECs, reducing the reliance on external electricity and enhancing the overall sustainability of hydrogen production. Such integrated systems have the potential to create decentralized energy solutions that are both efficient and environmentally friendly (Rosenbaum & Franks, 2014). The environmental impact and economic feasibility of bioelectrochemical processes have been extensively studied. Life cycle assessments and techno-economic analyses have shown that BESs have the potential to significantly reduce greenhouse gas emissions and provide a cost-effective solution for hydrogen production, particularly when integrated with waste treatment and renewable energy systems. However, challenges remain in scaling up these technologies and reducing the costs associated with advanced materials and system components (Chin et al., 2022). Table 9.1 presents some results found in literature. In fact, food waste, sewage sludge, and microalgae seem to be easier to convert to hydrogen, while agricultural residues are more difficult. This is mainly due to the recalcitrance of the lignocellulosic matrix which require pretreatment to facilitate the hydrolysis phase (Li & Chen, 2007). Overall, optimization of operating conditions is crucial to maximize hydrogen yield. These optimal conditions will depend on the fermentative strain and the composition of the substrate used. Compared with electrolysis, biological pathways such as DF and PF consume less energy and can use biomass waste or wastewater, making them environmentally friendly. However, these biological processes generally have lower hydrogen yields and slower production rates than electrolysis. While electrochemical methods are better suited to large-scale (Jia et al., 2016) and produce high-purity hydrogen, biological methods offer sustainable solutions using residual resources, potentially reducing environmental impact and improving energy security. Combining these methods could exploit the advantages of creating a more robust hydrogen production system.

9.2.3 Thermochemical processes

Various thermochemical pathways, including pyrolysis, gasification, and thermal liquefaction, are utilized to produce biohydrogen from biomass (Kaur et al., 2019).

Pyrolysis and gasification are particularly notable for their high efficiency in generating substantial amounts of biohydrogen. These processes result in the formation of products such as biocrude, water, charcoal, tar, and permanent gases (CO, CO_2, H_2, and CH_4). The specific products and their quantities are influenced by factors such as the reaction environment, final temperature, heating rate, heat source, and reactor type employed (Kaur et al., 2019).

9.2.3.1 Gasification

Biomass gasification is a promising technological approach for producing biohydrogen due to its efficiency in converting various biomass feedstocks into syngas, which contains a high concentration of hydrogen (Speight, 2023). It is a process where raw material molecules are decomposed into their elemental components under high temperatures and atmospheric or elevated pressures. This intricate procedure requires the presence of a gasifying agent such as steam, air, oxygen (O_2), carbon dioxide (CO_2), or a combination of these gases. This process involves the partial oxidation of carbonaceous materials at high temperatures (800°C–1000°C) and pressures (up to 33 bar) (Couto et al., 2013), resulting in a mixture of gases including methane (CH_4), hydrogen (H_2), higher hydrocarbons, carbon dioxide (CO_2), carbon monoxide (CO), and nitrogen (N_2) (Dai et al., 2023; Demirbas, 2006). The content of hydrogen is 15%–20% and could be increased up to 30%–40% if the air replaced by steam (Kaur et al., 2019) (Table 9.2).

The biomass gasification reaction could be presented as follows:

$$\textit{Biomass} + \textit{Air} \rightarrow H_2 + CO + CO_2 + N_2 + CnHm + tar + Char + H_2O \quad (9.1)$$

Numerous parameters affected the biohydrogen yield, including biomass type and its biochemical structure, temperature, particle size, type of catalyst, and the steam/biomass ratio. The generated syngas can subsequently undergo additional processing to enhance its hydrogen content, particularly using a water–gas shift reactor (Gao et al., 2019). The gasification process can be performed with or without catalysts, achieving higher efficiencies when catalysts are utilized (Chee et al., 2024). Various reactor types are employed in gasification, including fixed bed (Wojtacha-Rychter et al., 2024), fluidized bed (Erdem et al., 2024), and entrained flow reactors (Fawad et al., 2024). Introducing steam and/or oxygen into the gasification process increases the hydrogen yield in the syngas (Shen, 2024). However, a significant challenge remains in managing the disposal of char and tar which is complex mixture of high-molecular-weight aromatic hydrocarbons produced during gasification (Levin & Chahine, 2010).

Another type of gasification used for the conversion of biomass to biohydrogen is plasma gasification. Plasma consists of high-temperature ionized gases, enabling rapid and efficient heat transfer. This characteristic of plasma allows it to deliver thermal energy to incoming organic raw materials, simultaneously facilitating their pyrolysis and supporting

TABLE 9.2 Brief description, advantages, and disadvantages of different types of gasification technologies (Elhambakhsh et al., 2023; Rambhujun et al., 2020; Shen, 2024).

Gasification types	Description	Advantages	Disadvantages
Steam gasification	Produces hydrogen-rich syngas using steam and air as gasification agents, enhancing H_2 content in the syngas.	• High H_2 content (40%–60%) in the syngas • Abundant availability of gasification agents • Suitable for large-scale industrial applications	• High CO_2 emissions • Significant tar production
Oxygen gasification	Utilizes pure oxygen as the gasifying agent, leading to efficient syngas production.	• High concentration of combustible gases in the output • Low tar content in the produced syngas	• High operational costs • Energy-intensive oxygen generation required
Supercritical water gasification	Uses supercritical steam as both a gasifying agent and catalyst for conversion.	• Efficient conversion with high H_2 yield • H_2 production without coke and tar formation	• Challenges in recycling alkaline catalysts • Strict operating conditions needed • High-energy consumption for feedstock pumping
Plasma gasification	Uses high-temperature plasma to convert biomass into H_2-rich syngas, achieving efficient breakdown of feedstock with minimal residue.	• High productivity rates • Capable of processing diverse feedstocks • Strong potential to	• Electrode deactivation issues • Incomplete feedstock conversion • High-energy requirements

(Continued)

TABLE 9.2 (Continued)

Gasification types	Description	Advantages	Disadvantages
		generate value-added products alongside H_2.	• Need for robust and durable equipment • Production of unwanted by-products

endothermic vapor reforming reactions (Oliveira et al., 2022). This dual advantage has been effectively implemented in various systems, such as those described in US Patent No. 5,666,891, which features diverse and particularly useful configurations for arc electrodes. In these systems, organic compounds present in waste are decomposed through pyrolysis, driven by the high temperatures of the plasma, which break the molecular bonds of the organic materials (Oliveira et al., 2022). Plasma gasification can overcome some of the challenges of gasification process. When steam is introduced into the process chamber, these decomposed constituents undergo steam reforming reactions, converting into synthesis gas a clean-burning fuel composed primarily of carbon monoxide (CO), carbon dioxide (CO_2), and hydrogen (H_2). Non-volatile waste components are transformed into a molten state, which subsequently cools and solidifies into a stable vitrified glass. By carefully controlling the vitrification process, the resulting glass exhibits high chemical and environmental stability, along with significant resistance to leaching of hazardous components, offering intriguing potential for various downstream applications. Plasma gasification offers a solution to some of these challenges by subjecting the feedstock to extremely high temperatures (approximately 2000°C–4000°C) using plasma, the fourth state of matter formed by ionizing a gas (Ibrahimoglu & Yilmazoglu, 2020). This method boasts high-intensity nonionizing radiation, which aids in the thermal cracking of tar and allows for the gasification of feed materials with high moisture content, without stringent requirements on material size and structure (Farzad et al., 2016; Ibrahimoglu & Yilmazoglu, 2020). Supercritical water gasification (SCWG) is a biomass conversion method that operates at conditions above 374°C and 22.1 MPa, utilizing supercritical water as both a reactant and gasification medium. Unlike conventional thermochemical processes that use inert gases or steam, SCWG benefits from water's supercritical state, which increases the density and promotes the formation of H^+ and OH^- ions, facilitating hydrolysis and pyrolysis reactions. At higher temperatures, the formation of free radicals further enhances biomass degradation (Madenoğlu et al., 2011). A key advantage of SCWG is its efficient hydrogen (H_2) production, independent of the biomass's moisture content, which reduces energy compression costs for storage. Under supercritical conditions, lignocellulosic biomass, composed of cellulose and hemicellulose, dissolves and breaks down into simple sugars, which are gasified as the temperature and pressure rise. Lignin, another biomass component, is converted into phenolic compounds that are subsequently reformed into simple gases like H_2, CO, CO_2, and CH_4. This makes SCWG a highly efficient process for biomass gasification, leveraging the unique properties of supercritical water to enhance both the decomposition and gasification stages.

9.2.3.2 *Pyrolysis*

Pyrolysis involves the chemical transformation of materials through thermal energy in the absence of air or oxygen, or by utilizing heat and catalysts. This process results in the breaking of molecular bonds and the formation of smaller molecules. Pyrolysis typically occurs at temperatures ranging from 400 to 1000 °C (Tayibi et al., 2021). Depending on the rate of pyrolysis, the process yields various products, including gases, bio-oil, and char. Pyrolysis can be categorized into three main subclasses, conventional pyrolysis, flash pyrolysis, and fast pyrolysis, as shown in Fig. 9.3 (Czernik & Bridgwater, 2004). Table 9.3 demonstrates how different pyrolysis conditions influence biohydrogen production from various biomass feedstocks. The highest hydrogen yields were generally achieved at elevated temperatures, as evidenced by the slow pyrolysis of sugarcane bagasse which produced 28.8 vol% hydrogen at 780°C compared to just 9.6 vol% at 480°C (Al Arni, 2018). This substantial increase highlights how higher temperatures promote thermal cracking and reforming reactions that enhance hydrogen generation. However, temperature effects can vary by feedstock, as shown by fast pyrolysis of Bambara groundnut shells where 800°C yielded either 3 vol% (Mohammed et al., 2016) or 13.8 vol % H_2 from beach wood (Wang et al., 2005), suggesting other factors like heating rate and biomass composition play significant roles. The type of pyrolysis process also critically impacts results. Intermediate pyrolysis of rice husk at a slow 2°C/min heating rate generated only 8.6 vol% hydrogen (Weldekidan et al., 2018), while flash pyrolysis of bamboo at 900°C achieved 26.6 vol% yield (Maliutina et al., 2017), demonstrating how faster heating generally favors H_2 production. Catalytic approaches showed particular promise, with Fe-catalyzed pyrolysis of chestnut shells increasing hydrogen yields to 32% compared to 18.3% without catalyst (Xia et al., 2019). Even more

TABLE 9.3 Summary of different biomass to biohydrogen thermochemical technologies highlighting the operating conditions and the biohydrogen yield production.

Process	Process type	Biomass	Operating conditions	Biohydrogen yield production	References
Gasification	**Steam gasification** in atmospheric fluidized bed reactor	Corn stalk	• **T:** 800°C	450 mL/g	Li et al. (2017)
	Steam gasification in bubbling fluidized bed gasifier	Pine needles	• **T:** 793°C	41.85 g/kg biomass	Kumari and Mohanty (2022)
	Air-steam gasification in bubbling fluidized bed reactor	Straw (from Jiangsu Province, China)	• **T:** 800°C	62 g/kg biomass	Shen et al. (2008)
	Air-steam gasification in bubbling fluidized bed reactor	Citrus peels (from orange fruit industries)	• **T:** 850°C	0.60 m^3/kg biomass	Maisano et al. (2019)
	Supercritical water gasification using Ru/α-Al_2O_3/CaO	Sawdust	• **T:** 550°C • **P:** 36 MPa • **RT:** 10 min • **FC:** 6.7 wt.%	10.4 mol/kg	Onwudili and Williams (2013)

(Continued)

TABLE 9.3 (Continued)

Process	Process type	Biomass	Operating conditions	Biohydrogen yield production	References
	Supercritical water gasification	Biomass *(cauliflower residue, tomatoes residue, acorn, extracted acorn, and hazelnut shell.)*	• **T:** 600°C • **P:** 35 MPa • **RT:** 18 s	4.9-20.2 mol/kg	Madenoğlu et al. (2011)
	Supercritical water gasification using K_2CO_3 as catalyst			18.5-32.1 mol/kg	
	Supercritical water gasification using Trona as catalyst			18.9-31.1 mol/kg	
Pyrolysis	**Slow pyrolysis**	Sugarcane bagasse	• **T:** 480°C • **HR:** 20°C/min • **FR:** 1000 mL He/min	9.6 vol% (from 25 wt.% of total gas)	Al Arni (2018)
			• **T:** 780°C • **HR:** 20°C/min • **FR:** 1000 mL He/min	28.8 vol% (from 36 wt.% of total gas)	
	Intermediate pyrolysis	Rice husk	• **T:** 800°C • **HR:** 2°C/min • **FR:** 85 mL Ar/min	8.6 vol% (from 25.5 wt.% of total gas)	Weldekidan et al. (2018)

		Bambara groundnut landraces shell	• **T:** 500°C • **HR:** 50 °C/min • **FR:** 5 L N_2 /min	3 vol% (from 34 wt.% of total gas)	Mohammed et al. (2016)
	Fast pyrolysis	Beach wood	• **T:** 800°C • **HR:** 1000 °C/min • **FR:** 5 L N_2/min	13.8 vol% (from 56.9 wt.% of total gas)	Wang et al. (2005)
		Bamboo	• **T:** 500°C • **HR:** 1000 °C/min • **FR:** 5 L N_2/min	1.5 vol% (from 19 wt.% of total gas)	
	Flash pyrolysis	Palm Kernel shell	• **T:** 500°C • **FR:** 5 L N_2/min	18.5 vol% (from 5 wt.% of total gas)	Maliutina et al. (2017)
		Palm Kernel shell	• **T:** 900°C • **FR:** 5 L N_2/min	26.6 vol% (from 32 wt.% of total gas)	
	Catalytic pyrolysis using Fe as catalyst	Chestnut shells	• **T:** 400-800 °C • **HR:** 281 K/min	6.34%–32% with a total gas yield varied 23.3%–42.5% (for non-catalytic essay, the H_2 was 0-18.3% with a total gas yield of 14.5%–35%).	Xia et al. (2019)

(*Continued*)

TABLE 9.3 (Continued)

Process	Process type	Biomass	Operating conditions	Biohydrogen yield production	References
	Catalytic pyrolysis using K_2CO_3 as catalyst (10% Fe)	Sawdust	• **T:** 900°C	36% with a total gas yield of 54% (for non-catalytic essay, the H_2 20% with a total gas yield of 41%)	Zhou et al. (2018)
	Catalytic pyrolysis using Ni/CaO as catalyst (10 wt%)	Herb residue	• **T:** 650°C • **HR:** 283 K/min	78 vol%	Zhao et al. (2020)
Hydrothermal liquefaction	**Hydrothermal liquefaction** SS316 PARR reactor, using Ni $(NO_3)_2$ as catalyst	Pinewood slurry	• **T:** 275°C • **B:S ratio:** 1:10 • **Stirring:** 1300 rpm.	12.26 mol% H_2 (biocrude: 18.3 wt% and residue: 56.1 wt%)	Tungal and Shende (2014)

T, Temperature production; *HR*, heating rate; *RT*, residence time; *FR*, flow rate; *FC*, feed concentration; *B:S ratio*, biomass to solvent ratio.

impressive was sawdust pyrolysis using K_2CO_3 with 10% Fe catalyst, reaching 36% hydrogen yield versus 20% non-catalytic (Zhou et al., 2018).

9.2.3.3 Hydrothermal liquefaction

Hydrothermal liquefaction (HTL) has emerged as an efficient process for converting wet biomass into biohydrogen, overcoming key limitations of conventional thermochemical methods like pyrolysis. Unlike pyrolysis which requires dry feedstocks, HTL utilizes water as both reaction medium and hydrogen donor, making it ideal for aquatic biomass such as microalgae while eliminating energy-intensive drying steps. The process operates at moderate temperatures (250-400°C) that reduce energy demands compared to pyrolysis, while effectively suppressing tar formation and enhancing hydrogen yields through optimized reaction pathways. A study by (Tungal and Shende, 2014) demonstrated the effectiveness of HTL with pinewood, achieving significant production of hydrogen (12.26 mol%) along with bio-crude and bio-oil at optimal conditions. A critical advantage of HTL is its ability to address the inherent challenge of high oxygen content (35-46 wt.%) in biomass through efficient dehydration and decarboxylation reactions, significantly improving hydrogen production compared to fast pyrolysis. The integration of catalysts (Ni, Co, Ru-based) further boosts performance by promoting key reactions like water-gas shift and steam reforming (Fig. 9.4). However, scaling up HTL faces challenges including catalyst optimization to minimize coke formation, mitigation of reactor corrosion under subcritical conditions, and development of efficient hydrogen separation methods. Ongoing advances in catalytic HTL and supercritical water gasification continue to enhance the viability of this process as a sustainable route for biohydrogen production, with the potential to become a competitive alternative to fossil-based hydrogen generation (Peterson et al., 2008; Toor et al., 2011; Gollakota et al., 2018).

9.3 Biohydrogen storage and transportation technologies

9.3.1 Biohydrogen storage technologies

Storing biohydrogen is crucial for the advancement of biohydrogen applications across stationary power, transportation, and portable power systems. Achieving a clean hydrogen economy hinge on the development of highly efficient and dependable storage technologies. Currently, various methods are employed for hydrogen storage, including compressed gas, cryogenic liquefaction, cryo-compressed storage, and solid-state storage solutions. Each approach offers distinct advantages and challenges, contributing to ongoing research and innovation in the field.

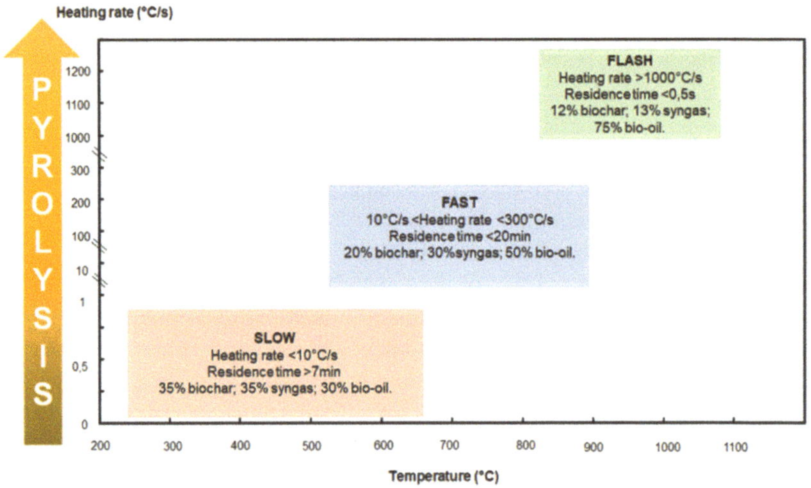

FIGURE 9.3 Illustration of the different pyrolysis types, their operational condition, and the generated product yields.
Adapted from Tayibi, S., Monlau, F., Bargaz, A., Jimenez, R., & Barakat, A. (2021). Synergy of anaerobic digestion and pyrolysis processes for sustainable waste management: A critical review and future perspectives. Renewable and Sustainable Energy Reviews, 152, https://doi.org/10.1016/j.rser.2021.111603. 111603.

9.3.1.1 High-pressure gaseous storage of biohydrogen

High-pressure hydrogen storage involves using specially designed gas cylinders to store hydrogen gas under significant pressure. This method offers benefits such as rapid refueling, established technology, operation at ambient temperatures, and relatively low costs. However, it also presents challenges, including low-energy density, the energy required for compression, larger storage volumes compared with conventional fuels like gasoline, and stringent safety requirements for storage containers (Zhang & Jiang, 2023; Zou et al., 2022). High-pressure hydrogen storage cylinders are used in various applications, including refueling stations and vehicles. For refueling stations, the technology has been widely adopted, with key suppliers producing cylinders that can handle increasing refueling pressures. This has led to the development of products supporting pressures up to 100 MPa, with advancements aimed at increasing capacities and diameters (Cao et al., 2021). In the automotive sector, the primary types of hydrogen storage cylinders are aluminum-lined fiber-wound (Type III) and plastic-lined fiber-wound (Type IV) cylinders. Type III cylinders have mature technology and are cost-effective (Zou et al., 2022). In contrast, Type IV cylinders are lighter, capable of withstanding higher pressures, have a higher hydrogen storage mass ratio, and offer longer service lives. Consequently, they are the focus of future development efforts,

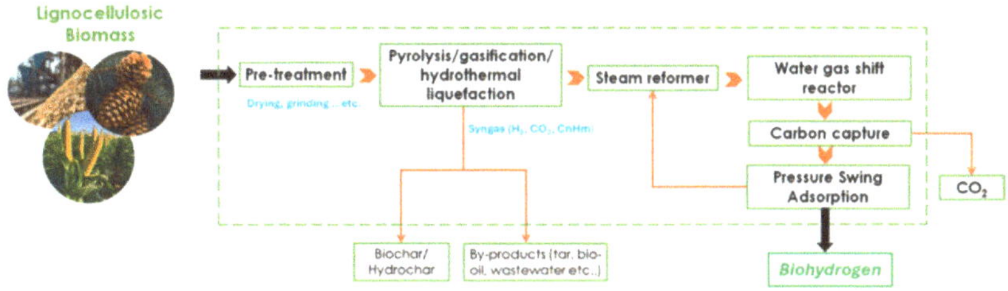

FIGURE 9.4 Thermochemical conversion routes of biomass into biohydrogen.

with many companies already working on Type IV cylinder products (Zou et al., 2022). Additionally, biohydrogen storage via high-pressure compressed steel cylinders is a well-established technology, typically operating at 200 bars (Faye et al., 2022). Innovations have led to the creation of lightweight composite cylinders capable of withstanding pressures up to 800 bar. This advancement aims to increase the volumetric density of hydrogen to 36 kg/m³, nearly half the density of liquid hydrogen at its standard boiling point of −252.87°C (Faye et al., 2022). High-pressure gaseous hydrogen storage remains the most mature and widely adopted method, and it is expected to continue being the primary hydrogen storage technology in the near future.

9.3.1.2 *Cryogenic liquid storage of biohydrogen*

Liquid hydrogen storage is a promising method for hydrogen containment due to its high storage density, achieving approximately 70 kg/m³ at atmospheric pressure and 1 bar (Barthelemy et al., 2017). This process involves cooling hydrogen gas to below its boiling point of 20 K, transforming it into a liquid state. The liquid is then stored in cryogenic tanks that are typically double-walled, metallic, and thermally insulated, often maintained in a vacuum to minimize heat transfer (Mori & Hirose, 2009; Zhang et al., 2015). Despite these advantages, such as non-corrosive properties and compact storage containers, the technology faces significant challenges. The energy-intensive process required to liquefy hydrogen, coupled with the high cost of tank materials, evaporation losses, and safety concerns, limits its broad application, particularly in large-scale vehicle use (Pudukudy et al., 2014). However, advancements in this field are gradually overcoming these barriers. With the growing interest in hydrogen as a clean energy source, investments in liquid hydrogen storage technologies have surged, fostering the development of more efficient storage systems and the establishment of standardized protocols (Zhang & Jiang, 2023). Although primarily utilized in aerospace, the technology is expected to expand into broader civilian applications as these technical challenges are addressed (Zhang & Jiang, 2023).

9.3.1.3 Cryo-compress storage of biohydrogen

Cryo-compressed hydrogen (CcH_2) is an advanced method of storing hydrogen that combines the benefits of compressed hydrogen and liquid hydrogen. In this technique, hydrogen is stored at cryogenic temperatures (20–50 K) and high pressures (typically between 250–350 atm or 35 MPa) (Aceves et al., 2010; Yan et al., 2021). Unlike traditional cryogenic vessels that store liquid hydrogen at near-ambient pressures, cryo-compressed hydrogen can exist in different phases, including liquid, supercritical, or a two-phase state (saturated liquid and vapor) (Aceves et al., 2010). One of the main advantages of CcH_2 is its high hydrogen density, which ranges from 60.0–71.5 kg/m³ under conditions of 35–110 K and 5–70 MPa. This density is close to that of liquid hydrogen at 71.0 kg/m³ (Yan et al., 2021). Additionally, insulated pressure vessels for CcH2 significantly prolong the storage period by allowing higher internal pressures, increasing the vessel's heat receptivity. This reduces the risk of vehicles being stranded due to evaporative losses as venting stops when the tank reaches ambient temperature at the allowable pressure (> 250 atm), maintaining 30% of the initial liquid density (Aceves et al., 2010). However, the cryo-compression process requires more energy and incurs higher costs compared withtraditional hydrogen liquefiers to achieve similar hydrogen densities (Wang et al., 2022). Despite this, CcH_2 offers several operational advantages, including long non-destructive periods, high-pressure resistance, reduced hydrogen loss due to boil-off, and the ability to facilitate rapid hydrogen release and filling (Barthelemy et al., 2017; Chen et al., 2020; Moradi & Groth, 2019; Wang et al., 2022). A challenge associated with CcH_2 is the higher heat leakage compared with compressed hydrogen and liquid nitrogen (Yanxing et al., 2019). Additionally, storage vessels must be specifically designed to resist hydrogen embrittlement, which adds complexity to the design and operation (Barthelemy et al., 2017). While CcH_2 offers distinct benefits, such as the potential to eliminate ullage space due to hydrogen becoming a supercritical fluid above 13 atm, it is important to carefully consider these advantages in light of its higher energy requirements and potential design challenges (Aceves et al., 2006). Overall, CcH_2 is a promising technology for hydrogen storage, offering flexibility for extended range fueling with liquid H_2 and cost-effective solutions for short trips with compressed hydrogen (Aceves et al., 2006). Nonetheless, careful consideration of its energy demands and technical challenges is essential for practical application.

9.3.1.4 Solid-state storage of biohydrogen

Solid-state hydrogen storage technology leverages materials like metal hydrides, which can absorb and release hydrogen through reversible chemical reactions. This method is advantageous for its safety and practicality, as it does not require extreme temperatures or pressures, making it well-suited for portable applications and hydrogen-powered vehicles (Barthelemy et al., 2017). The technology operates by enabling hydrogen to adsorb and desorb materials with high surface areas, such as activated carbon, carbon nanotubes, and various carbon-based substances. Additionally, chemical hydrogen storage can be achieved through the reversible adsorption and release of hydrogen using metal hydrides, including alloys based on magnesium, iron, lanthanum-nickel-rare earth, titanium, and zirconium (Zhang & Jiang, 2023). Solid-state storage offers significant benefits, including high hydrogen storage density, elimination of the need for high-pressure containers, provision of high-purity hydrogen, and enhanced safety with lower energy consumption (Zhang & Jiang, 2023). However, challenges persist, such as the relatively immature state of the technology and the current limitations in hydrogen charging and discharging efficiency. Efforts to optimize storage capacity and efficiency continue with hybrid approaches like cryogenic compressed hydrogen storage emerging. These approaches aim to combine the advantages of both liquid and compressed gas storage while mitigating energy losses, thereby advancing the integration of hydrogen as a clean energy solution across various sectors (Zhang & Jiang, 2023).

9.3.2 Biohydrogen transportation technologies

The transportation of hydrogen is a critical factor that affects its overall cost, emissions, and energy consumption, making it crucial in the realization of a hydrogen economy (Witkowski et al., 2017). Hydrogen can be transported in different states: gaseous, liquid, and solid, each with its own technical and economic implications (Li et al., 2021; Ma et al., 2008). Gaseous hydrogen is typically transported through pipelines, which are considered the most cost-effective method for large-scale transportation (Lahnaoui et al., 2017). However, the initial cost of constructing these pipelines is significantly high (Fekete et al., 2015). Pipeline transportation works best for large-scale operations but relies heavily on high utilization rates to be cost-effective. For example, when the pipeline utilization rate is 100% and the transportation distance is 100 km, the cost of hydrogen transportation is 0.17 $/kg. However, at a distance of 500 km, the cost increases to 0.42 $/kg, and if the utilization rate drops below 20%, the cost advantage diminishes significantly (Zhang & Jiang, 2023). Taking the example of China, where the hydrogen pipeline network is still in its early stages, high-pressure gaseous hydrogen trailers are currently the predominant mode of transport (Zhang & Jiang, 2023). These trailers are cost-effective during the early stages of market development when the transportation radius and demand for hydrogen are relatively small. For example, long pipe trailers operating at 20 MPa are suitable for short-distance transportation within a 200 km radius. However, domestic regulations currently limit the maximum working pressure to 20 MPa, while internationally, 50 MPa trailers are already in use. If domestic standards were relaxed to allow for higher pressures, the transportation costs could be significantly reduced. For a 100 km journey, the cost for 20 MPa trailers is 1.22 $/kg, while for 50 MPa trailers, it is reduced to 0.79 $/kg, approximately 65% of the former (Zhang & Jiang, 2023). It is expected that 50 MPa trailers will be widely used in China by 2025. On the contrary, liquid hydrogen transportation is known for its high density and purity, typically accomplished using cryogenic tanks through trucks or ships. Despite its advantages, this method requires substantial energy consumption and incurs high costs due to the complexities of liquefaction (Dauletbay, 2024). However, liquid hydrogen transportation is highly efficient and well-suited for large-scale, long-distance transportation, remaining cost-effective for distances up to 2000 km. For instance, while a tank truck can transport a few tons of liquid hydrogen per trip, a liquid hydrogen transport ship can carry hundreds of tons, making it a more efficient and cost-effective option for long-distance sea transportation (Dauletbay, 2024; Zhang & Jiang, 2023). In addition to these methods, some studies have suggested that combining hydrogen with carriers, such as biomass, and installing reverse conversion facilities at usage sites (e.g., hydrogenation and energy stations), could lower the overall transportation costs compared with direct hydrogen transport (Zhang & Jiang, 2023). This highlights the potential of using biomass hydrogen carriers as a cost-effective and sustainable alternative in hydrogen logistics. Looking toward the future, as the hydrogen economy continues to expand, there will be a shift toward high-density and high-safety hydrogen storage technologies. A comprehensive hydrogen pipeline network, capable of large-scale, long-distance, and low-cost transportation, will likely become the backbone of hydrogen distribution (Li et al., 2021; Ma et al., 2008). Additionally, the development of standards for solid-state and organic liquid storage, as well as pipeline distribution, will be crucial to support the evolving hydrogen infrastructure.l

9.4 Biohydrogen applications

9.4.1 Fuels cells for electricity production and transport fields

As a renewable and environmentally friendly energy carrier, biohydrogen has the potential to revolutionize the energy and transport sectors by offering a sustainable alternative to fossil fuels. In the field of electricity production, hydrogen fuel cells, which convert hydrogen into electricity with water and heat as by-products, are particularly promising. These fuel cells can be used at a variety of scales, providing reliable and efficient electricity for residential, commercial, and industrial use, particularly in off-grid locations. Indeed, when integrated with renewable energy sources such as solar and wind, hydrogen fuel cells can improve the stability of the electricity supply by storing excess energy produced by electrolysis. In addition, fuel cell vehicles also offer greater range and shorter fueling times than battery electric vehicles, making them an attractive option. In fact, major car manufacturers have already developed and marketed hydrogen fuel cell vehicles, and the hydrogen infrastructure is developing in countries such as Japan, USA, and parts of Europe (Ferraren-De Cagalitan & Abundo, 2021).

For heavy transport, hydrogen fuel cells are particularly promising, offering a viable solution for long-distance road, air, and sea transport. Projects such as Airbus' ZEROe concept aim to introduce hydrogen-powered aircraft by 2035, while the maritime industry is exploring hydrogen fuel cells to power ships. In addition, hydrogen can be blended with natural gas to create a cleaner fuel for power stations and heating applications to reduce carbon emissions. The development of hydrogen infrastructure, including production facilities, storage systems, distribution networks, and fueling stations, is crucial to the widespread adoption of biohydrogen. The scalability and versatility of biohydrogen make it a key player in the future energy landscape, with its feasible production from a variety of feedstocks through processes such as AD and bioelectrochemical systems (Sharma & Ghoshal, 2015). Nowadays, advances in hydrogen production technologies are improving efficiency and reducing costs, making biohydrogen increasingly competitive. Continued research and development, as well as collaborative efforts and political support, are essential to overcome the challenges and realize the full potential of biohydrogen in transforming the energy sector and achieving global climate goals (Sharma & Ghoshal, 2015) (Fig. 9.5).

9.4.2 Industry fields

9.4.2.1 Chemical industries

The chemical industry is the world's largest producer and consumer of hydrogen (Fig. 9.5). Hydrogen fulfills various roles along the chemical value chain (Ausfelder & Bazzanella, 2016). First and foremost, hydrogen is an integral part of many of the largest volume chemicals, both of organic and inorganic nature. It can be supplied as a reactant either in molecular form or bound within a molecule to produce the desired product mainly ammonia and methanol (Fig. 9.6). Hydrogen could be used in obtaining low temperature for the operation of physical installation, in soldering of pure metals (Bondarenko et al., 2022) and fabrication of porous metals (Nakajima, 2007). For instance, in Russia and China, the main industry that consumes hydrogen is primary the production of ammonia (NH_3) and methanol (CH_3OH). Indeed, it was reported that 28 companies in Russia use 2.46 million tons of hydrogen per year, whereas in methanol sector 0.6 million tons of hydrogen was consumed in 2013 (Bondarenko et al., 2022).

Ammonia is considered as the base chemical for nitrogen-based fertilizers that can be produced by hydrogen. One of the main nitrogen-based fertilizers produced from ammonia through reacting with carbon dioxide is urea. Indeed, and by considering a conversion efficiency of 100%, 10 kg of urea can be produced by 1 kg of hydrogen (Mali et al., 2021). Moreover, the ammonia technology allows the production most of the nitrogen and phosphorus-based fertilizers, including ammonium sulfates, ammonium nitrate, and ammonium chloride (Mali et al., 2021) (Fig. 9.6).

9.5 Challenges of production of biohydrogen from biomass

Biohydrogen from lignocellulosic biomass has been demonstrated its technological potential and environmentally friendly pathway for energy production (Chen et al., 2013; Dutta et al., 2023; Osman et al., n.d.; Rani et al., 2024). However, discussion on the biohydrogen production from biomass has limitations in several aspects including production efficiency, technological, economic, and environmental limitations. One of the main challenges is achieving high yields of hydrogen production from lignocellulosic biomass. The main hydrogen production process is a biological conversion of lignocellulosic material. Many biological processes have inherent limitations on the amount of hydrogen that can be produced, often due to competing microbial reactions or inefficiencies in the conversion process (Feng et al., 2023; Monlau, Trably, et al., 2013; Zhao et al., 2024). On the contrary, the main technical challenge for biohydrogen production from lignocellulosic biomass is the efficient breakdown of the complex structure of lignocellulose into fermentable sugars, as lignocellulosic biomass material is typically composed of cellulose, hemicellulose, and lignin (Bhatia et al., 2021; Chen et al., 2021). To utilize this lignocellulosic biomass as resource for biohydrogen, a fractionation process is strictly needed to isolate and separate each biomolecule individually. Lignin–carbohydrate complex is associated together

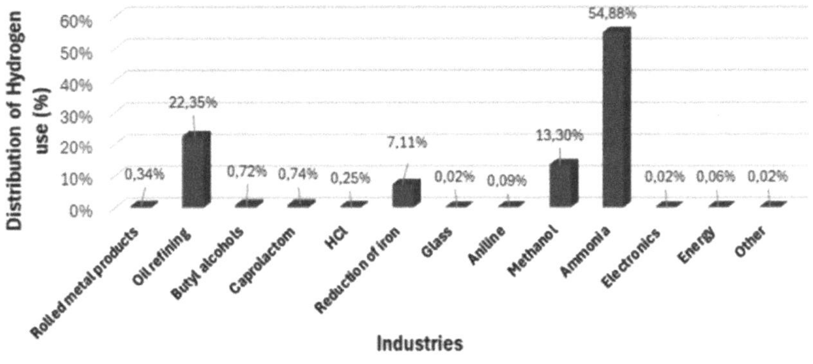

FIGURE 9.5 Hydrogen application in different industries and sectors.
Adapted from Ausfelder, F., & Bazzanella, A. (2016). Hydrogen in the chemical industry. In Hydrogen Science and Engineering: Materials, Processes, Systems and Technology (pp. 19–40). John Wiley & Sons, Ltd. https://doi.org/10.1002/9783527674268.ch02.

with diversity chemical linkages such as ester bonds, hydrogen bonds, and intermolecular bonds which make lignocellulosic material hard for alteration (Gao et al., 2021; Pang et al., 2019; Shakeel et al., 2024). The fractionation process categorizes into several processes, including physical, chemical, physicochemical, biological, and integration processes. The fractionation process is generally responsible for altering the lignocellulosic structure to release more fermentable sugars and reactive biomolecules for further processing. Additionally, the efficient conversion of these sugars into hydrogen gas by hydrogen-producing microorganisms can be challenging due to factors such as substrate inhibition, low hydrogen yields, and competition from other microbial pathways. Therefore, the design of bioreactors for optimal conditions, the selection of suitable hydrogen-producing strains and the overall process optimization are all critical technical challenges that researchers are working to overcome to make biohydrogen production from lignocellulosic biomass more economically viable and sustainable (Deivayanai et al., 2022; Prabakar et al., 2018; Kannah et al., 2021).

9.5.1 Technical and operational challenges

Several types of organic biomass could be converted into biohydrogen such as agro-waste, microalgae, seaweeds, and municipal solid wastes. However, the main technological limitation of hydrogen production from biomass is how to achieve high hydrogen productivity. Hydrogen production via fermentation process is still growing from primary stage with the gradual change in the

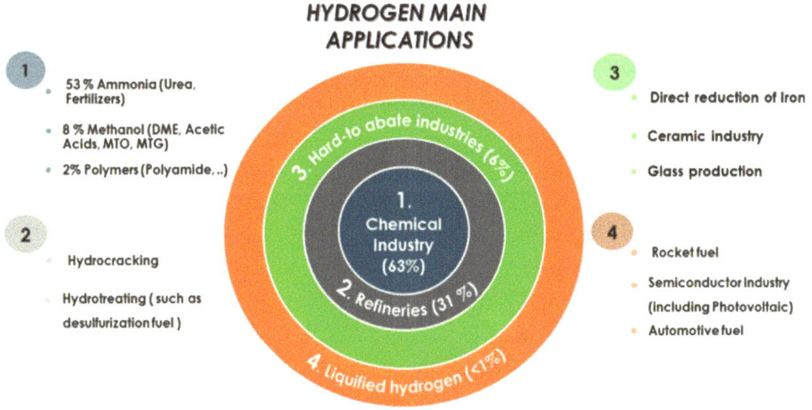

FIGURE 9.6 Distribution of hydrogen application in different industry fields.
Based on Bondarenko, V.L., Ilyinskaya, D.N., Kazakova, A.A., Kozlovtsev, P.S., Lavrov, N.A., & Razenko, E.A. (2022). Hydrogen, its unique properties and applications in industry. Chemical and Petroleum Engineering, 57(11–12), 1015–1019. https://doi.org/10.1007/s10556-022-01039-7.

technology. Several studies have demonstrated that the pretreatment and usage of energy in the process affect the overall cost. In terms of process efficiency, an improvement of hydrogen production processes efficiency is essential to increase yields and reduce energy consumption. Technologies like DF offer advantages in terms of simplicity and lower energy intensity (Bhatia et al., 2021; Chen et al., 2021; Prabakar et al., 2018; Rathi et al., 2024). Continuous advancements in biotechnology, process efficiency, and feedstock conversion are needed to make biohydrogen production more practical and economically viable. Another technical barrier of DF is the pretreatment process before fermentation is a major challenge since pretreatment process adjusts to different types of lignocellulosic biomass. Moreover, the development and operation of functional bioreactors and the thermodynamics limitations on the hydrogen production yield by microbial fermentation. Furthermore, the conventional hydrogen yield from DF is restricted to 4 mol H_2/ mol glucose, and it is the major technical obstacle for its utilization. Various parameters such as substrate and inhibitor concentrations, temperature, pH, and reactor type directly affect H_2 production yield by DF. Modeling of the H_2 production is very important to ameliorate, analyze, and predict H_2 production during fermentation. Several studies of mathematical models, including the kinetic of cell growth and product(s) formation, substrate utilization, and inhibition, are studied for H_2 production improvement. In addition, some models are developed to explain the effect of pH, temperature, and dilution rate on H_2 production. The obtained model kinetic constants can be used in the design, operation, and optimization of the fermentative H_2 production process (Ayodele et al., 2023; Zhao et al., 2024). Another technical barrier of hydrogen production is the type of reactor. The different types of reactors are required for different types of H_2 production. In case of fermentative process, packed bed and fluidized bed reactor are required. For photobiological H_2 production, tubular, vertical helical, and flat panel bioreactors are typically applied. To effectively produce H_2 with high production yield, the design and operational conditions of reactors are strictly needed for economically viability (Rani et al., 2024). Regarding feedstock availability, the hydrogen production must be ensuring a consistent and sustainable supply of feedstock for hydrogen production. Lignocellulosic biomass, such as agricultural residues (wheat straw, rice husk), forestry waste, and energy crops, is a promising feedstock for biohydrogen production due to its abundance and sustainability. Another feedstock from food industry wastes is vastly available for H_2 production (Osman et al., 2023). Food waste from the food processing industry, such as wastewater, can serve as feedstock for biohydrogen production. These wastes can provide high hydrogen yields without extensive pretreatment. Moreover, industrial wastewater from sectors like distilleries can be utilized as a feedstock for biohydrogen production. Studies have shown successful biohydrogen generation from distillery wastes. Other biomass sources like algae, organic waste, and agricultural by-products can also be used as feedstock for biohydrogen production. Organic components of municipal solid waste can be also converted into biohydrogen through biological processes, contributing to waste management and clean energy production. In addition, crops specifically grown for bioenergy purposes, such as switchgrass and miscanthus, can serve as feedstock for biohydrogen production (Osman et al., 2023). Several technical factors affect H_2 productivity directly, including process efficiency, type of reactor, and feedstock availability. To overcome these technical obstacles, the integration of other energy-generating systems with DF enhances the energy recovery from the substrate. Besides, an integration of DF and another process such as PF, AD will decrease feedback inhibition. Beyond sustainable technologies, process scaling up is a crucial key for hydrogen production on a commercial scale. The challenges of process scaling up are to transform from laboratory-scale experiments to industrial-scale production, which poses challenges in maintaining efficiency and cost-effectiveness. Real-time data-driven insights and automation can enhance process efficiency at an industrial scale. For industrial implementations, the production of hydrogen needs to be technologically viable, economically feasible and should decrease environmental impacts. Certain key technical achievements including in-depth understanding mechanism preparation feedstock and reactions must be required (Prabakar et al., 2018; Rathi et al., 2024).

9.5.2 Economic challenges

Development of biohydrogen economy is a promising strategy to reutilize organic waste and achieve carbon neutrality. Therefore, economic analysis to reveal commercial feasibility is crucial to developing the biohydrogen economy and attracting investors (Ayodele et al., 2023; Feng et al., 2023). The cost of hydrogen production is a significant challenge, especially when compared to traditional fossil fuel-based methods. Technologies need to be developed to make hydrogen production more cost-effective. In fermentative process of H_2 via fermentation, DF is costly compared with other technologies. The high operating cost is associated with feedstocks such as microbial fermentation, substrate, and feedstock preparation. The cost of acquiring and processing feedstock, especially lignocellulosic biomass, can be a significant economic barrier. Lignocellulosic biomass may require significant processing and pretreatment, leading to higher feedstock costs compared with traditional fossil fuels (Feng et al., 2023). Ensuring a cost-effective and sustainable feedstock supply is crucial for reducing production costs. Besides, the substrate cost is the major problem that influences the biohydrogen cost. Another economic barrier is technological development costs. Research and development costs for improving biohydrogen production technologies, optimizing processes, and scaling-up production can be substantial, hindering widespread adoption. Moreover, establishing the necessary infrastructure for biomass collection, processing, and hydrogen production requires substantial investment, adding to the overall production costs. On the contrary, scaling-up hydrogen production from lignocellulosic biomass from laboratory-scale to industrial-scale operations can be challenging and may require additional investments in equipment and processes. The market competition with conventional H_2 production is critical. Hydrogen produced from lignocellulosic biomass faces competition from conventional fossil fuels, which are currently more economically viable. Overcoming market competition and establishing a competitive pricing strategy is crucial for economic success. In terms of policy and regulatory challenge, inconsistent policies, lack of incentives, and regulatory barriers can impede the economic viability of biohydrogen production. Clear and supportive policies are needed to promote investment and growth in the biohydrogen sector.

9.5.3 Storage challenges

Despite the several advantages of biohydrogen production from lignocellulosic biomass, there are some drawbacks, which affect impediments in industrial scale. Hydrogen storage, compressor and distribution networks, and lack of durable fuel cell technologies are all technological constraints that need to be overcome. Effective biohydrogen storage is an important point for a biohydrogen-based economy as it enables a reliable energy supply system and the efficient utilization of renewable energy resources. Several types of biohydrogen storage include physical-based storage (liquid hydrogen, compressed hydrogen, cryo-compressed hydrogen) and material-based methods. Despite the widespread of using compressed hydrogen, high hydrogen pressure increases the risks of explosion. Therefore, lightweight pressure vessels made from low-cost materials that can resist hydrogen diffusion and embrittlement caused by hydrogen storage are required (Hren et al., 2023). Concerning liquid hydrogen, developing more effective and energy-efficient hydrogen liquefaction methods is necessary to reduce cost and energy demand. On the contrary, the liquid hydrogen boil-off phenomenon, which occurs due to outer energy input, can result in a loss of about 1.5%–3% of the stored hydrogen per day (Feng et al., 2023). To address this issue, it is necessary to modify the storage tank design to prevent hydrogen evaporation and develop systems that can capture or reuse the vaporized hydrogen. Regarding material-based hydrogen storage, it is a strategy for storing hydrogen reversibly by either adsorption or absorption. This method is still immature, and further research is necessary to improve storage capacity and hydrogenation/dehydrogenation kinetics, identify safer materials, and investigate ambient storage conditions.

9.5.4 Environmental challenges

Traditional hydrogen production methods using fossil fuels contribute to environmental pollution. Developing eco-friendly production methods, such as biohydrogen production from biological pathways, is crucial for sustainability.

Addressing the environmental challenges requires the development and adoption of sustainable biohydrogen production technologies, efficient resource management practices, and adherence to environmental regulations to minimize negative environmental impacts and promote the transition toward a cleaner and more sustainable energy future. CO_2 emissions are considered primary GHG sources because of their detrimental impacts on human health and the environment. Therefore, the shift to a carbon-neutral industry and economy with sustainable energy production is one of the most important issues for biohydrogen production. Biohydrogen has the potential to considerably reduce CO_2 emissions, water contamination, and fossil resource utilization when derived from environmentally friendly sources and renewable resources. Basically, the global warming potential (GWP) is a measurement of CO_2 releases corresponding to the amount of CO_2 emitted and the magnitude of its effect on the atmospheric radiation forcing that contributes to climate change. GWP values of the processes were in the range of 0.91–14.6 kg-CO_2 eq. The DF process is considered the most environmentally harmful technique with the highest GWP value of 14.6 kg-CO_2 eq. among other processes due to their superior electricity and heat requirements by the production process (Goren et al., 2023). Another environmental parameter such as ozone depletion, ionizing radiation, and photochemical oxidant formation is typically evaluated for the environmental impacts of biohydrogen production process. Ecotoxicity and eutrophication are other important environmental impact parameters that must be evaluated. The average values of these parameters for H_2 production processes are different depending on the biohydrogen production technology. For example, DF process has significantly the highest potential of freshwater eutrophication (0.0098 kg Peq.), mainly contributed by the biological and chemical oxygen demand of the biomass hydrolysate coming from the utilization of anaerobic sludge or/and wastewater and the formation of organic substances in the biohydrogen production method. Furthermore, the ecotoxicity values of the processes are another environmental impact, considering their terrestrial (TE-FE), freshwater (FE), and marine ecotoxicities (ME). The highest TE-FE (0.12 kg 1,4 DCB eq.), FE (0.61 kg 1,4 DCB eq.), and ME (0.82 kg 1,4 DCB eq.) values were obtained for the DF process (Goren et al., 2023). Overall discussion on environmental impacts of biohydrogen production from organic wastes in particular lignocellulosic material has been established several environmental assessments for sustainability biohydrogen production. The comprehensive assessment is to fill the gap by assessing the practically feasible biohydrogen production methods and make a comparative evaluation by considering various environmental impact criteria. However, considering the harmful environmental impacts of conventional biohydrogen production methods, it is important to spread the biological methods to reduce carbon and water footprint.

9.6 Conclusion and perspectives

Biohydrogen is a green and sustainable fuel that plays a pivotal role in environmental protection and reducing fossil fuel dependence. Transitioning to a biohydrogen economy requires advancements in three main areas: production, storage, and delivery. Despite progress, several challenges remain. In biohydrogen production, low yields and the need for large-scale experiments are primary obstacles. Multiple pathways exist, including fermentation, MECs, and thermochemical processes using biomass, each with unique benefits and challenges. Fermentation converts organic matter into biohydrogen but faces efficiency challenges. MECs use electric current to drive microbial biohydrogen production, showing promise for higher biohydrogen rates and the ability to use various organic compounds as substrates. Thermochemical processes involve converting biomass into biohydrogen through methods like gasification and pyrolysis, which can offer higher yields but require significant energy input and optimization to be economically viable. Improving biohydrogen yield involves optimizing conditions, using additives, genetic engineering, and developing hybrid systems (Feng et al., 2022). Further research is necessary to understand existing pathways better and discover new methods. Scaling up from laboratory-scale to large-scale production involves changing operational parameters and addressing increased maintenance costs, making large-scale production complex and requiring extensive research and cost analysis. Biohydrogen storage methods currently in use are often unsafe or cost prohibitive. Compressed biohydrogen and liquid biohydrogen are well-established but expensive due to strict storage conditions and high vessel requirements. Enhancing safety in compressed biohydrogen storage involves developing better-structured vessels with improved materials. Liquid biohydrogen, which is prone to boil-off and has high liquefaction costs, is unsuitable for long-term, large-scale storage. Addressing these issues requires cost-effective insulation systems and methods to reduce liquefaction costs (Aceves et al., 2010; Yan et al., 2021). Material-based biohydrogen storage offers a safer, long-term alternative, but it is still under development. This method needs more effective biohydrogen carriers, longer carrier lifetimes, and solutions to slow kinetic rates and high-energy input during dehydrogenation and hydrogenation (Yanxing et al., 2019). BioHydrogen delivery also presents significant challenges. The current infrastructure is unreliable, as biohydrogen can easily escape from pipelines, causing environmental harm and pipeline embrittlement. While blending biohydrogen with natural gas is possible, new infrastructure is essential to fully implement a biohydrogen economy and reduce fossil fuel usage. Key research directions include developing materials resistant to embrittlement and constructing systems to prevent biohydrogen leakage (Liu et al., 2023; Okonkwo et al., 2023). Existing transportation methods rely on fossil fuels, which are unsustainable. Therefore, biohydrogen-powered transportation systems should be developed, and designing vehicles to use evaporated biohydrogen as fuel during liquid biohydrogen transportation is a promising approach to reduce costs (Aziz, 2021; Sherif et al., 1997). Applications of biohydrogen are vast and diverse, ranging from its use in fuel cells for electric vehicles to industrial processes requiring clean biohydrogen. In transportation, biohydrogen can be used in fuel cell vehicles, offering a zero-emission alternative to fossil fuels. In industrial applications, biohydrogen serves as a clean energy source for processes like ammonia production and metal refining, reducing carbon footprints. Additionally, biohydrogen can be integrated into existing natural gas pipelines, providing a cleaner energy mix and gradually transitioning to a biohydrogen economy (Bertasini et al., 2023). To fully harness biohydrogen's potential as a clean energy source, integrating production processes with advanced bioreactor designs, genetic modifications of organisms, and sustainable feedstock usage is crucial. Continued innovation in these areas, combined with strategic planning and investment, is essential. A thorough review of current methods highlights the need for more research and technological improvements. Fermentation and MECs must overcome yield limitations, while thermochemical processes using biomass need optimization for scalability and economic viability (Abdalla et al., 2018; Jeje et al., 2024; Moradi & Groth, 2019). Collaboration between industry, academia, and government entities is vital to overcoming technical obstacles and paving the way for a greener energy future. With ongoing innovation and strategic efforts, biohydrogen can significantly mitigate climate change and advance sustainable energy solutions. The future of biohydrogen is promising, contingent on sustained research, development, and cross-sector cooperation to realize its full potential.

References

Abdalla, A. M., Hossain, S., Nisfindy, O. B., Azad, A. T., Dawood, M., & Azad, A. K. (2018). Hydrogen production, storage, transportation and key challenges with applications: A review. *Energy Conversion and Management, 165*, 602–627. https://doi.org/10.1016/j.enconman.2018.03.088.

Aceves, S. M., Berry, G. D., Martinez-Frias, J., & Espinosa-Loza, F. (2006). Vehicular storage of hydrogen in insulated pressure vessels. *International Journal of Hydrogen Energy, 31*(15), 2274–2283. https://doi.org/10.1016/j.ijhydene.2006.02.019.

Aceves, S. M., Espinosa-Loza, F., Ledesma-Orozco, E., Ross, T. O., Weisberg, A. H., Brunner, T. C., & Kircher, O. (2010). High-density automotive hydrogen storage with cryogenic capable pressure vessels. *International Journal of Hydrogen Energy, 35*(3), 1219–1226. https://doi.org/10.1016/j.ijhydene.2009.11.069.

Aceves-Lara, C.-A., Latrille, E., Bernet, N., Buffière, P., & Steyer, J.-P. (2008). A pseudo-stoichiometric dynamic model of anaerobic hydrogen production from molasses. *Water Research, 42*(10), 2539–2550. https://doi.org/10.1016/j.watres.2008.02.018.

Ahmed, S. F., Rafa, N., Mofijur, M., Badruddin, I. A., Inayat, A., Ali, M. S., Farrok, O., & Khan, T. M. Y. (2021). Biohydrogen production from biomass sources: Metabolic pathways and economic analysis. *Frontiers in Energy Research, 9*. https://doi.org/10.3389/fenrg.2021.753878.

Akhlaghi, N., & Najafpour-Darzi, G. (2020). A comprehensive review on biological hydrogen production. *International Journal of Hydrogen Energy, 45*(43), 22492–22512. https://doi.org/10.1016/j.ijhydene.2020.06.182 Elsevier Ltd.

Al Arni, S. (2018). Comparison of slow and fast pyrolysis for converting biomass into fuel. *Renewable Energy, 124*, 197–201. https://doi.org/10.1016/j.renene.2017.04.060.

Al-Janabi, S. K., Barron, A. R., Shabbani, H. J. K., Othman, M. R., & Kim, J. (2024). Advances in hydrogen production from sustainable recourses through biological and thermochemical pathways: Review and bibliometric analysis. *International Journal of Hydrogen Energy, 60*, 28–45. https://doi.org/10.1016/j.ijhydene.2024.02.054 Elsevier Ltd.

Ali, S. S., Al-Tohamy, R., Elsamahy, T., & Sun, J. (2024). Harnessing recalcitrant lignocellulosic biomass for enhanced biohydrogen production: Recent advances, challenges, and future perspective. *Biotechnology Advances, 72*. https://doi.org/10.1016/j.biotechadv.2024.108344 108344.

Arun, J., SundarRajan, P., Pavithra, K. G., Priyadharsini, P., Shyam, S., Goutham, R., Le, Q. H., & Pugazhendhi, A. (2024). New insights into microbial electrolysis cells (MEC) and microbial fuel cells (MFC) for simultaneous wastewater treatment and green fuel (hydrogen) generation. *Fuel, 355*. https://doi.org/10.1016/j.fuel.2023.129530 129530.

Ausfelder, F., & Bazzanella, A. (2016). Hydrogen in the chemical industry. In *Hydrogen science and engineering: Materials, processes, systems and technology*. John Wiley & Sons, Ltd, 19–40. https://doi.org/10.1002/9783527674268.ch02.

Ayodele, D. T., Ogunbiyi, O. D., Akamo, D. O., Otun, K. O., Akinpelu, D. A., Adegoke, J. A., Fapojuwo, D. P., & Oladoye, P. O. (2023). Factors affecting biohydrogen production: Overview and perspectives. *International Journal of Hydrogen Energy, 48*(71), 27513–27539. https://doi.org/10.1016/j.ijhydene.2023.04.001.

Aziz, M. (2021). Liquid hydrogen: A review on liquefaction, storage, transportation, and safety. *Energies, 14*(18). https://doi.org/10.3390/en14185917 MDPI.

Azwar, M. Y., Hussain, M. A., & Abdul-Wahab, A. K. (2014). Development of biohydrogen production by photobiological, fermentation and electrochemical processes: A review. *Renewable and Sustainable Energy Reviews, 31*, 158–173. https://doi.org/10.1016/j.rser.2013.11.022.

Barthelemy, H., Weber, M., & Barbier, F. (2017). Hydrogen storage: Recent improvements and industrial perspectives. *International Journal of Hydrogen Energy, 42*(11), 7254–7262. https://doi.org/10.1016/j.ijhydene.2016.03.178.

Bastola, J. K., & Hiloidhari, M. (2024). Biogas and biohydrogen for net zero cooking and transportation in Bhutan. *Sustainable Production and Consumption, 45*, 79–90. https://doi.org/10.1016/j.spc.2023.12.024.

Bertasini, D., Battista, F., Mancini, R., Frison, N., & Bolzonella, D. (2024). Hydrogen and methane production through two stage anaerobic digestion of straw residues. *Environmental Research, 247*. https://doi.org/10.1016/j.envres.2024.118101 118101.

Bertasini, D., Battista, F., Rizzioli, F., Frison, N., & Bolzonella, D. (2023). Decarbonization of the European natural gas grid using hydrogen and methane biologically produced from organic waste: A critical overview. *Renewable Energy, 206*, 386–396. https://doi.org/10.1016/j.renene.2023.02.029.

Bhatia, S. K., Jagtap, S. S., Bedekar, A. A., Bhatia, R. K., Rajendran, K., Pugazhendhi, A., Rao, C. V., Atabani, A. E., Kumar, G., & Yang, Y. H. (2021). Renewable biohydrogen production from lignocellulosic biomass using fermentation and integration of systems with other energy generation technologies. *Science of the Total Environment, 765*. https://doi.org/10.1016/j.scitotenv.2020.144429 Elsevier B.V.

Bondarenko, V. L., Ilyinskaya, D. N., Kazakova, A. A., Kozlovtsev, P. S., Lavrov, N. A., & Razenko, E. A. (2022). Hydrogen, its unique properties and applications in industry. *Chemical and Petroleum Engineering, 57*(11–12), 1015–1019. https://doi.org/10.1007/s10556-022-01039-7.

Brentner, L. B., Jordan, P. A., & Zimmerman, J. B. (2010). Challenges in developing biohydrogen as a sustainable energy source: Implications for a research agenda. *Environmental Science and Technology, 44*(7), 2243–2254. https://doi.org/10.1021/es9030613.

Cao, J., Qin, X., Geng, G., Zhang, W., & Yu, B. (2021). Current status and prospects of hydrogen storage and transportation technology. *Acta Petrolei Sinica (Petroleum Processing Section), 37*(6), 1461–1478.

Chee, A. L. K., Chin, B. L. F., Sulaiman, S. A., Chai, Y. H., Saptoro, A., Umar, H. A., Lock, S. S. M., & Yiin, C. L. (2024). Syngas-enriched hydrogen production via catalytic gasification of water hyacinth using renewable palm kernel shell hydrochar. *Fuel, 362*. https://doi.org/10.1016/j.fuel.2023.130811.

Chen, C.-C., Chuang, Y.-S., Lin, C.-Y., Lay, C.-H., & Sen, B. (2012). Thermophilic dark fermentation of untreated rice straw using mixed cultures for hydrogen production. *International Journal of Hydrogen Energy, 37*(20), 15540–15546. https://doi.org/10.1016/j.ijhydene.2012.01.036.

Chen, C. Y., Saratale, G. D., Lee, C. M., Chen, P. C., & Chang, J. S. (2008). Phototrophic hydrogen production in photobioreactors coupled with solar-energy-excited optical fibers. *International Journal of Hydrogen Energy, 33*(23), 6886–6895. https://doi.org/10.1016/j.ijhydene.2008.09.014.

Chen, L., Xiao, R., Cheng, C., Tian, G., Chen, S., & Hou, Y. (2020). Thermodynamic analysis of the para-to-ortho hydrogen conversion in cryo-compressed hydrogen vessels for automotive applications. *International Journal of Hydrogen Energy, 45*(46), 24928–24937. https://doi.org/10.1016/j.ijhydene.2020.05.252.

Chen, R., Wang, Y. Z., Liao, Q., Zhu, X., & Xu, T. F. (2013). Hydrolysates of lignocellulosic materials for biohydrogen production. *Biochemistry and Molecular Biology Reports, 46*(5), 244–251. https://doi.org/10.5483/BMBRep.2013.46.5.038.

Chen, W. H., Farooq, W., Shahbaz, M., Naqvi, S. R., Ali, I., Al-Ansari, T., & Amin, N. A. S. (2021). Current status of biohydrogen production from lignocellulosic biomass, technical challenges and commercial potential through pyrolysis process. *Energy, 226*(8), 120433. https://doi.org/10.1016/j.energy.2021.120433 Elsevier Ltd.

Chen, Y., Zhao, S., Wang, X., Peng, Q., Lin, R., Wang, Y., Shen, R., Cao, X., Zhang, L., Zhou, G., Li, J., Xia, A., & Li, Y. (2016). Synergetic integration of $Cu_{1.94}S–Zn_xCd_{1-x}S$ heteronanorods for enhanced visible-light-driven photocatalytic hydrogen production. *Journal of the American Chemical Society, 138*(13), 4286–4289. https://doi.org/10.1021/jacs.5b12666.

Cheng, J., Su, H., Zhou, J., Song, W., & Cen, K. (2011). Hydrogen production by mixed bacteria through dark and photo fermentation. *International Journal of Hydrogen Energy, 36*(1), 450–457. https://doi.org/10.1016/j.ijhydene.2010.10.007.

Chin, M. Y., Phuang, Z. X., Woon, K. S., Hanafiah, M. M., Zhang, Z., & Liu, X. (2022). Life cycle assessment of bioelectrochemical and integrated microbial fuel cell systems for sustainable wastewater treatment and resource recovery. *Journal of Environmental Management, 320*. https://doi.org/10.1016/j.jenvman.2022.115778 115778.

Chookaew, T., Prasertsan, P., & Ren, Z. J. (2014). Two-stage conversion of crude glycerol to energy using dark fermentation linked with microbial fuel cell or microbial electrolysis cell. *New Biotechnology, 31*(2), 179–184. https://doi.org/10.1016/j.nbt.2013.12.004.

Couto, N., Rouboa, A., Silva, V., Monteiro, E., & Bouziane, K. (2013). Influence of the biomass gasification processes on the final composition of syngas. *Energy Procedia, 36*, 596–606. https://doi.org/10.1016/j.egypro.2013.07.068.

Czernik, S., & Bridgwater, A. V. (2004). Overview of Applications of Biomass Fast Pyrolysis Oil. *Energy & Fuels, 18*(2), 590–598. https://doi.org/10.1021/ef034067u.

Dai, F., Zhang, S., Luo, Y., Wang, K., Liu, Y., & Ji, X. (2023). Recent progress on hydrogen-rich syngas production from coal gasification. *Multidisciplinary Digital Publishing Institute (MDPI) – Processes, 11*(6), 1765. https://doi.org/10.3390/pr11061765.

Dauletbay, A. (2024). Transportation of hydrogen: Hydrogen usage. In Z. Abdallah, & N. Aldoumani (Eds.). *Hydrogen Technologies*, IntechOpen. https://doi.org/10.5772/intechopen.1005066.

Deivayanai, V. C., Yaashikaa, P. R., Kumar, P. S., & Rangasamy, G. (2022). A comprehensive review on the biological conversion of lignocellulosic biomass into hydrogen: Pretreatment strategy, technology advances and perspectives. *Bioresource Technology, 365*. https://doi.org/10.1016/j.biortech.2022.128166 Elsevier Ltd.

Demirbas, M. F. (2006). Hydrogen from various biomass species via pyrolysis and steam gasification processes. *Energy Sources, Part A: Recovery, Utilization and Environmental Effects, 28*(3), 245–252. https://doi.org/10.1080/009083190890003.

Dutta, N., Usman, M., Ashraf, M. A., Luo, G., El-Din, M. G., & Zhang, S. (2023). Methods to convert lignocellulosic waste into biohydrogen, biogas, bioethanol, biodiesel and value-added chemicals: A review. *Environmental Chemistry Letters, 21*(2), 803–820. https://doi.org/10.1007/s10311-022-01511-z Springer Science and Business Media Deutschland GmbH.

Elhambakhsh, A., Van Duc Long, N., Lamichhane, P., & Hessel, V. (2023). Recent progress and future directions in plasma-assisted biomass conversion to hydrogen. *Renewable Energy, 218*. https://doi.org/10.1016/j.renene.2023.119307 119307.

Elsayad, R. M., Sharshir, S. W., Khalil, A., & Basha, A. M. (2024). Recent advancements in wastewater treatment via anaerobic fermentation process: A systematic review. *Journal of Environmental Management, 366*. https://doi.org/10.1016/j.jenvman.2024.121724 121724.

Erdem, K., Han, D. G., & Midilli, A. (2024). A parametric study on hydrogen production by fluidized bed co-gasification of biomass and waste plastics. *International Journal of Hydrogen Energy, 52*, 1434–1444. https://doi.org/10.1016/j.ijhydene.2023.10.115.

Fang, H. H. P., Zhu, H., & Zhang, T. (2006). Phototrophic hydrogen production from glucose by pure and co-cultures of *Clostridium butyricum* and *Rhodobacter sphaeroides*. *International Journal of Hydrogen Energy, 31*(15), 2223–2230. https://doi.org/10.1016/j.ijhydene.2006.03.005.

Farzad, S., Mandegari, M. A., & Görgens, J. F. (2016). A critical review on biomass gasification, co-gasification, and their environmental assessments. *Biofuel Research Journal, 3*(4), 483–495. https://doi.org/10.18331/BRJ2016.3.4.3 Green Wave Publishing of Canada.

Fawad, A., Qyyum, M. A., & Ul Jabbar, A. (2024). A comprehensive simulation study of integrated gasification, enrichment, and separation processes for biohydrogen production from sugarcane bagasse. *International Journal of Hydrogen Energy*. https://doi.org/10.1016/j.ijhydene.2024.05.332.

Faye, O., Szpunar, J., & Eduok, U. (2022). A critical review on the current technologies for the generation, storage, and transportation of hydrogen. *International Journal of Hydrogen Energy, 47*(29), 13771–13802. https://doi.org/10.1016/j.ijhydene.2022.02.112 Elsevier Ltd.

Fekete, J. R., Sowards, J. W., & Amaro, R. L. (2015). Economic impact of applying high strength steels in hydrogen gas pipelines. *International Journal of Hydrogen Energy, 40*(33), 10547–10558. https://doi.org/10.1016/j.ijhydene.2015.06.090.

Feng, S., Ngo, H. H., Guo, W., Chang, S. W., Nguyen, D. D., Liu, Y., Zhang, X., Bui, X. T., Varjani, S., & Hoang, B. N. (2022). Wastewater-derived biohydrogen: Critical analysis of related enzymatic processes at the research and large scales. *Science of The Total Environment, 851*. https://doi.org/10.1016/j.scitotenv.2022.158112 158112.

Feng, S., Ngo, H. H., Guo, W., Chang, S. W., Nguyen, D. D., Bui, X. T., Zhang, X., Ma, X. Y., & Hoang, B. N. (2023). Biohydrogen production, storage, and delivery: A comprehensive overview of current strategies and limitations. *Chemical Engineering Journal, 471*. https://doi.org/10.1016/j.cej.2023.144669 144669.

Ferraren-De Cagalitan, D. D. T., & Abundo, M. L. S. (2021). A review of biohydrogen production technology for application towards hydrogen fuel cells. *Renewable and Sustainable Energy Reviews, 151*. https://doi.org/10.1016/j.rser.2021.111413 111413.

Gao, C., Yang, J., & Han, L. (2021). Systematic comparison for effects of different scale mechanical-NaOH coupling treatments on lignocellulosic components, micromorphology and cellulose crystal structure of wheat straw. *Bioresource Technology, 326*. https://doi.org/10.1016/j.biortech.2021.124786.

Gao, W., Zhou, T., Gao, Y., & Wang, Q. (2019). Enhanced water gas shift processes for carbon dioxide capture and hydrogen production. *Applied Energy, 254*. https://doi.org/10.1016/j.apenergy.2019.113700.

Ghimire, A., Frunzo, L., Pirozzi, F., Trably, E., Escudie, R., Lens, P. N. L., & Esposito, G. (2015). A review on dark fermentative biohydrogen production from organic biomass: Process parameters and use of by-products. *Applied Energy, 144*, 73–95. https://doi.org/10.1016/j.apenergy.2015.01.045.

Gollakota, A. R. K., Kishore, N., & Gu, S. (2018). A review on hydrothermal liquefaction of biomass. *Renewable and Sustainable Energy Reviews, 81*(1), 1378–1392. https://doi.org/10.1016/j.rser.2017.05.178.

De Gioannis, G., Dell'Era, A., Muntoni, A., Pasquali, M., Polettini, A., Pomi, R., Rossi, A., & Zonfa, T. (2023). Bio-electrochemical production of hydrogen and electricity from organic waste: preliminary assessment. *Clean Technologies and Environmental Policy, 25*(1), 269–280. https://doi.org/10.1007/s10098-022-02305-1.

De Gioannis, G., Friargiu, M., Massi, E., Muntoni, A., Polettini, A., Pomi, R., & Spiga, D. (2014). Biohydrogen production from dark fermentation of cheese whey: Influence of pH. *International Journal of Hydrogen Energy, 39*(36), 20930–20941. https://doi.org/10.1016/j.ijhydene.2014.10.046.

Goren, A. Y., Dincer, I., & Khalvati, A. (2023). Comparative environmental sustainability assessment of biohydrogen production methods. *Science of the Total Environment, 904*. https://doi.org/10.1016/j.scitotenv.2023.166613 Elsevier B.V.

Gupta, R., Zhang, L., Hou, J., Zhang, Z., Liu, H., You, S., Sik Ok, Y., & Li, W. (2023). Review of explainable machine learning for anaerobic digestion. *Bioresource Technology, 369*. https://doi.org/10.1016/j.biortech.2022.128468 128468.

Gupta, S., Fernandes, A., Lopes, A., Grasa, L., & Salafranca, J. (2024). Photo-fermentative bacteria used for hydrogen production. *Applied Sciences (Switzerland), 14*(3). https://doi.org/10.3390/app14031191 Multidisciplinary Digital Publishing Institute (MDPI).

Hassan, Q., Nassar, A. K., Al-Jiboory, A. K., Viktor, P., Telba, A. A., Awwad, E. M., Amjad, A., Fakhruldeen, H. F., Algburi, S., Mashkoor, S. C., Jaszczur, M., Sameen, A. Z., & Barakat, M. (2024). Mapping Europe renewable energy landscape: Insights into solar, wind, hydro, and green hydrogen production. *Technology in Society, 77*. https://doi.org/10.1016/j.techsoc.2024.102535.

Hovorukha, V., Havryliuk, O., Gladka, G., Tashyrev, O., Kalinichenko, A., Sporek, M., & Dołhańczuk-Śródka, A. (2021). Hydrogen dark fermentation for degradation of solid and liquid food waste. *Energies, 14*(7). https://doi.org/10.3390/en14071831.

Hren, R., Vujanović, A., Van Fan, Y., Klemeš, J. J., Krajnc, D., & Čuček, L. (2023). Hydrogen production, storage and transport for renewable energy and chemicals: An environmental footprint assessment. *Renewable and Sustainable Energy Reviews, 173*. https://doi.org/10.1016/j.rser.2022.113113 113113.

Ibrahimoglu, B., & Yilmazoglu, M. Z. (2020). Numerical modeling of a downdraft plasma coal gasifier with plasma reactions. *International Journal of Hydrogen Energy, 45*(5), 3532–3548. https://doi.org/10.1016/j.ijhydene.2018.12.198.

Jeje, S. O., Marazani, T., Obiko, J. O., & Shongwe, M. B. (2024). Advancing the hydrogen production economy: A comprehensive review of technologies, sustainability, and future prospects. *International Journal of Hydrogen Energy, 78*, 642–661. https://doi.org/10.1016/j.ijhydene.2024.06.344.

Kannah, R. Y., Kavitha, S., Preethi, Karthikeyan, O. P., Kumar, G., Dai-Viet, N. V., & Banu, J. R. (2021). Techno-economic assessment of various hydrogen production methods—A review. *Bioresource Technology, 319*. https://doi.org/10.1016/j.biortech.2020.124175 Elsevier Ltd.

Jia, J., Seitz, L. C., Benck, J. D., Huo, Y., Chen, Y., Ng, J. W. D., Bilir, T., Harris, J. S., & Jaramillo, T. F. (2016). Solar water splitting by photovoltaic-electrolysis with a solar-to-hydrogen efficiency over 30. *Nature Communications, 7*(1). https://doi.org/10.1038/ncomms13237 13237.

Kapdan, I. K., & Kargi, F. (2006). Bio-hydrogen production from waste materials. *Enzyme and Microbial Technology, 38*(5), 569–582. https://doi.org/10.1016/j.enzmictec.2005.09.015.

Kaur, R., Gera, P., Jha, M. K., & Bhaskar, T. (2019). Thermochemical route for biohydrogen production. In *Biomass, biofuels, biochemicals: Biohydrogen, second edition*. Elsevier, 187–218. https://doi.org/10.1016/B978-0-444-64203-5.00008-3.

Khetkorn, W., Khanna, N., Incharoensakdi, A., & Lindblad, P. (2013). Metabolic and genetic engineering of cyanobacteria for enhanced hydrogen production. *Biofuels, 4*(5), 535–561. https://doi.org/10.4155/bfs.13.41.

Khan, M. A., Ngo, H. H., Guo, W., Liu, Y., Zhang, X., Guo, J., Chang, S. W., Nguyen, D. D., & Wang, J. (2018). Biohydrogen production from anaerobic digestion and its potential as renewable energy. *Renewable Energy, 129*, 754–768. https://doi.org/10.1016/j.renene.2017.04.029.

Kumar, P., Pal, N., & Sharma, H. (2021). Techno-economic analysis of solar photo-voltaic/diesel generator hybrid system using different energy storage technologies for isolated islands of India. *Journal of Energy Storage, 41*. https://doi.org/10.1016/j.est.2021.102965 102965.

Kumari, P., & Mohanty, B. (2022). Maximization of hydrogen production from pine needles steam gasification based on response surface methodology. *Biomass Conversion and Biorefinery, 12*(7), 2335–2348. https://doi.org/10.1007/s13399-020-00761-7.

Lahnaoui, A., Wulf, C., & Dalmazzone, D. (2017). Building an optimal hydrogen transportation system for mobility, focus on minimizing the cost of transportation via truck. *Energy Procedia, 142*, 2072–2079. https://doi.org/10.1016/j.egypro.2017.12.579.

Levin, D. B., & Chahine, R. (2010). Challenges for renewable hydrogen production from biomass. *International Journal of Hydrogen Energy, 35*(10), 4962–4969. https://doi.org/10.1016/j.ijhydene.2009.08.067.

Li, B., Yang, H., Wei, L., Shao, J., Wang, X., & Chen, H. (2017). Hydrogen production from agricultural biomass wastes gasification in a fluidized bed with calcium oxide enhancing. *International Journal of Hydrogen Energy, 42*(8), 4832–4839. https://doi.org/10.1016/j.ijhydene.2017.01.138.

Li, D., & Chen, H. (2007). Biological hydrogen production from steam-exploded straw by simultaneous saccharification and fermentation. *International Journal of Hydrogen Energy, 32*(12), 1742–1748. https://doi.org/10.1016/j.ijhydene.2006.12.011.

Li, J., Li, G., Ma, S., & Song, J. (2021). An overview on hydrogen energy storage and transportation technology and its typical application in power system. *Modern Electric Power, 38*(5), 535–545. https://doi.org/10.19725/j.cnki.1007-2322.2021.0023 overview on hydrogen energy storage and transportation technology and its typical application in power system.

Liu, J., Zhao, M., & Rong, L. (2023). Overview of hydrogen-resistant alloys for high-pressure hydrogen environment: On the hydrogen energy structural materials. *Clean Energy, 7*(1), 99–115. https://doi.org/10.1093/ce/zkad009 Oxford University Press.

Loss, R. A., Fontes, M. L., Reginatto, V., & Antônio, R. V. (2013). Biohydrogen production by a mixed photoheterotrophic culture obtained from a Winogradsky column prepared from the sediment of a southern Brazilian lagoon. *Renewable Energy, 50*, 648–654. https://doi.org/10.1016/j.renene.2012.07.011.

Ma, J., Liu, S., Zhou, W., & Pan, X. (2008). Comparison of hydrogen transportation methods for hydrogen refueling station. *Tongji Daxue Xuebao/Journal of Tongji University, 36*(5), 615–619. https://www.scopus.com/inward/record.uri?eid=2-s2.0-45449094225&partnerID=40&md5=1fe06e0765052384249dc060bf003735.

Madenoğlu, T. G., Boukis, N., Sağlam, M., & Yüksel, M. (2011). Supercritical water gasification of real biomass feedstocks in continuous flow system. *International Journal of Hydrogen Energy, 36*(22), 14408–14415. https://doi.org/10.1016/j.ijhydene.2011.08.047.

Maisano, S., Urbani, F., Cipitì, F., Freni, F., & Chiodo, V. (2019). Syngas production by BFB gasification: Experimental comparison of different biomasses. *International Journal of Hydrogen Energy, 44*(9), 4414–4422. https://doi.org/10.1016/j.ijhydene.2018.11.148.

Mali, B., Niraula, D., Kafle, R., & Bhusal, A. (2021). Green hydrogen: Production methodology, applications and challenges in Nepal. *7th International Conference on Engineering, Applied Sciences and Technology (ICEAST)*, 68–76. https://doi.org/10.1109/ICEAST52143.2021.9426300.

Maliutina, K., Tahmasebi, A., Yu, J., & Saltykov, S. N. (2017). Comparative study on flash pyrolysis characteristics of microalgal and lignocellulosic biomass in entrained-flow reactor. *Energy Conversion and Management, 151*, 426–438. https://doi.org/10.1016/j.enconman.2017.09.013.

Mirza, S. S., Qazi, J. I., Liang, Y., & Chen, S. (2019). Growth characteristics and photofermentative biohydrogen production potential of purple non sulfur bacteria from sugar cane bagasse. *Fuel, 255*. https://doi.org/10.1016/j.fuel.2019.115805 115805.

Mohammed, I. Y., Abakr, Y. A., Musa, M., Yusup, S., Singh, A., & Kazi, F. K. (2016). Valorization of Bambara groundnut shell via intermediate pyrolysis: Products distribution and characterization. *Journal of Cleaner Production, 139*, 717–728. https://doi.org/10.1016/j.jclepro.2016.08.090.

Monlau, F., Barakat, A., Trably, E., Dumas, C., Steyer, J. P., & Carrère, H. (2013). Lignocellulosic materials into biohydrogen and biomethane: Impact of structural features and pretreatment. *Critical Reviews in Environmental Science and Technology, 43*(3), 260–322. https://doi.org/10.1080/10643389.2011.604258.

Monlau, F., Trably, E., Barakat, A., Hamelin, J., Steyer, J.-P., & Carrere, H. (2013). Two-stage alkaline-enzymatic pretreatments to enhance biohydrogen production from sunflower stalks. *Environmental Science & Technology, 47*(21), 12591–12599. https://doi.org/10.1021/es402863v.

Moradi, R., & Groth, K. M. (2019). Hydrogen storage and delivery: Review of the state of the art technologies and risk and reliability analysis. *International Journal of Hydrogen Energy, 44*(23), 12254–12269. https://doi.org/10.1016/j.ijhydene.2019.03.041.

Mori, D., & Hirose, K. (2009). Recent challenges of hydrogen storage technologies for fuel cell vehicles. *International Journal of Hydrogen Energy, 34*(10), 4569–4574. https://doi.org/10.1016/j.ijhydene.2008.07.115.

Nakajima, H. (2007). Fabrication, properties and application of porous metals with directional pores. *Progress in Materials Science, 52*(7), 1091–1173. https://doi.org/10.1016/j.pmatsci.2006.09.001.

Okonkwo, P. C., Barhoumi, E. M., Belgacem, I. B., Mansir, I. B., Aliyu, M., Emori, W., Uzoma, P. C., Beitelmal, W. H., Akyüz, E., Radwan, A. B., & Shakoor, R. A. (2023). A focused review of the hydrogen storage tank embrittlement mechanism process. *International Journal of Hydrogen Energy, 48*(35), 12935–12948. https://doi.org/10.1016/j.ijhydene.2022.12.252.

Oliveira, M., Ramos, A., Ismail, T. M., Monteiro, E., & Rouboa, A. (2022). A review on plasma gasification of solid residues: Recent advances and developments. *Energies, 15*(4). https://doi.org/10.3390/en15041475 MDPI.

Onwudili, J. A., & Williams, P. T. (2013). Hydrogen and methane selectivity during alkaline supercritical water gasification of biomass with ruthenium-alumina catalyst. *Applied Catalysis B: Environmental, 132–133*, 70–79. https://doi.org/10.1016/j.apcatb.2012.11.033.

Osman, A. I., Deka, T. J., Baruah, D. C., & Rooney, D. W. Critical challenges in biohydrogen production processes from the organic feedstocks. https://doi.org/10.1007/s13399-020-00965-x/Published.

Osman, A. I., Deka, T. J., Baruah, D. C., & Rooney, D. W. (2023). Critical challenges in biohydrogen production processes from the organic feedstocks. *Biomass Conversion and Biorefinery, 13*(10), 8383–8401. https://doi.org/10.1007/s13399-020-00965-x.

Pang, J., Zheng, M., Li, X., Sebastian, J., Jiang, Y., Zhao, Y., Wang, A., & Zhang, T. (2019). Unlock the compact structure of lignocellulosic biomass by mild ball milling for ethylene glycol production. *ACS Sustainable Chemistry & Engineering, 7*(1), 679–687. https://doi.org/10.1021/acssuschemeng.8b04262.

Peterson, A. A., Vogel, F., Lachance, R. P., Fröling, M., Antal, M. J., & Tester, J. W. (2008). Thermochemical biofuel production in hydrothermal media: A review of sub- and supercritical water technologies. *Energy & Environmental Science, 1*(1), 32–65. https://doi.org/10.1039/B810100K.

Prabakar, D., Manimudi, V. T., Suvetha K, S., Sampath, S., Mahapatra, D. M., Rajendran, K., & Pugazhendhi, A. (2018). Advanced biohydrogen production using pretreated industrial waste: Outlook and prospects. *Renewable and Sustainable Energy Reviews, 96*, 306–324. https://doi.org/10.1016/j.rser.2018.08.006 Elsevier Ltd.

Pudukudy, M., Yaakob, Z., Mohammad, M., Narayanan, B., & Sopian, K. (2014). Renewable hydrogen economy in Asia—Opportunities and challenges: An overview. *Renewable and Sustainable Energy Reviews, 30*, 743–757. https://doi.org/10.1016/j.rser.2013.11.015 Elsevier Ltd.

Rambhujun, N., Salman, M. S., Wang, T., Pratthana, C., Sapkota, P., Costalin, M., Lai, Q., & Aguey-Zinsou, K.-F. (2020). Renewable hydrogen for the chemical industry. *MRS Energy & Sustainability, 7*, E33. https://doi.org/10.1557/mre.2020.33.

Rani, P., Yadav, D. K., Yadav, A., Bishnoi, N. R., Kumar, V., Ram, C., Pugazhendhi, A., & Kumar, S. S. (2024). Frontier in dark fermentative biohydrogen production from lignocellulosic biomass: Challenges and future prospects. *Fuel, 366*. https://doi.org/10.1016/j.fuel.2024.131187.

Rathi, B. S., Kumar, P. S., Rangasamy, G., & Rajendran, S. (2024). A critical review on biohydrogen generation from biomass. *International Journal of Hydrogen Energy, 52*, 115–138. https://doi.org/10.1016/j.ijhydene.2022.10.182.

Rincón-Pérez, J., Razo-Flores, E., Morales, M., Alatriste-Mondragón, F., & Celis, L. B. (2020). Improving the biodegradability of scenedesmus obtusiusculus by thermochemical pretreatment to produce hydrogen and methane. *BioEnergy Research, 13*(2), 477–486. https://doi.org/10.1007/s12155-019-10067-w.

Rivera-Tinoco, R., & Bouallou, C. (2010). Using biomass as an energy source with low CO_2 emissions. *Clean Technologies and Environmental Policy, 12*(2), 171–175. https://doi.org/10.1007/s10098-009-0241-4.

Rosenbaum, M. A., & Franks, A. E. (2014). Microbial catalysis in bioelectrochemical technologies: Status quo, challenges and perspectives. *Applied Microbiology and Biotechnology, 98*(2), 509–518. https://doi.org/10.1007/s00253-013-5396-6.

Sahrin, N. T., Khoo, K. S., Lim, J. W., Shamsuddin, R., Ardo, F. M., Rawindran, H., Hassan, M., Kiatkittipong, W., Abdelfattah, E. A., Da Oh, W., & Cheng, C. K. (2022). Current perspectives, future challenges and key technologies of biohydrogen production for building a carbon–neutral future: A review. *Bioresource Technology, 364*. https://doi.org/10.1016/j.biortech.2022.128088 Elsevier Ltd.

Saravanan, A., Kumar, P. S., Khoo, K. S., Show, P. L., Carolin, C. F., Jackulin, C. F., Jeevanantham, S., Karishma, S., Show, K. Y., Lee, D. J., & Chang, J. S. (2021). Biohydrogen from organic wastes as a clean and environment-friendly energy source: Production pathways, feedstock types, and future prospects. *Bioresource Technology, 342*. https://doi.org/10.1016/j.biortech.2021.126021 Elsevier Ltd.

Shakeel, U., Zhang, Y., Topakas, E., Wang, W., Liang, C., & Qi, W. (2024). Unraveling interplay between lignocellulosic structures caused by chemical pretreatments in enhancing enzymatic hydrolysis. *Carbohydrate Polymers*. https://doi.org/10.1016/j.carbpol.2024.122037 122037.

Sharma, S., & Ghoshal, S. K. (2015). Hydrogen the future transportation fuel: From production to applications. *Renewable and Sustainable Energy Reviews, 43*, 1151–1158. https://doi.org/10.1016/j.rser.2014.11.093.

Shen, L., Gao, Y., & Xiao, J. (2008). Simulation of hydrogen production from biomass gasification in interconnected fluidized beds. *Biomass and Bioenergy, 32*(2), 120–127. https://doi.org/10.1016/j.biombioe.2007.08.002.

Shen, Y. (2024). Biomass pretreatment for steam gasification toward H2-rich syngas production—An overview. *International Journal of Hydrogen Energy, 66*, 90–102. https://doi.org/10.1016/j.ijhydene.2024.04.096 Elsevier Ltd.

Sherif S.A., Zeytinoglu N., & Veziroglug T.N. (1997). Liquid hydrogen: Potential, problems, and a proposed research program (22, 7).

Show, K.-Y., Lee, D.-J., & Chang, J.-S. (2011). Bioreactor and process design for biohydrogen production. *Bioresource Technology, 102*(18), 8524–8533. https://doi.org/10.1016/j.biortech.2011.04.055.

Silva, F. M. S., Mahler, C. F., Oliveira, L. B., & Bassin, J. P. (2018). Hydrogen and methane production in a two-stage anaerobic digestion system by co-digestion of food waste, sewage sludge and glycerol. *Waste Management, 76*, 339–349. https://doi.org/10.1016/j.wasman.2018.02.039.

Speight, J.G. (2023). Production of bio-syngas and bio-hydrogen by gasification. In R. Luque, C. Sze Ki Lin, K. Wilson, & C. Du (Eds.), Handbook of biofuels production (pp. 419–448).

Srivastava, R. K., Shetti, N. P., Reddy, K. R., & Aminabhavi, T. M. (2020). Biofuels, biodiesel and biohydrogen production using bioprocesses. A review. *Environmental Chemistry Letters, 18*(4), 1049–1072. https://doi.org/10.1007/s10311-020-00999-7 Springer.

Sun, Q., Xiao, W., Xi, D., Shi, J., Yan, X., & Zhou, Z. (2010). Statistical optimization of biohydrogen production from sucrose by a co-culture of *Clostridium acidisoli* and *Rhodobacter sphaeroides*. *International Journal of Hydrogen Energy, 35*(9), 4076–4084. https://doi.org/10.1016/j.ijhydene.2010.01.145.

Surti, P., Kailasa, S. K., & Mungray, A. K. (2021). Genetic engineering strategies for performance enhancement of bioelectrochemical systems: A review. *Sustainable Energy Technologies and Assessments, 47*. https://doi.org/10.1016/j.seta.2021.101332 101332.

Tayibi, S., Monlau, F., Bargaz, A., Jimenez, R., & Barakat, A. (2021). Synergy of anaerobic digestion and pyrolysis processes for sustainable waste management: A critical review and future perspectives. *Renewable and Sustainable Energy Reviews, 152*. https://doi.org/10.1016/j.rser.2021.111603 111603.

Toor, S. S., Rosendahl, L., & Rudolf, A. (2011). Hydrothermal liquefaction of biomass: A review of subcritical water technologies. *Energy, 36*(5), 2328–2342. ISSN 0360-5442 https://doi.org/10.1016/j.energy.2011.03.013.

Tungal, R., & Shende, R. V. (2014). Hydrothermal liquefaction of pinewood (Pinus ponderosa) for H_2, biocrude and bio-oil generation. *Applied Energy, 134*, 401–412. https://doi.org/10.1016/j.apenergy.2014.07.060.

Wang, H., Zhao, Y., Dong, X., Yang, J., Guo, H., & Gong, M. (2022). Thermodynamic analysis of low-temperature and high-pressure (cryo-compressed) hydrogen storage processes cooled by mixed-refrigerants. *International Journal of Hydrogen Energy, 47*(67), 28932–28944. https://doi.org/10.1016/j.ijhydene.2022.06.193.

Wang, X., Kersten, S. R. A., Prins, W., & van Swaaij, W. P. M. (2005). Biomass pyrolysis in a fluidized bed reactor. Part 2: experimental validation of model results. *Industrial & Engineering Chemistry Research, 44*(23), 8786–8795. https://doi.org/10.1021/ie050486y.

Weldekidan, H., Strezov, V., Town, G., & Kan, T. (2018). Production and analysis of fuels and chemicals obtained from rice husk pyrolysis with concentrated solar radiation. *Fuel, 233*, 396–403. https://doi.org/10.1016/j.fuel.2018.06.061.

Witkowski, A., Rusin, A., Majkut, M., & Stolecka, K. (2017). Comprehensive analysis of hydrogen compression and pipeline transportation from thermodynamics and safety aspects. *Energy, 141*, 2508–2518. https://doi.org/10.1016/j.energy.2017.05.141.

Wojtacha-Rychter, K., Howaniec, N., & Smoliński, A. (2024). Investigation of co-gasification characteristics of coal with wood biomass and rubber seals in a fixed bed gasifier. *Renewable Energy, 220*. https://doi.org/10.1016/j.renene.2023.119725.

Xia, S., Li, K., Xiao, H., Cai, N., Dong, Z., xu, C., Chen, Y., Yang, H., Tu, X., & Chen, H. (2019). Pyrolysis of Chinese chestnut shells: Effects of temperature and Fe presence on product composition. *Bioresource Technology, 287*. https://doi.org/10.1016/j.biortech.2019.121444 121444.

Xu, X., Zhou, Q., & Yu, D. (2022). The future of hydrogen energy: Bio-hydrogen production technology. *International Journal of Hydrogen Energy, 47*(79), 33677–33698. https://doi.org/10.1016/j.ijhydene.2022.07.261 Elsevier Ltd.

Yan, Y., Zhao, X., Xu, Z., Wei, W., & Ni, Z. (2021). Loading procedure for testing the cryogenic performance of cryo-compressed vessel for fuel cell vehicles. *Applied Thermal Engineering, 183*. https://doi.org/10.1016/j.applthermaleng.2020.115798 115798.

Yanxing, Z., Maoqiong, G., Yuan, Z., Xueqiang, D., & Jun, S. (2019). Thermodynamics analysis of hydrogen storage based on compressed gaseous hydrogen, liquid hydrogen and cryo-compressed hydrogen. *International Journal of Hydrogen Energy, 44*(31), 16833–16840. https://doi.org/10.1016/j.ijhydene.2019.04.207.

Yuan, H., Lu, Y., Abu-Reesh, I. M., & He, Z. (2015). Bioelectrochemical production of hydrogen in an innovative pressure-retarded osmosis/microbial electrolysis cell system: Experiments and modeling. *Biotechnology for Biofuels, 8*(1), 116. https://doi.org/10.1186/s13068-015-0305-0.

Zhang, G., & Jiang, Z. (2023). Overview of hydrogen storage and transportation technology in China. *Unconventional Resources, 3*, 291–296. https://doi.org/10.1016/j.uncres.2023.07.001.

Zhang, J., Xue, D., Wang, C., Fang, D., Cao, L., & Gong, C. (2023). Genetic engineering for biohydrogen production from microalgae. *IScience*. https://doi.org/10.1016/j.isci.

Zhang, T., Jiang, D., Zhang, H., Jing, Y., Tahir, N., Zhang, Y., & Zhang, Q. (2020). Comparative study on bio-hydrogen production from corn stover: Photo-fermentation, dark-fermentation and dark-photo co-fermentation. *International Journal of Hydrogen Energy, 45*(6), 3807–3814. https://doi.org/10.1016/j.ijhydene.2019.04.170.

Zhang, Y.-huan, Jia, Z.-chao, & Yuan, Z.-ming (2015). Development and application of hydrogen storage. *Journal of Iron and Steel Research International, 22*(9), 757–770.

Zhao, B., Song, G., Zhou, W., Chen, L., Sun, L., Yang, S., Guan, H., Zhu, D., Chen, G., Ding, W., Wang, J., & Yang, H. (2020). Catalytic pyrolysis of herb residues for the preparation of hydrogen-rich gas. *Energy & Fuels, 34*(2), 1131–1136. https://doi.org/10.1021/acs.energyfuels.9b02177.

Zhao, Z. T., Ding, J., Wang, B. Y., Bao, M. Y., Liu, B. F., Pang, J. W., Ren, N. Q., & Yang, S. S. (2024). Advances in the biomass valorization in dark fermentation systems: A sustainable approach for biohydrogen production. *Chemical Engineering Journal, 481*. https://doi.org/10.1016/j.cej.2023.148444 Elsevier B.V.

Zhou, W., Bai, B., Chen, G., Ma, L., Jing, D., & Yan, B. (2018). Study on catalytic properties of potassium carbonate during the process of sawdust pyrolysis. *International Journal of Hydrogen Energy, 43*(30), 13829–13841. https://doi.org/10.1016/j.ijhydene.2018.02.002.

Zou, C., Li, J., Zhang, X., Jin, X., Xiong, B., Yu, H., Liu, X., Wang, S., Li, Y., Zhang, L., Miao, S., Zheng, D., Zhou, H., Song, J., & Pan, S. (2022). Industrial status, technological progress, challenges, and prospects of hydrogen energy. *Natural Gas Industry B, 9*(5), 427–447. https://doi.org/10.1016/j.ngib.2022.04.006.

Chapter 10

Life cycle assessment and cost–benefit analysis of biogas and biohydrogen production

Tugba Keskin[1,2,3], Enes Icer[3], and Merve Asli Ergun[3]
[1]*Department of Industrial Engineering, Faculty of Engineering, Izmir Democracy University, Izmir, Turkey,* [2]*Research and Application Center for Sustainable Environmental Studies, Izmir Democracy University, Izmir, Turkey,* [3]*Department of Operations Research, Graduate School of Natural and Applied Sciences, Izmir Democracy University, Izmir, Türkiye*

10.1 Introduction

Climate change is increasingly becoming a subject of worldwide concern and comprehension. The emphasis has shifted from simply safeguarding natural resources for future generations to improving the ability to withstand and adjust to challenges. Sustainability concerns persistently influence the formulation of policies and the management of resources. The concept of circular economy has become essential for many businesses (Aziz et al., 2019). An important aspect that contributes to the global sustainability issue is the dependence on technologies that rely on depleting fossil resources. The increase in scientific investigation and growing enthusiasm for sustainable energy sources have accelerated the transition away from traditional technologies. The increasing awareness of environmental contamination and the depletion of fossil fuels have led researchers and the industry to focus on the utilization of renewable energy, particularly biomass, as part of a circular economy. To ensure the long-term sustainability of the world's population, which is estimated to reach 9 billion by 2050, it is essential to increase energy resources by at least 50% and reduce CO_2 emissions by 80% (Filippa et al., 2020). In 2015, the United Nations Conference on Climate Change established an agreement to restrict the rise in temperature to 1.5°C. The initial phase of this reduction strategy involves the utilization of renewable technologies. Numerous researchers have concentrated on various forms of renewable technologies, including photovoltaic cells, wind energy, solar systems, and biofuels. Waste reuse and waste management are key priorities in

Biogas in the Circular Economy. DOI: https://doi.org/10.1016/B978-0-443-29230-9.00007-3

environmental efforts aimed at ensuring sustainability (Roy et al., 2009). From this perspective, technologies that offer waste management using renewable energy sources are highly significant (Sinsuw et al., 2024). Biogas and biohydrogen systems are ecologically beneficial technologies that are intensively investigated and considered significant among several renewable energy sources, including solar energy, wind energy, and biomass energy (Daş et al., 2024).

Utilizing organic wastes, by-products, and biomass in renewable energy generation technologies offers benefits in terms of sustainability, climate change mitigation, and circular economy. However, the crucial aspect in this situation is to interact using numerical values. When analyzing numerical data, we may identify the environmental effects of activities and spot areas with significant consequences, known as hot spots. Furthermore, it is crucial to analyze numerical data to accurately understand the environmental effects of activities, including their contribution to global warming and their impact on land and aquatic life. The procedure by which all of these activities can be carried out is referred to as Life Cycle Analysis/Assessment (LCA) (Hijazi et al., 2016).

LCA is an effective tool for evaluating the environmental impact of waste treatment technologies (Zaman, 2010). LCA comprehensively assesses all aspects of life cycles associated with a product, service, or system, as well as the interrelationships between them. It provides a comprehensive overview of the environmental effects that can arise over the whole life cycle of the product. The LCA technique considers stages like raw material acquisition, final disposal, and shipment, which are typically not included in standard environmental impact assessments (Banar et al., 2009). LCA relies on ISO 14040 to 14043 (Environmental Management) that have been published by International Organization for Standardization (ISO). Finally, these standards were combined, and a new standard was published (ISO 14044). This standard is a guideline for LCA explaining all steps in detail (Hauschild, 2017). When conducting LCA studies, it is crucial to consider variables such as defining the aim and scope, presenting the results, and tailoring the communication depending on their level of interaction.

The applications of LCA studies are as outlined by Alpaydın (2014).

- In the field of industry, product design, the process of renewing and developing products, as well as strategic planning.
- Development of a eco-labeling program to address the environmental impact of a product.
- Making determinations about matters for research initiatives, funding for development, and administrative frameworks within the public sector.

LCA has four main steps:

Goal scope definition: The essential phase of LCA study. During this stage, the system boundaries will be established and the input–output mass and energy balances can be determined. For the LCA study to be successful, it is crucial to have a clear and precise definition of the product system during the goal and scope phase. A product system is a set of unit processes that involve elementary and product flows and provide one or more specific roles. It represents the life cycle of a product. Regardless of whether it is in the product, manufacturing, or service sectors, it is essential to clearly outline the phases of the system in great detail. This stage allows us to comprehend the system and helps in making decisions regarding the actions to be made to minimize environmental effects during the conclusion evaluation phase. Another important step is to select a proper functional unit. The functional unit is crucial for gaining a comprehensive understanding of a process and for making comparisons with other systems (Ramírez-Arpide et al., 2018). A functional unit is a standardized measure of the performance of a product system that is used as a reference. Another important definition for determining the goal and scope is input and output. The input refers to a product, material, or energy flow that enters a unit process. Examples of products and materials include raw resources, intermediary products, and co-products. The output of a unit process consists of products, materials, or energy flows. These can include raw materials, intermediate products, co-products, and releases. Therefore, a process is all the activities of the system from input to output, including energy flow and elementary flow. Once the goal and scope have been established, the system boundaries should be properly defined. Establishing the system boundaries is essential throughout the inventory data collection phase.

System boundaries can be defined as follows:

- Cradle to Grave: From the resource extraction point to disposal into the environment.
- Cradle to Gate: From resource extraction to gate of factory (before transportation).
- Cradle to Cradle: The final product is recycled after disposal.
- Gate to Gate: Analyzing only one product or process without using data from resources or disposal.

Inventory analysis: At this step, it is crucial to identify all the specific aspects of the operation, ranging from the minimal amount of water required to the substantial electricity usage involved. During this stage, all data will be gathered and stored in a product system, based on the functional unit. Inventory analysis should cover all the aspects of the bill of materials and also resource inputs such as electricity and water. The quality and accuracy of all the data are very important for better results; therefore for inventory analysis, different databases are developed such as EcoInvent, Agribalyse, and Worldsteel (Klöpffer, 1997).

Inventory analysis step is directly related to the goal and scope step. The scope can be defined according to Greenhouse Gas Protocol which is released on 2004. In GHG protocol, the scope of LCA studies is divided into three subcategories:

Scope 1 inventory: The activities of an organization that directly causes GHG emissions such as direct emissions and direct electricity usage amount.
Scope 2 inventory: The purchased activities GHG emissions such as steam and electricity.
Scope 3 inventory: The indirect emissions other than Scope 2.

The collected data for inventory analysis can be primary data (directly taken from the system) or secondary data, which can be obtained from published literature, databases, estimations, or national inventories.

Impact analysis: The many sources and emissions forming the Life Cycle Inventory (LCI) are converted into environmental effect categories during the Life Cycle Impact Assessment (LCIA) phase. The LCI flow is partitioned into one or more categories. Characterization factors are used to aggregate flows within each group. These factors are determined by the physical and chemical qualities of the substances responsible for the impact, as well as the way that flows from the product system into the environment. The combined amount is referred to as the "potential impact" and is often expressed in kilograms comparable to a particular reference material for the relevant category. In this step, using all the data in inventory analysis, classification, characterization, and evaluation of all impacts of the system to the environment will be quantified. Different impact methods have different analysis categories such as global warming, acidification, and eutrophication. Commonly used impact methods are as follows: CML, Recepi, and ILCD.

Interpretation: All steps of LCA can be developed with interpretation of results. After completion of interpretations in all steps, a final assessment can be performed. According to these steps and ISO 14044 – *Environmental management—LCA—requirements and guidelines* regulations, some software has been developed that includes all databases and impact assessment methods. Common softwares used are as follows: *SimaPro*, *Gabi*, *Open LCA*, and *CCalc*. Biogas and biohydrogen production are the two most known technologies for generating energy from biowaste and utilizing biomass. Various technologies combine different reactor types, operating conditions, and system designs to handle different forms of waste. Gaining a comprehensive understanding of the distinctions between these systems is a crucial method for finding the most suitable design parameters or system. This section of the chapter will analyze the necessary characteristics and system design units for conducting LCA studies, focusing on biogas and biohydrogen technologies (Fig. 10.1).

10.2 Life cycle assessment for biogas production

Biogas production is a biofuel production method that has successfully transitioned from the pilot-size phase to large-scale industrial production. Given the consideration of large-scale production, there have been studies in

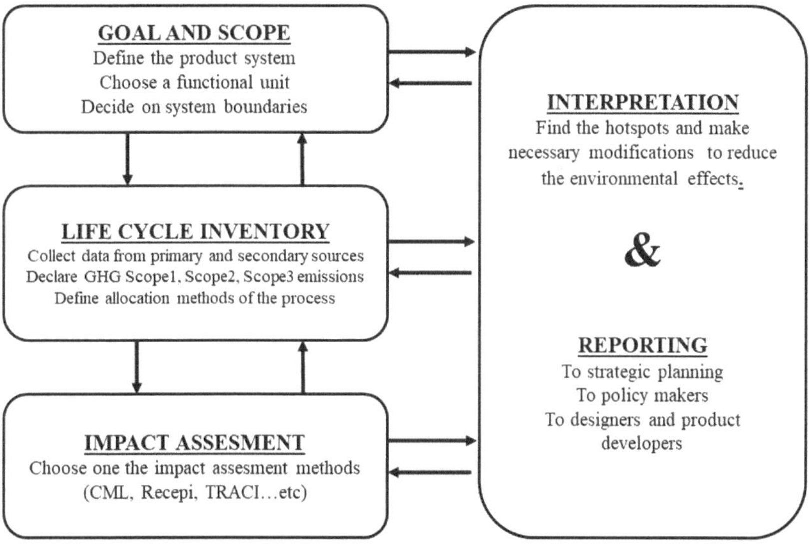

FIGURE 10.1 ISO 14044 LCA methodology. *LCA*, Life Cycle Assessment.

the literature that examine biogas production systems and conduct life cycle analyses to identify their environmental consequences.

10.2.1 Goal and scope definition for biogas production

The primary objective of LCA for biogas energy systems is to assess the environmental impacts associated with the production and utilization of biogas as an energy source. This includes evaluating the energy and material inputs required, as well as the emissions released into both the living and non-living components of the environment. Various LCA studies on biogas have distinct objectives, which may include one or more of the following (Fig. 10.1):

- Calculate the environmental impact of biogas system
- Compare the effect of different substrate sources
- Compare the effects of different disposal scenarios on environmental impact
- To make decision on process parameters to reduce environmental impact

The determination of system boundaries is based on the primary objective of the investigation. To implement a gate-to-gate strategy, data collection should start from the point of waste collection at the plant. The transportation of waste should be excluded from the analysis, and the boundary should be defined to conclude at the biogas production stage. When using the cradle-to-gate strategy, it is important to analyze the effects of waste transportation to

TABLE 10.1 LCA studies on biogas production.

Waste type	Reactor type	System boundaries	LCIA	GHG emissions (CO_2 eq.)	Software	Functional unit	References
Organic household waste, grease trap removal sludge, and ley crops	Anaerobic digester	Cradle to grave	CML	23.7 g	—	g CO_2	Shinde et al. (2021)
OFMSW (organic fraction of municipal solid waste)	Anaerobic digester	Cradle to grave	CML-IA	−5.67E +05 1.34E+06 kg	SimaPro	T biogas/t suspended soilds	Pasciucco et al. (2023)
MOW (municipal organic waste) and animal manure	CSTR	Cradle to grave	ILCD	49 g	SimaPro	1 Mj compressed biogas	van den Oever et al. (2021)
Animal waste, pig manure (c-bp), and cow manure (s-bp)	CSTR	Cradle to grave	CML	1.982 ton for c-bp and 5.67 ton for s-bp	SimaPro	Ton CO_2	Sinsuw et al. (2024)
Wastewater	Anaerobic digester	Cradle to grave	ReCiPe 2016	28.18– 18.05 kg	SimaPro	1 m^3 of treated water	Shih et al. (2021)
Wastewater	—	Cradle to gate	—	0.092 — 0.257 kg	SimaPro	1 t biomass	García et al. (2017)
Extraction of raw materials, production of inputs to the biofuel plant, processing and transportation of materials	Anaerobic digester	Cradle to gate	NS	~0.085 kg	SimaPro	105.263 tonnes of biomass and 1 MJ energy produced	Khoshnevisan et al. (2016)
Conventional farming system; open digestate storage	Anaerobic digester	Gate to gate	CML 2001	5.75E-03– 7.65E-03 kg	OpenLCA	1 MJ of bioenergy	Ramírez-Arpide et al. (2018)

Integrated industrial biogas plant	Anaerobic digester	Cradle to gate	Danish EDIP 2003	−291,591.44 kg	SimaPro	30 tons of cattle manure /Day 5 tons of poultry manure 15 tons of industrial biogas plant whey	Taylan et al. (2018)
Agro-industrial waste	Anaerobic digester (CSTR)	Cradle to gate	ReCiPe 2016	5.10E-03– 6.5E-03 kg	SimaPro	1 MJ biogas	Ugwu et al. (2022)
Palm oil mill effluent (POME)	Anaerobic digester	Cradle to gate	ReCiPe 2016	2.49E-08–1.42E-03kg	SimaPro	1 ton POME	Aziz et al. (2019)
Sewage sludge	Anaerobic digester	Gate to gate	ReCiPe 2016	−0,214 kg	SimaPro	1 m^3 biogas	Singh et al. (2020)

LCA, Life cycle assessment.

the plant. To eliminate the limitations on access, it is necessary to incorporate the electricity production stage using combined heat and power (CHP) or the regeneration stage. The scope of LCA studies can be expanded to include the cradle stage by incorporating the actions involved in utilizing the digestate as fertilizer (Esteves et al., 2019).

Biogas production systems can be initiated by implementing waste management practices. Both solid and liquid waste should be delivered to a biogas-producing plant. The biogas plant can be located within the production plant where waste is generated, as indicated by Hollas et al. (2021). Every LCA study on biogas production covers a goal step including anaerobic digestion. Different reactor types can be used for biogas production (Table 10.1).

A generalized schematic view of biogas production systems is presented. As it can be seen in the figure, the boundaries can be expanded from gate (biogas production) to usage of electricity such as CHP unit addition to produce electricity or biogas upgrading systems. For manure studies, the effect of storage time can be taken into consideration because manure has its own greenhouse gas-affecting character. Hollas et al. (2021) studied on the environmental effects of swine manure storage using three different scenarios. The first scenario was focused on direct biogas production in a covered lagoon, and second and third scenarios were based on liquid–solid separation and the biogas production in an anaerobic CSTR-type digester. The difference between the second and third scenarios was the hydraulic retention times (30 and 5 min). To calculate the environmental effect of reactor type on biogas production systems (pilot plant system using slaughter waste and commercial plant system using a mixture of pig manure and sludge), a cradle-to-grave approach was chosen (Sinsuw et al., 2024). Transport of manure to biogas system and transport of biogas is not taken into account. The biogas production system has two important outcomes: methane production and electricity production. Generally, these two functional units can be used to calculate the environmental impact: Biogas produced in metric tone or cubic meters or energy generation as MJ and kWh (Ramírez-Arpide et al., 2018). The functional unit can also be selected as the organic substrate used (Fig. 10.2).

10.2.2 Life cycle inventory for biogas production

This phase is of great importance as well. For inventory analysis, it is preferable to utilize the primary local data. However, if necessary, global (GLO) or area-based (RER—rest of the world) inventory data can also be obtained from software and databases like the Ecoinvent databases. One can extract the required data from national statistics databases (Cherubini et al., 2018).

The used raw material is one of the important data for inventory stage. Because biogas can be produced from a very wide range of waste materials,

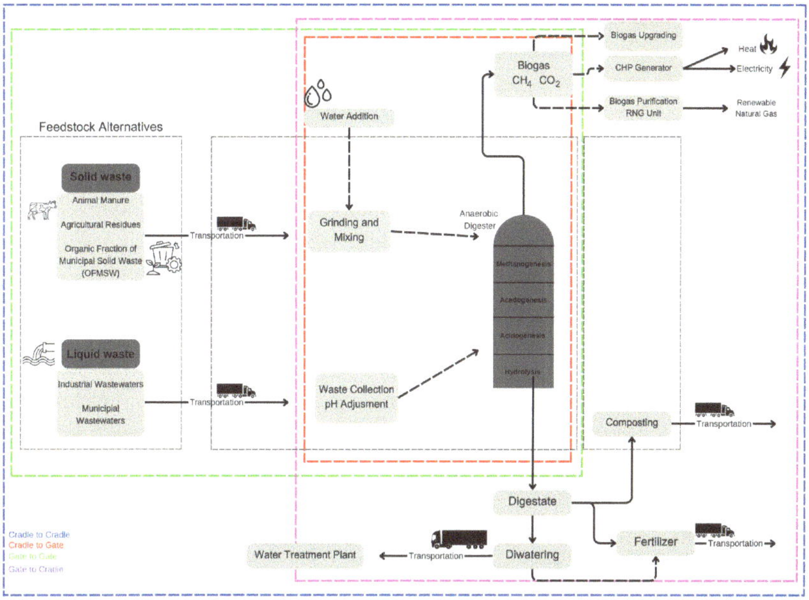

FIGURE 10.2 System boundaries for biogas production.

and every waste has different collection and transportation characteristics. The collection of food waste (Guven et al., 2018), sewage treatment wastes (Mainardis et al., 2021; Zhang & Xu, 2020), animal manure (Xue et al., 2019), or agricultural wastes (Xue et al., 2019) is different from each other in terms of inventory data. Some studies also use co-digestion processes and transporting wastes from different areas that are affecting the LCI data characterization. The reliability of LCI data is essential. The quality of data, as per the US LCI database development criteria, is dependent upon the collection time, its age, the source, the collecting methodology, and its representativeness. Statistical methods effectively resolve uncertainties and gaps in the data (Cooper & Kahn, 2012). In some biogas plants, the digestate is hygenized at plant area, and in some biogas plants, the digestate is transported to another plant to be converted into fertilizer (Hamelin et al., 2010). Some biogas plants directly burn the produced gas at flame, while othersconvert into electricity using a CHP generator, and in some cases, biogas is further processed for regeneration (Fig. 10.2). Different processes come up with different inventory data and affect final LCA environmental impact values. Three different kinds of transportation should be taken into account: transport of wastes to plant, transport of biogas to end user, or transport of digestate to fertilizer-producing plants. Biogas reactor is another important parameter for LCI. Anaerobic digesters can have different properties and capacities: batch-type reactors (Hollas et al., 2021; Mainardis et al., 2021), two-stage anaerobic digester (Sinsuw et al., 2024), and continuous stirred tank reactor (CSTR) (Li et al., 2021; Singh et al., 2020; Ugwu et al., 2022). UASB-type reactors and hybrid reactors can be used. The last step of inventory analysis covers the final use of biogas. Biogas can be converted into electricity and heat using CHP electricity generator or can be upgraded to use as natural gas. There are many biogas upgrading methods, and different upgrading methods have different environmental impact values (Starr et al., 2012). Another important output of the system to be considered is the use of digestate. Digestate can be directly send to a fertilizer production plant, hygenized at plant, or pyrolyzed to convert into biochar. These different processes have different energy needs in terms of electricity consumption. The final usage can also impact the environmental impact value. Utilizing fertilizers or charcoal can enhance the system analysis in a cradle-to-cradle approach and reduce the environmental impact values.

10.2.3 Life cycle impact assessment for biogas production

LCIA can be performed by using different methodologies. The generally used impact categories are as follows global warming, ozone depletion, acidification, eutrophification, photochemical oxidation potential, and water depletion. Different studies examine different environmental impacts. In literature, CML 2001 method is the mostly used method. This impact methodology covers long-term effects of the system (Acero et al., 2014). Another impact methodology is developed by Intergovernmental Panel on Climate Change (IPCC). Recepi method is also one of the mostly used method. It is a combination of eco-indicator and CML methods. The different point of Recepi methodology is having two levels of calculation: midpoint and endpoint. The midpoint method analyzes cause–effect chain and reaches out to the effect on human health, damage to ecosystem, and damage to resources. Generally, midpoint method is used, but some studies can calculate endpoint impact to analyze actual damages. The common impact category for all biogas production studies is global warming potential (in CO_2 eq) (Esteves et al., 2019).

10.2.3.1 General concept for life cycle analysis in biogas studies

In recent years, in LCA studies on biogas production, system boundaries have expanded from cradle to grave with increasing technological opportunities and the use of digestate in fertilizer or biochar production (Pasciucco et al., 2023; Shih et al., 2021; Shinde et al., 2021; Sinsuw et al., 2024; van den Oever et al., 2021). Different global warming results were found depending on the type of waste used in the processes and the anaerobic reactor difference (1.42×10^{-3} kg to 49 g). As it can be seen in Table 10.1, the most popular software in recent years is *Sima Pro*. One study used IPCC calculation methods (Shinde et al., 2021), and one study used open LCA (Ramírez-Arpide et al., 2018). Recepi and CML are the most used impact assessment methods. Some studies used the organic substrate as a functional unit (Aziz et al., 2019; García et al., 2017; Khoshnevisan et al., 2016; Shih et al., 2021; Taylan et al., 2018), and some studies used energy values in MJ (Singh et al., 2020; Ugwu et al., 2022). Looking at the GHG emission values in the table, the CO_2 value of 818 kg for biohydrogen production from palm oil has been reduced to 189 kg with improvements made in the process. In wheat straw, there was a CO_2 production of 5.60 and 5.32 kg from sweet sorghum stalk and 5.18 kg from steam potato. In solid wastes such as municipal waste, CO_2 emissions between 2.13 and 1.12×10^4 kg have been calculated (Table 10.2).

10.3 Life cycle assessment for biohydrogen production

Hydrogen shows great potential as an energy carrier, particularly for use in heating and transportation systems. Hydrogen production technologies are experiencing rapid global growth. Hydrogen technologies have significant potential in achieving the net zero emission target (Amaya-Santos et al., 2021). Biohydrogen derived from biomass is gaining more recognition due to its energy-efficient, eco-friendly, and carbon-neutral characteristics in comparison with conventional methods of hydrogen production, such as water electrolysis and hydrocarbon fuel reformation. The present methods for utilizing biomass consist of thermochemical techniques such as pyrolysis and gasification, as well as biological methods such as biophotolysis, biological water gas shift reaction [lower calorific value (LCV): 3.00 kWh/Nm^3, sand higher calorific value (HCV) 3.55 kWh/Nm^3, of H_2 compared with CH_4], and dark fermentation (Ni et al., 2006). Biological processes are often conducted under normal temperatures and pressures, making them less energy-intensive and more environmentally appealing compared with thermochemical processes (Chelvam et al., 2024). Despite the benefits of utilizing waste biomass and occurring at ambient temperatures, biohydrogen synthesis suffers from the drawback of low yield improvements (3.26–1544) (El Bari et al., 2022). Although the majority of hydrogen technologies have reached the TRL 7–9 level, biohydrogen technologies are now at the TRL 3–4 level. Implementing strategies to enhance low yields will improve the practical use of the technology (Wilkinson et al., 2023).

Biohydrogen production can be divided into subsections such as dark fermentation, photofermentation, biophotolysis, and hybridization of these technologies (Barghash et al., 2022; Morya et al., 2022). Among various biohydrogen production technologies, this book chapter focuses on technologies that convert organic waste into energy carriers (see Chapter 9).

TABLE 10.2 LCA studies on biohydrogen production.

Waste type	Reactor type	System boundaries	LCIA	GHG emissions (CO_2 eq)	Software	Functional unit	References
Palm oil wastewater	UASFF reactor (upflow anaerobic sludge filter fermenter)	Cradle to grave	ReCiPe 2016	818 to 189 kg	SimaPro	1 kg biohydrogen production	Akhbari et al. (2021)
Apple processing plant wastes	Batch reactor	Cradle to gate	ReCiPe 2016	—	SimaPro	1 m^3 biohydrogen production	Baş et al. (2023)
Biowaste and biomass	Rotating cone reactor	Gate to gate	CML-IA,2001	2.13E+00–1.54E+00	SimaPro	1 kg biohydrogen production	Arfan et al. (2023)
Wheat straw (WS-H_2), sweet sorghum stalk (SSS-H_2), and steam potato peels (SPP-H_2)	Bioreactor	Gate to gate	CML 2002+	5.60 kg for WS-H_2, 5.32 kg for SSS-H_2, and 5.18 kg for SPP-H_2	SimaPro	460 tons of biohydrogen production in 1 year	Djomo and Blumberga (2011)

Solid waste (MSW)	Gas shift reactors	Cradle to gate	Environmental Footprint 3.0	–284,16 to 294,78 kg	Gabi	1 MWh is the production of transport class H2	Amaya-Santos et al. (2021)
Solid waste and waste wood	Anaerobic reactor	Gate to gate	CML 2001	4.99 kg for solid waste 4.11 kg for waste wood	Gabi	1 kg biohydrogen production	Lui et al. (2022)
Microalgal biomass cultivated	Anaerobic fermentation reactor	Cradle to gate	—	5–6 kg	SimaPro	1 MJ of biohydrogen produced	Ferreira et al., 2013
Sugar beet molasses (SBM), cheese whey (CW), wine vinasses mixed with wastewater treatment plant sludge (WVS)	Anaerobic membrane bioreactor	Cradle to gate and Gate to gate	ReCiPe Midpoint and Endpoint	3.56 kg for SBM, 39.74 kg for CW, and 9.13 kg for WVS	SimaPro	1 Nm^3 biohydrogen production	Camacho et al. (2022)
Sugarcane and potato peels	Thermo reactor and photoreactor	Cradle to cradle	—	24–31 g for sugarcane and 60–77 g for potato peels	SimaPro	1 MJ of biohydrogen produced	Ferreira et al. (2011)
Sludge and treated effluent (TE)	Batch reactor	Gate to cradle	ReCiPe 2016	1.12×10^4 kg	OpenLCA	1 m^3 biohydrogen production	Barghash et al. (2022)

(Continued)

TABLE 10.2 (Continued)

Waste type	Reactor type	System boundaries	LCIA	GHG emissions (CO_2 eq)	Software	Functional unit	References
Potato steam peels	Two-stage bioreactor	Gate to gate	Impact 2002+ v.2.1	1000 to 1500 g CO2 per kilogram of hydrogen	SimaPro	1 kg of biohydrogen produced	(Djomo et al., 2008)
Rice husk	Anaerobic reactor	Gate to gate	CML 2001	50 kg	OpenLCA	1 kg of biohydrogen produced	Reaño (2020)

LCA, Life cycle assessment.

10.3.1 Goal and scope definition for biohydrogen production

LCA studies for biohydrogen production also used ISO 14400 Standards. Due to the restricted application of pilot scale studies, the primary objective of LCA studies was to assess the environmental impact to gain insights into potential chances for scaling up. The environmental impact of biohydrogen technology was assessed, and areas of notable impact were identified. In general, laboratory-scale reactors were used to conduct LCA studies, with gate-to-gate system boundaries (Djomo & Blumberga, 2011). As an example of a pilot scale application, Akhbari et al. (2021) used a cradle-to-grave approach for biohydrogen production from palm oil effluent. Technologies for biohydrogen production from waste materials are still in the process of being developed. Biohydrogen production systems continue to utilize various reactor designs and process parameters (see Chapter 9). Another potential usage involves the utilization of catalytic materials and adsorbents. When conducting LCA calculations, it is important to consider the impact of the materials used. Seo et al. (2019) conducted a LCA of zeolite A, Has-Clay, and Kanuma clay, evaluated their environmental impact, and found that the Klauma Clay has the lowest environmental impact. Dark fermentation technologies are highly efficient in converting waste into biohydrogen, and they have the advantage of requiring minimal energy. Utilizing end products (namely organic acids) as substrates in photofermentative hydrogen production during dark fermentation technologies has the dual benefit of improving hydrogen yields and effectively utilizing waste, hence reducing environmental impacts (Fig. 10.3). Scope definitions in biohydrogen production generally include calculations within the system boundaries, including the impact of the materials used on a laboratory scale, the characteristic features of the waste material used, and the energy and water needs of the system (from cradle to gate) (Fig. 10.3).

10.3.2 Life cycle inventory for biohydrogen production

The LCA procedures for biogas or biohydrogen production, which rely on organic substrates and microbial fermentation, are analogous. The disposal of the substrate source and the residual material in the bioreactor postfermentation is crucial in these processes. The initial stage to be examined in the inventory analysis of biohydrogen generation is the waste material phase. Dark fermentation offers the opportunity to engage in various biohydrogen conversions, particularly those involving waste materials. Typically, high biohydrogen conversion efficiencies are attained when using wastes that contain accessible carbohydrates, proteins, and lipids. Several of these organic wastes are as follows: apple processing waste (Baş et al., 2023), potato steam peels (Djomo et al., 2008), sugar beet molasses and cheese whey (Camacho et al., 2022), sugarcane and potato peels(Ferreira et al., 2011), and microalgal biomass (Ferreira et al., 2013). Various processes are employed to generate hydrogen from lignocellulosic wastes, which are the most abundant forms of organic waste globally. These technologies typically include thermochemical processes that have a high energy content. Thermochemical processes have undergone the most advanced technological verifications, regardless of their enormous energy content. There are examples reported in the literature when lignocellulosic waste materials are utilized to produce biohydrogen. Some examples are as follows: biohydrogen production from solid waste (Amaya-Santos et al., 2021), waste wood (Lui et al., 2022), and rice husk (Reaño, 2020).

Reactor selection is a crucial factor to consider when calculating inventory, as it has a direct impact on energy usage. Although anaerobic fermenters and batch reactors are the most frequently utilized types of reactors, other bioreactor options, such as rotating cone reactors (Arfan et al., 2023) and upflow reactors (Akhbari et al., 2021), can also be employed. The inventory data for each of these items vary and have an impact on LCA calculations.

Various databases can be utilized for inventory calculations in biohydrogen generation. Inventory values can be sourced either from local databases or obtained from the databases of existing software. The most popular software utilized in studies on biohydrogen production is *Sima Pro*; however, the literature also includes work with *Gabi* and *OpenLCA* (Table 10.2).

10.3.3 Life cycle impact assessment for biohydrogen production

Global warming is the typical impact category examined in LCA research on biohydrogen production and in other biofuel studies. In addition to this, acidification, ecotoxicity in freshwater, eutrophication in freshwater, photochemical ozone production, and impacts on human health are other frequently observed areas of environmental effect (Amaya-Santos et al., 2021). The impact category to be examined also varies depending on the process flow, raw material, and inventory. For example, in a pilot scale study conducted with real palm oil effluent, in addition to the above impact categories, land use, terrestrial ecotoxicity, human noncarcinogenic activity, and mineral resource scarcity categories should also be examined (Akhbari et al., 2021). The study investigated the influence of various organic wastes on resource depletion, human health, ecosystem quality, and water usage (Djomo & Blumberga, 2011). When performing a study on biohydrogen production using electrical energy from a PV panel, it is important to examine the effects of metal depletion and ionic radiation (Barghash et al., 2022). These and several comparable instances can be found in the literature. The crucial aspect in selecting the impact category is to ensure a clear definition of the process and the ability to accurately forecast its environmental impacts.

10.4 Cost and benefit analysis for biogas and biohydrogen systems

While biogas and biohydrogen technologies offer ecologically acceptable solutions by utilizing waste as a substrate to address the climate crisis, their affordability must be carefully assessed during the investment decision-making process. It is important to thoroughly analyze parameters such as economic risks, government incentives, profit rates, and payback time (Menind & Olt, 2009).

For a successful cost analysis, engineering first CAPEX (Construction and Infrastructure Costs) and OPEX (Operational Costs) should be calculated.

CAPEX components can be property purchase, land usage, design and installation cost of anaerobic digester, automation level, type of reactor, setting up a laboratory infrastructure, etc.

The formula for CAPEX (Sánchez-Martín et al., 2023) is as follows:

$$CAPEX\ Flow\frac{Nm^3}{h}x\ Investment\ cost\frac{€}{Nm^3/h}$$

OPEX components that are related to the administration are labor cost to follow-up analysis and the operation, maintenance of equipment, consumables such as basal medium chemicals, plastic and glass equipment that are used for analysis, transport of feedstock to plant (sometimes fees for purchasing feedstock can be added if needed), insurance, administrative costs, plant maintenance, etc. (Gwavuya et al., 2012). Total biogas produced, net biogas production, and avoided CO_2 values will be used to calculate the total revenue values (Rasheed et al., 2016). Biogas can be converted directly into electricity using a CHP generator, and the following formula can directly be used to

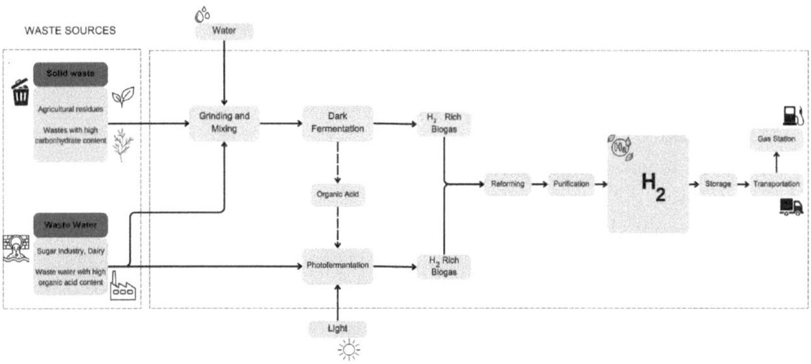

FIGURE 10.3 LCA for biohydrogen systems (system boundaries). *LCA*, Life cycle assessment.

calculate the electricity benefit. For biohydrogen production, the electricity conversion using a fuel cell can be taken into account.

The biogas production can be calculated as follows (Sánchez-Martín et al., 2023):

$$Tot.\ biogas\ prod. \left(\frac{m^3}{year}\right) = Volatile\ solid\ (VS)\left(\frac{kg}{year}\right) x\ Biogas\ yield \left(\frac{m^3}{kgVS}\right)$$

Energy benefit calculation formula (Sánchez-Martín et al., 2023) is as follows:

$$Electricity\ from\ biogas \left(\frac{kwh}{year}\right) = Tot.\ biogas\ prod \left(\frac{m^3}{year}\right) x\ Calorific\ value \left(\frac{kwh}{year}\right)$$

These plants are using waste and converting it into sustainable energy. Therefore, we can calculate the avoided CO_2 as follows (Sánchez-Martín et al., 2023):

$$CO_{2\,avoided}\left(\frac{t}{year}\right) = Emission\ factor \left(\frac{kg\ CO_2}{kwh}\right) xTot.\ biogas\ prod. \left(\frac{kwh}{year}\right)$$

Emission factor can be dependent on country's regulations and from different databases, but an average value for emission factor is 0.182 kg CO_2/kWh. The avoided CO_2 value can be used to calculate the CO_2 revenue value. The price of CO_2 will depend again on countries regulations (Sánchez-Martín et al., 2023).

$$CO_{2\,avoided}\left(\frac{€}{year}\right) = CO_{2\,avoided}\left(\frac{t}{year}\right) x\ CO_{2\,cost}\left(\frac{€}{t}\right)$$

The difference between these costs and revenues will be the net profit value. For a feasibility study, OPEX costs can be calculated by 1%–10% of CAPEX (Sánchez-Martín et al., 2023). Investment costs can also be calculated from COD removal ratio (Wattanasilp et al., 2021).

Cost–benefit analysis is a useful tool for forecasting the behavior of economic agents and is considered a positive attribute. The normative property of something refers to its ability to prescribe or dictate behavior. One of the most significant benefits for decision-makers is having prior knowledge of the time required to payback the initial investment. When every component is thoroughly evaluated in a cost–benefit analysis, it will serve as a viable implementation strategy for governments and investors alike (Renard, 1988).

Simply cost–benefit can be calculated by using the formula below (Wattanasilp et al., 2021):

$$Benefit\ Cost\ Ratio\,(BCR) = \frac{Value\ of\ expected\ benefits}{Value\ of\ expected\ costs}$$

If the BCR is greater than 1, it indicates that the project's cash flow will exceed its costs, which is indicative of an appropriate financial decision. The general components of a biogas of biohydrogen plant are listed in Table 10.3

Gittinger (1972) suggested to use of net present value (NPV) and internal rate of return (IRR) for cost–benefit analysis of a plant. IRR is a calculation method considering the discounted flow rate (Rasheed et al., 2016).

$$NPV = \sum_{t=1}^{n} \frac{Bt - Ct}{(1 + i)^t} \tag{10.1}$$

$$IRR\ (\%):\ 0 = \sum_{t=1}^{n} \frac{Bt - Ct}{(1 + i)^t}$$

where Bt: benefit during a year; Ct: cost during a year; i: discount value and t: lifetime of the biogas plant (can be assumed at least 20 years). NPV is the value of positive and negative cash flows of the investment life cycle. It is a common method used to analyze capital budgets. The advantage of NPV over IRR is the detailed calculations of all aspects of cost components. In case to select one project from several options, it is better to use NPV because IRR cannot analyze multiple discount rates. For short-term projects, IRR is better, while NPV is more useful for longer projects. IRR can be used to calculate the breakeven point, and NPV shows a longer-time profit.

When investing in a plant, it is crucial to thoroughly evaluate the potential profits that can be obtained from the establishment and perform meticulous calculations. Country regulations and laws also influence the profit calculations associated with plant investments. Effective cost–benefit analyses are an essential requirement for making profitable investments (Rasheed et al., 2016; Sánchez-Martín et al., 2023; Wattanasilp et al., 2021; Xue et al., 2019).

10.5 Conclusions and future prospects

Biogas production has achieved a technology readiness level (TRL) of 9, indicating that it is fully developed and already commercialized. On the contrary, biohydrogen production is still at a TRL of 3–4, indicating that it is in the early stages of development and not yet ready for widespread use. Thorough and accurate LCA studies have been carried out on the generation of biogas. Nevertheless, LCA studies on biohydrogen production have primarily been carried out on a small scale in laboratory settings. The environmental effect values derived from LCA study in biogas production plants vary. The variation in substrates employed across different research, and

TABLE 10.3 Components of a cost–benefit analysis.

Costs for a plant	Revenues from plant
Capital costs Design and engineering Property purchases Plant installation (bioreactors, pumps, compressors, scrubbers, etc.) Preliminary supplies	Revenues from heat sales Revenues from biogas sales Revenues from digestate sales Revenues from electricity sales Revenues from carbon credit sales
Operating costs Maintenance costs Labor Salaries Feedstock transportation Overhead (management, administration)	

the differences in system boundaries during calculations account for this difference. The selection of the waste substrate, the closeness of the waste to the plant, and the utilization of the biogas end product for heating or electricity are the key factors in minimizing the environmental impact of biogas production. Any enhancements made to the estimations of environmental effects will have a beneficial impact on the cost analysis. For instance, incorporating a renewable energy source, like a solar panel, into a system that relies on biogas plants for power production can effectively decrease both the carbon footprint and the expenses. Similarly, in LCA calculations that begin at the waste's origin, the primary environmental impact arises from the emission of greenhouse gases throughout the waste's transportation to the facility. Implementing a sustainable transportation alternative will significantly contribute to decreasing the overall environmental effect and further reduce the related expenses. Increased biohydrogen production yields, improved technology, and reduced costs will result in both lower environmental impact and improved cost benefits. Considering that anaerobic technologies can be used in both production processes, the end product in biogas production still contains CO_2 (Filippa et al., 2020). Therefore, biohydrogen is a much more advantageous energy source in terms of environmental impact. Integrating the well-established technology of biogas into biohydrogen technology for future research will offer significant advantages in terms of environmental effects, LCA calculations, and costs. Climate change is a crucial concern within the United Nations Development Goals and sustainability goals. Due to the increasing industrialization and rapid consumption patterns, there has been a substantial rise in greenhouse gas emissions. Currently, relying solely on renewable energy sources is not enough. Measuring our environmental effects using LCA and implementing the required modifications in our production systems to consistently decrease them till achieving carbon neutrality will be a crucial measure to ensure a clean environment for future generations. Another important factor is to understand clearly the cost effect of these technologies by making a detailed cost and benefit analysis. A biogas plant investment has several advantages, but a cost–benefit analysis should be used to clearly understand the profitability. Profitability could depend on many factors, such as national regulations, labor options, and feedstock availability.

LCA analysis and cost–benefit analysis should be used together to understand and quantify the numerical impacts of biogas or biohydrogen plants because for sustainability, economic, environmental, and social factors are always must be in harmony with each other.

AI disclosure

During the preparation of this work, the author(s) used Quillbot.com (premium) to improve English language by using paraphrasing tool. After using this tool/service, the author(s) reviewed and edited the content as needed and take(s) full responsibility for the content of the publication.

References

Acero, A. P., Rodríguez, C., & Ciroth, A. (2014). *LCIA methods impact assessment methods in life cycle assessment and their impact categories*. GreenDelta GmbH.

Akhbari, A., Onn, C. C., & Ibrahim, S. (2021). Analysis of biohydrogen production from palm oil mill effluent using a pilot-scale up-flow anaerobic sludge blanket fixed-film reactor in life cycle perspective. *International Journal of Hydrogen Energy, 46*(68), 34059–34072. https://doi.org/10.1016/j.ijhydene.2021.07.186, http://www.journals.elsevier.com/international-journal-of-hydrogen-energy/.

Alpaydın, Ö. (2014). Gaziantep ili için entegre katı atık yönetiminde yaşam döngüsü değerlendirmesi uygulanması.

Amaya-Santos, G., Chari, S., Sebastiani, A., Grimaldi, F., Lettieri, P., & Materazzi, M. (2021). Biohydrogen: A life cycle assessment and comparison with alternative low-carbon production routes in UK. *Journal of Cleaner Production, 319*, 128886. https://doi.org/10.1016/j.jclepro.2021.128886.

Arfan, M., Eriksson, O., Wang, Z., & Soam, S. (2023). Life cycle assessment and life cycle costing of hydrogen production from biowaste and biomass in Sweden. *Energy Conversion and Management, 291*. https://doi.org/10.1016/j.enconman.2023.117262 117262.

Aziz, N. I. H. A., Hanafiah, M. M., & Gheewala, S. H. (2019). A review on life cycle assessment of biogas production: Challenges and future perspectives in Malaysia. *Biomass and Bioenergy, 122*, 361–374. https://doi.org/10.1016/j.biombioe.2019.01.047, http://www.journals.elsevier.com/biomass-and-bioenergy/.

Banar, M., Cokaygil, Z., & Ozkan, A. (2009). Life cycle assessment of solid waste management options for Eskisehir, Turkey. *Waste Management, 29*(1), 54–62. https://doi.org/10.1016/j.wasman.2007.12.006.

Barghash, H., AlRashdi, Z., Okedu, K. E., & Desmond, P. (2022). Life-cycle assessment study for bio-hydrogen gas production from sewage treatment plants using solar PVs. *Energies, 15*(21). https://doi.org/10.3390/en15218056, http://www.mdpi.com/journal/energies/.

El Bari, H., Lahboubi, N., Habchi, S., Rachidi, S., Bayssi, O., Nabil, N., Mortezaei, Y., & Villa, R. (2022). Biohydrogen production from fermentation of organic waste, storage and applications. *Cleaner Waste Systems, 3*. https://doi.org/10.1016/j.clwas.2022.100043.

Baş, F., Gökçek, Ö. B., Muratçobanoğlu, H., & Demirel, S. (2023). *Life Cycle Assessment (LCA) of the Effect of Magnetite Addition on Biohydrogen Production* (2nd International Conference on Engineering.).

Camacho, C. I., Estévez, S., Conde, J. J., Feijoo, G., & Moreira, M. T. (2022). Dark fermentation as an environmentally sustainable win-win solution for bioenergy production. *Journal of Cleaner Production, 374*. https://doi.org/10.1016/j.jclepro.2022.134026, https://www.journals.elsevier.com/journal-of-cleaner-production.

Chelvam, K., Hanafiah, M. M., Woon, K. S., & Ali, K. A. (2024). A review on the environmental performance of various hydrogen production technologies: An approach towards hydrogen economy. *Energy Reports, 11*, 369–383. https://doi.org/10.1016/j.egyr.2023.11.060.

Cherubini, E., Franco, D., Zanghelini, G. M., & Soares, S. R. (2018). Uncertainty in LCA case study due to allocation approaches and life cycle impact assessment methods. *International Journal of Life Cycle Assessment, 23*(10), 2055–2070. https://doi.org/10.1007/s11367-017-1432-6, http://www.springerlink.com/content/0948-3349.

Cooper, J. S., & Kahn, E. (2012). Commentary on issues in data quality analysis in life cycle assessment. *International Journal of Life Cycle Assessment, 17*(4), 499–503. https://doi.org/10.1007/s11367-011-0371-x.

Daş, İ. T. O., Özmıhçı, S., & Büyükkamacı, N. (2024). Environmental impact analysis of different wastes to biohydrogen, biogas and biohytane processes. *International Journal of Hydrogen Energy, 56*, 1446–1463. https://doi.org/10.1016/j.ijhydene.2023.12.184, https://www.sciencedirect.com/science/journal/03603199.

Djomo, S. N., & Blumberga, D. (2011). Comparative life cycle assessment of three biohydrogen pathways. *Bioresource Technology, 102*(3), 2684–2694. https://doi.org/10.1016/j.biortech.2010.10.139.

Djomo, S. N., Humbert, S., & Blumberga, D. (2008). Life cycle assessment of hydrogen produced from potato steam peels. *International Journal of Hydrogen Energy, 33*(12), 3067–3072. https://doi.org/10.1016/j.ijhydene.2008.02.006.

Esteves, E. M. M., Herrera, A. M. N., Esteves, V. P. P., & Morgado, Cd. R. V. (2019). Life cycle assessment of manure biogas production: A review. *Journal of Cleaner Production, 219*, 411–423. https://doi.org/10.1016/j.jclepro.2019.02.091, https://www.journals.elsevier.com/journal-of-cleaner-production.

Ferreira, A. F., Ortigueira, J., Alves, L., Gouveia, L., Moura, P., & Silva, C. (2013). Biohydrogen production from microalgal biomass: Energy requirement, CO_2 emissions and scale-up scenarios. *Bioresource Technology, 144*, 156–164. https://doi.org/10.1016/j.biortech.2013.06.079, http://www.elsevier.com/locate/biortech.

Ferreira, A. F., Ribau, J. P., & Silva, C. M. (2011). Energy consumption and CO_2 emissions of potato peel and sugarcane biohydrogen production pathways, applied to Portuguese road transportation. *International Journal of Hydrogen Energy, 36*(21), 13547–13558. https://doi.org/10.1016/j.ijhydene.2011.08.008.

Filippa, F., Panara, F., Leonardi, D., Arcioni, L., & Calderini, O. (2020). Life cycle assessment analysis of alfalfa and corn for biogas production in a farm case study. *Processes, 8*(10), 1285. https://doi.org/10.3390/pr8101285.

García, C. A., Morales, M., Quintero, J., Aroca, G., & Cardona, C. A. (2017). Environmental assessment of hydrogen production based on Pinus patula plantations in Colombia. *Elsevier Ltd, Colombia Energy, 139*, 606–616. https://doi.org/10.1016/j.energy.2017.08.012, http://www.elsevier.com/inca/publications/store/4/8/3/.

Gittinger, J. P. (1972). *Economic Analysis of Agricultural Projects*. Baltimore, Maryland, London, UK: Johns Hopkins University Press. https://www.cabidigitallibrary.org/doi/full/10.5555/19721800195.

Guven, H., Eriksson, O., Wang, Z., & Ozturk, I. (2018). Life cycle assessment of upgrading options of a preliminary wastewater treatment plant including food waste addition. *Water Research, 145*, 518–530. https://doi.org/10.1016/j.watres.2018.08.061, http://www.elsevier.com/locate/watres.

Gwavuya, S. G., Abele, S., Barfuss, I., Zeller, M., & Müller, J. (2012). Household energy economics in rural Ethiopia: A cost-benefit analysis of biogas energy. *Renewable Energy, 48*, 202–209. https://doi.org/10.1016/j.renene.2012.04.042.

Hamelin, L, Wesnæs, M., Wenzel, H., & Molt, B. (2010). Environmental project life cycle assessment of biogas from separated slurry. https://www2.mst.dk/udgiv/publications/2010/978-87-92668-03-5/pdf/978-87-92668-04-2.pdf.

Hauschild, M. Z. (2017). Introduction to LCA methodology. In *Life cycle assessment: Theory and practice*. Denmark: Springer International Publishing, 59–66. https://doi.org/10.1007/978-3-319-56475-3_6.

Hijazi, O., Munro, S., Zerhusen, B., & Effenberger, M. (2016). Review of life cycle assessment for biogas production in Europe. *Renewable and Sustainable Energy Reviews, 54*, 1291–1300. https://doi.org/10.1016/j.rser.2015.10.013, https://www.journals.elsevier.com/renewable-and-sustainable-energy-reviews.

Hollas, C. E., Bolsan, A. C., Chini, A., Venturin, B., Bonassa, G., Cândido, D., Antes, F. G., Steinmetz, R. L. R., Prado, N. V., & Kunz, A. (2021). Effects of swine manure storage time on solid-liquid separation and biogas production: A life-cycle assessment approach. *Renewable and Sustainable Energy Reviews, 150*. https://doi.org/10.1016/j.rser.2021.111472 111472.

Khoshnevisan, B., Shafiei, M., Rajaeifar, M. A., & Tabatabaei, M. (2016). Biogas and bioethanol production from pinewood pre-treated with steam explosion and N-methylmorpholine-N-oxide (NMMO): A comparative life cycle assessment approach. *Energy, 114*, 935–950. Available from: https://doi.org/10.1016/j.energy.2016.08.024, http://www.elsevier.com/inca/publications/store/4/8/3/.

Klöpffer, W. (1997). Life cycle assessment: From the beginning to the current state. *Environmental Science and Pollution Research, 4*(4), 223–228. https://doi.org/10.1007/BF02986351, http://www.springerlink.com/content/0944-1344.

Li, J., Xiong, F., & Chen, Z. (2021). An integrated life cycle and water footprint assessment of nonfood crops based bioenergy production. *Scientific Reports, 11*(1). https://doi.org/10.1038/s41598-021-83061-y, http://www.nature.com/srep/index.html.

Lui, J., Sloan, W., Paul, M. C., Flynn, D., & You, S. (2022). Life cycle assessment of waste-to-hydrogen systems for fuel cell electric buses in Glasgow, Scotland. *Bioresource Technology, 359*. https://doi.org/10.1016/j.biortech.2022.127464 127464.

Mainardis, M., Buttazzoni, M., Gievers, F., Vance, C., Magnolo, F., Murphy, F., & Goi, D. (2021). Life cycle assessment of sewage sludge pretreatment for biogas production: From laboratory tests to full-scale applicability. *Journal of Cleaner Production, 322*. https://doi.org/10.1016/j.jclepro.2021.129056 129056.

Menind, A., & Olt, J. (2009). Biogas plant investment analysis, cost benefit and main factors. In *Conference Proceeding*. Engineering For Rural Development.

Morya, R., Raj, T., Lee, Y., Pandey, A. K., Kumar, D., Singhania, R. R., Singh, S., Verma, J. P., & Kim, S. H. (2022). Recent updates in biohydrogen production strategies and life-cycle assessment for sustainable future. *Bioresource Technology, 366*. https://doi.org/10.1016/j.biortech.2022.128159, http://www.elsevier.com/locate/biortech.

Ni, M., Leung, D. Y. C., Leung, M. K. H., & Sumathy, K. (2006). An overview of hydrogen production from biomass. In: *Fuel Processing Technology, 87*(5), 461–472. https://doi.org/10.1016/j.fuproc.2005.11.003.

van den Oever, A. E. M., Cardellini, G., Sels, B. F., & Messagie, M. (2021). Life cycle environmental impacts of compressed biogas production through anaerobic digestion of manure and municipal organic waste. *Journal of Cleaner Production, 306*. https://doi.org/10.1016/j.jclepro.2021.127156, https://www.journals.elsevier.com/journal-of-cleaner-production.

Pasciucco, F., Francini, G., Pecorini, I., Baccioli, A., Lombardi, L., & Ferrari, L. (2023). Valorization of biogas from the anaerobic co-treatment of sewage sludge and organic waste: Life cycle assessment and life cycle costing of different recovery strategies. *Journal of Cleaner Production, 401*. https://doi.org/10.1016/j.jclepro.2023.136762 136762.

Ramírez-Arpide, F. R., Demirer, G. N., Gallegos-Vázquez, C., Hernández-Eugenio, G., Santoyo-Cortés, V. H., & Espinosa-Solares, T. (2018). Life cycle assessment of biogas production through anaerobic co-digestion of nopal cladodes and dairy cow manure. *Journal of Cleaner Production, 172*, 2313–2322. https://doi.org/10.1016/j.jclepro.2017.11.180.

Rasheed, R., Khan, N., Yasar, A., Su, Y., & Tabinda, A. B. (2016). Design and cost-benefit analysis of a novel anaerobic industrial bioenergy plant in Pakistan. *Renewable Energy, 90*, 242–247. https://doi.org/10.1016/j.renene.2016.01.008, http://www.journals.elsevier.com/renewable-and-sustainable-energy-reviews/.

Reaño, R. L. (2020). Assessment of environmental impact and energy performance of rice husk utilization in various biohydrogen production pathways. *Bioresource Technology, 299*. Available from: https://doi.org/10.1016/j.biortech.2019.122590, http://www.elsevier.com/locate/biortech.

Renard, R. (1988). The economic costs and benefits of biogas projects in developing countries. *Mircen Journal of Applied Microbiology and Biotechnology, 4*(1), 67–80. https://doi.org/10.1007/BF00936816.

Roy, P., Nei, D., Orikasa, T., Xu, Q., Okadome, H., Nakamura, N., & Shiina, T. (2009). A review of life cycle assessment (LCA) on some food products. *Journal of Food Engineering, 90*(1), 1–10. https://doi.org/10.1016/j.jfoodeng.2008.06.016.

Seo, Y., Suzuki, M., Takagi, T., & Dowaki, K. (2019). Life-cycle assessment of adsorbents for biohydrogen production. *Resources, 8*(1), 52. https://doi.org/10.3390/resources8010052.

Shih, M.-F., Lay, C.-H., Lin, C.-Y., & Chang, S.-H. (2021). Exploring the environmental and economic potential for biogas production from swine manure wastewater by life cycle assessment. *Clean Technologies and Environmental Policy*. https://doi.org/10.1007/s10098-021-02157-1.

Shinde, A. M., Dikshit, A. K., Odlare, M., Thorin, E., & Schwede, S. (2021). Life cycle assessment of bio-methane and biogas-based electricity production from organic waste for utilization as a vehicle fuel. *Clean Technologies and Environmental Policy, 23*(6), 1715–1725. https://doi.org/10.1007/s10098-021-02054-7, https://link.springer.com/journal/10098.

Singh, A. D., Upadhyay, A., Shrivastava, S., & Vivekanand, V. (2020). Life-cycle assessment of sewage sludge-based large-scale biogas plant. *Bioresource Technology, 309*. https://doi.org/10.1016/j.biortech.2020.123373, http://www.elsevier.com/locate/biortech.

Sinsuw, A. A. E., Chen, T.-H., Dokmaingam, P., Suriandjo, H. S., & Chu, C.-Y. (2024). Life cycle assessment of environmental impacts for two-stage anaerobic biogas plant between commercial and pilot scales. *International Journal of Hydrogen Energy, 52*, 58–70. https://doi.org/10.1016/j.ijhydene.2023.06.331.

Starr, K., Gabarrell, X., Villalba, G., Talens, L., & Lombardi, L. (2012). Life cycle assessment of biogas upgrading technologies. *Waste Management, 32*(5), 991–999. https://doi.org/10.1016/j.wasman.2011.12.016.

Sánchez-Martín, L., Ortega Romero, M., Llamas, B., Suárez Rodríguez, Md. C., & Mora, P. (2023). Cost model for biogas and biomethane production in anaerobic digestion and upgrading: Case study: Castile and Leon. *Materials, 16*(1). https://doi.org/10.3390/ma16010359, http://www.mdpi.com/journal/materials.

Taylan, O., Kaya, D., Bakhsh, A. A., & Demirbas, A. (2018). Bioenergy life cycle assessment and management in energy generation. *Energy Exploration and Exploitation, 36*(1), 166–181. https://doi.org/10.1177/0144598717725871, http://eea.sagepub.com/.

Ugwu, S. N., Harding, K., & Enweremadu, C. C. (2022). Comparative life cycle assessment of enhanced anaerobic digestion of agro-industrial waste for biogas production. *Journal of Cleaner Production, 345*. https://doi.org/10.1016/j.jclepro.2022.131178, https://www.journals.elsevier.com/journal-of-cleaner-production.

Wattanasilp, C., Songprakorp, R., Nopharatana, A., & Khompatraporn, C. (2021). Techno-cost-benefit analysis of biogas production from industrial cassava starch wastewater in Thailand for optimal utilization with energy storage. *Energies, 14*(2), 416. https://doi.org/10.3390/en14020416.

Wilkinson, J., Mays, T., & McManus, M. (2023). Review and meta-analysis of recent life cycle assessments of hydrogen production. *Cleaner Environmental Systems, 9*. https://doi.org/10.1016/j.cesys.2023.100116 100116.

Xue, Y.-nan, Luan, W.-xin, Wang, H., & Yang, Y.-jie (2019). Environmental and economic benefits of carbon emission reduction in animal husbandry via the circular economy: Case study of pig farming in Liaoning, China. *Journal of Cleaner Production, 238*. https://doi.org/10.1016/j.jclepro.2019.117968 117968.

Zaman, A. U. (2010). Comparative study of municipal solid waste treatment technologies using life cycle assessment method. *International Journal of Environmental Science and Technology, 7*(2), 225–234. https://doi.org/10.1007/BF03326132, http://www.ceers.org/ijest/issues/full/v7/n2/702002.pdf.

Zhang, C., & Xu, Y. (2020). Economic analysis of large-scale farm biogas power generation system considering environmental benefits based on LCA: A case study in China. *Journal of Cleaner Production, 258*. https://doi.org/10.1016/j.jclepro.2020.120985 120985.

Index

Note: Page numbers followed by "*f*" and "*t*" refer to figures and tables, respectively.

B

C

D

M